Advances in Comparative and Environmental Physiology 5

Volumes already published

Vol. 1: NaCl Transport in Epithelia
Edited by *R. Greger* (1988)

Vol. 2: From the Contents: Hibernation – Vapor Absorption
Nutrient Transport – Stable Pollutants (1988)

Vol. 3: Molecular and Cellular Basis of Social Behavior in Vertebrates
Edited by *J. Balthazart* (1989)

Vol. 4: Animal Adaptation to Cold
Edited by *L. C. H. Wang* (1989)

Vol. 5: From the Contents: Osmoregulation – Chemoreception
Temperature – Spectroscopy – Metalloproteins
Maternal-Fetal Relationships (1989)

Volumes in preparation

Vol. 6: Vertebrate Gas Exchange: From Environment to Cell
Edited by *R. G. Boutilier* (1990)

Vol. 7: Volume and Osmolality Control in Animal Cells
Edited by *R. Gilles, E. K. Hoffman,* and *L. Bolis*

Advances in

Comparative and Environmental Physiology 5

With Contributions by

M. Brouwer · W. E. S. Carr · W. Ross Ellington · D. W. Engel
R. A. Gleeson · B. Korsgaard · T. S. Moerland · T. Mustafa
D. J. Prior · B. D. Sidell · K. C. Srivastava · H. G. Trapido-Rosenthal · R. E. Weber · R. W. Wiseman

With 50 Figures

Springer-Verlag
Berlin Heidelberg New York
London Paris Tokyo Hong Kong

ISBN-13: 978-3-642-74512-6 e-ISBN-13: 978-3-642-74510-2
DOI: 10.1007/978-3-642-74510-2

Softcover reprint of the hardcover 1st edition 1989

2131/3145-543210 – Printed on acid-free paper

Foreword to the Series

The aim of the series is to provide comprehensive, integrated reviews giving sound, critical, and provocative summaries of our present knowledge in environmental and comparative physiology, from the molecular to the organismic level.

Living organisms have evolved a widespread range of basic solutions to cope with the different problems, both organismal and environmental, with which they are faced. A clear understanding of these solutions is of course of fundamental interest for all biologists, zoologically or medically oriented. They can be best comprehended in the framework of the environmental and/or comparative approaches. These approaches demand either wide surveys of animal forms or a knowledge of the specific adaptive features of the species considered. This diversity of requirements, both at the conceptual and technological level, together with the fact that physiology and biochemistry have long been mainly devoted to the service of medicine, can account for the fact that these approaches emerged only slowly amongst the other new, more rapidly growing disciplines of the biological sciences.

The field has now gained the international status it deserves and the organization of a series devoted to it appeared timely to me in view of its actual rapid development and of the interest it arouses for a growing number of biologists, physiologists, and biochemists, independently of their basic, major orientation.

Liège, Belgium *Raymond Gilles*

List of Editors

Those interested in contributing to this series should contact Professor R. Gilles or one of the Series Editors. Review proposals should include a short outline – and a Table of Contents – briefly defining the aims and scopes of the review.

Contents

Chapter 1

Neuronal Control of Osmoregulatory Responses in Gastropods

David J. Prior[1]

Contents

1 Introduction

Combining the experimental approaches of environmental physiology and cellular neurobiology can be particularly useful when studying the control of the physiological changes and behavioral modifications that occur in response to environmental stress.

Behavioral neurobiologists seek to understand the neuronal interactions that underly naturally occurring behavioral patterns and responses (see Camhi 1983). Similarly, environmental physiologists study the morphological, cellular, and molecular adaptations that enable organisms to make compensatory responses to environmental stress (see Prosser 1973). In that many regulatory responses are now known to be modulated by the nervous system, it is understandable that analyses of the control of such responses should involve approaches derived from both comparative physiology and cellular neurobiology.

Cellular excitability is one of the most labile of physiological functions. Both axonal and synaptic transmission are dependent upon the maintenance of appropriate ionic gradients. Variations in the ionic and osmotic composition of the environment can cause dramatic changes in neuronal function and thereby the integrative properties of the nervous system which underlie behavioral patterns. As such, it is natural that osmoregulation is an area of inquiry that blends environmental physiology and neurobiology.

[1] Department of Biological Sciences, Northern Arizona University, Flagstaff, AZ 86011, USA

Advances in Comparative and Environmental Physiology, Vol. 5

2 Invertebrate Neurons and Osmotic Stress

Much of the basic literature on the effects of osmotic or ionic stress on neuronal function comes from work on osmoconforming organisms such as the marine annelids *Sabella penicillus* (e.g., Treherne and Pichon 1978) and *Mercierella enigmatica* Fauvel (e.g., Carlson and Treherne 1977; Benson and Treherne 1978a, b), bivalve mollusks such as the freshwater *Anodonta cygnea* (Carlson and Treherne 1969) and the marine *Mya arenaria* (Beres and Pierce 1981), *Mytilus edulis* (Willmer 1978a, b, c), and certain marine gastropods, such as *Elysia chlorotica* (e.g., Parker and Pierce 1985; Quinn and Pierce 1987). In essentially all of these studies initiation and conduction of axonal action potentials has been the focus of the analysis.

A common finding among early studies of neuronal responses to osmotic stress was that following exposure to hyposmotic stress, axons swell and then return to their original dimensions within 2 to 3 h (e.g., Gilles 1979). However, during this period, the electrophysiological responses of different neuron types are quite different. Earlier reports on the effects of hyposmotic stress in marine annelids and bivalves indicated that, in general, neurons were initially hyperpolarized, followed by a progressive depolarization and reduction in excitability (see Treherne 1980 for a review). In contrast, in other marine species, such as the horseshoe crab, *Limulus polyphemus*, exposure of neurons to hyposmotic stress results in an immediate depolarization and an increase in cyclic synaptic activity (e.g., in the cardiac follower neurons; Prior and Pierce 1981). Studies of the effects of osmotic stress on basic membrane properties and patterned synaptic function in gastropods can be readily coupled with a regulatory biological approach in which mechanisms underlying the adaptive responses are examined.

The variations in hemolymph osmolality experienced by gastropods are quite remarkable and, as has been reported, follow changes in environmental concentrations in aquatic forms and body hydration in terrestrial forms (see Burton 1983; Riddle 1983; Prior 1985). Terrestrial gastropods are especially well-suited for analyses of the neural control of water regulatory responses. Due to the drying effects of their environment terrestrial slugs and snails are highly susceptible to dehydration. Of course, dehydration stress is especially crucial for slugs, which lack an external shell. Active slugs can lose 30%–40% of their initial body weight (IBW) in just 2 h due to evaporative water loss from the integument, lung, and the dilute mucous trail that is deposited during locomotion (Dainton 1954a, b; Burton 1964, 1966; Machin 1964a, 1966, 1975; Prior et al. 1983; see Riddle 1983 and Prior 1985 for reviews). Thus, even brief exposure to a drying environment can result in a dramatic loss of body water. Concomitant with this loss of body water is a proportional increase in hemolymph osmolality and ionic concentration (Burton 1964, 1965a, b, c, 1968a, b, 1971a, b; Prior et al. 1983; Prior 1984; see Burton 1983 for a review).

Within the Gastropoda, there are reports of widely varying responses to osmotic stress in identified neurons among different species and even within a single species. In early studies of the terrestrial slug *Arion* extracellular recording techniques were used with isolated preparations of the central nervous system to examine the responses of central neurons to variations in osmolality. In this case,

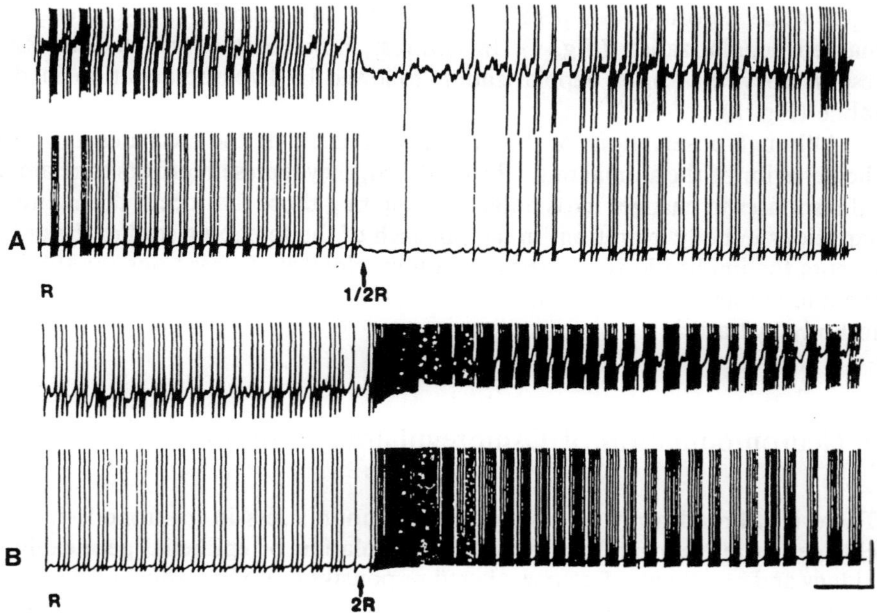

Fig. 1 A, B. Effects of dilution and concentration changes of the bathing medium on the intracellular recording of RAPN. Upper recordings in *A* and *B* are amplifications of the lower full spike recordings. *Arrows* indicate the changes of the bathing medium from normal Ringer solution to 1/2 R (*A*) and from normal Ringer solution to 2 R (*B*). All recordings were obtained from the same neuron. Calibration: upper recordings (12 mV), lower recordings (50 mV). Time: 1 min (Hasegawa and Takeda 1986)

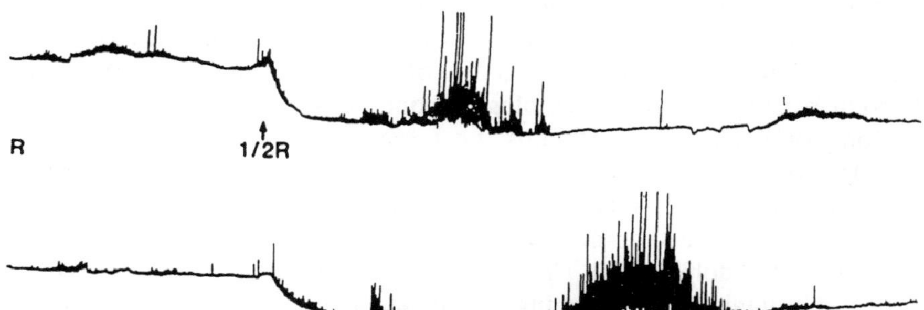

Fig. 2. Effects of dilution changes of the bathing medium on the intracellular recording of RAPN. *Arrows* indicate the change of the bathing medium from normal Ringer solution (*R*) to 1/2R. Calibration: 12 mV. Time: 1 min (Hasegawa and Takeda 1986)

neurons were found that were excited by exposure of the preparation to concentrated saline (1.2×Locke solution; Hughes and Kerkut 1956; Kerkut and Taylor 1956). But in a recent report on neuronal responses to osmotic stress in the terrestrial snail *Achatina fulica* (Hasegawa and Takeda 1986), most neurons examined became hyperpolarized in response to hyposmotic stress (85 mOsm kg^{-1}

H_2O) while exposure to hyperosmotic saline (340 mOsm kg^{-1} H_2O) resulted in marked depolarization (Fig. 1). In contrast, however, neurons such as RAPN responded with a brief depolarization followed by hyperpolarization in dilute saline (Fig. 2).

Although observations of this sort do not elucidate the mechanisms whereby the behavior of the animal might be modified, they represent efforts to relate neural responses to changes in osmolality simulating those that occur during normal variations in body hydration. In fact, in each of the foregoing studies the authors discuss the implications of their electrophysiological observations for studies of the alterations in locomotor activity that are known to occur during variations in body hydration.

3 Neuromodulation of Osmoregulatory Responses

There is a broad array of responses to osmotic stress in gastropods including changes in integumental permeability, modulation of epithelial permeability in kidney and heart and, of course, altered responsiveness of neuronal networks such as the central pattern generators underlying cyclic behaviors. In all of these responses neuromodulation has been either demonstrated or at least implicated.

3.1 Integumental Permeability

Modulation of epithelial permeability is crucial for many of the regulatory responses of animals. Osmoregulation is especially dependent upon the precise control of the movement of water and electrolytes across epithelia such as renal structures and integument. The permeability of such epithelia can be specifically altered in response to certain physiological conditions (e.g., Taylor and Palmer 1982; DiBona 1983; Hays 1983; Phillips and Lewis 1983).

Many of the early studies of epithelial permeability centered around the movement of water and electrolytes across the integument of intact amphibians. It was shown that dehydrated animals could rapidly absorb water across their skin (e.g., Bentley 1971; Adolph 1933). This "hydroosmotic" effect was not observed in fully hydrated animals, thus indicating that it was a specific dehydration-induced increase in integumental permeability.

It was shown by Brunn (1921) and others (e.g., Heller 1941; Sawyer 1951) that injection of fully hydrated frogs with mammalian neurohypophysial extracts also resulted in rapid water absorption. Similarly, injections of frog pituitary extract also caused the effect (Heller 1941). This "Brunn effect" (i.e., hormonally-mediated increase in integumental permeability) was subsequently shown to be due to the octapeptide arginine-vasotocin (Pickering and Heller 1959). Hormonally-dependent changes in permeability have subsequently been demonstrated in experiments on a number of epithelia (e.g., Bentley 1971; Phillips and Lewis 1983).

Measurements of electrical conductance have led to the general description of epithelia as being "leaky" (e.g., gall bladder, proximal nephron) or "tight" (e.g., amphibian skin, urinary bladder; Fromter and Diamond 1972). The degree of leakiness has been attributed to the relative resistance of the tight junctions separating the epithelial cells. The junctional resistance in leaky epithelia is about 100 $\Omega\ cm^{-2}$, while in tight epithelia it is often greater than 500 $\Omega\ cm^{-2}$ (Lewis 1983).

Permeability appears to be more precisely controlled in tight epithelia than in leaky epithelia. Leaky membranes are highly permeable to water, support neither large osmotic gradients nor transepithelial potentials, and seem unresponsive to hormonal stimulation (Lewis 1983). In contrast, the hydraulic permeability of tight epithelia has been shown to be under hormonal and, apparently, neural control (Taylor and Palmer 1982).

There are growing numbers of observations that suggest the existence of direct neural modulation of epithelial permeability. For example, it has been shown that in isolated toad skin there is an increase in water movement in response to epinephrine or the β-adrenergic agonists, isoproterenol and isoprenaline (Brown et al. 1980; DeSousa and Grosso 1982). It may be that neural modulation plays an important role in many epithelial osmoregulatory responses.

Terrestrial gastropods can rapidly recover from dehydration by absorbing water through the integument of their foot while in contact with a moist surface (e.g., Dainton 1954a). This process of contact-rehydration is reminiscent of the observations with amphibians; furthermore, it has been shown to be mediated by a well-regulated behavioral pattern (Prior 1982, 1984). When slugs (e.g., *Limax maximus*) that have been dehydrated to 60%–70% of their initial body weight (%IBW) are placed in a dry test chamber with a pad of moist tissue paper, they move directly onto the pad and assume a characteristic flattened posture (Fig. 3). They remain in this posture until they have absorbed enough water to rehydrate

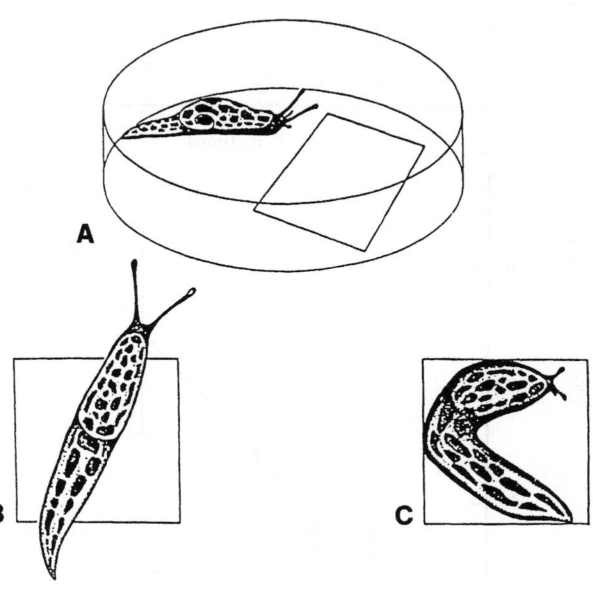

Fig. 3A–C. The chamber for testing contact-rehydration is shown in *A* and consists of a dry plastic petri dish containing a pad of moist tissue paper. A trial was inititated by placing a slug in the chamber about 2 cm from the pad; *B* Fully hydrated slug moving across the pad without stopping (note the fully extended superior tentacles); *C* Dehydrated slug in the characteristic flattened posture during contact-rehydration. The superior tentacles are half withdrawn and the foot fringe is expanded (Prior 1984)

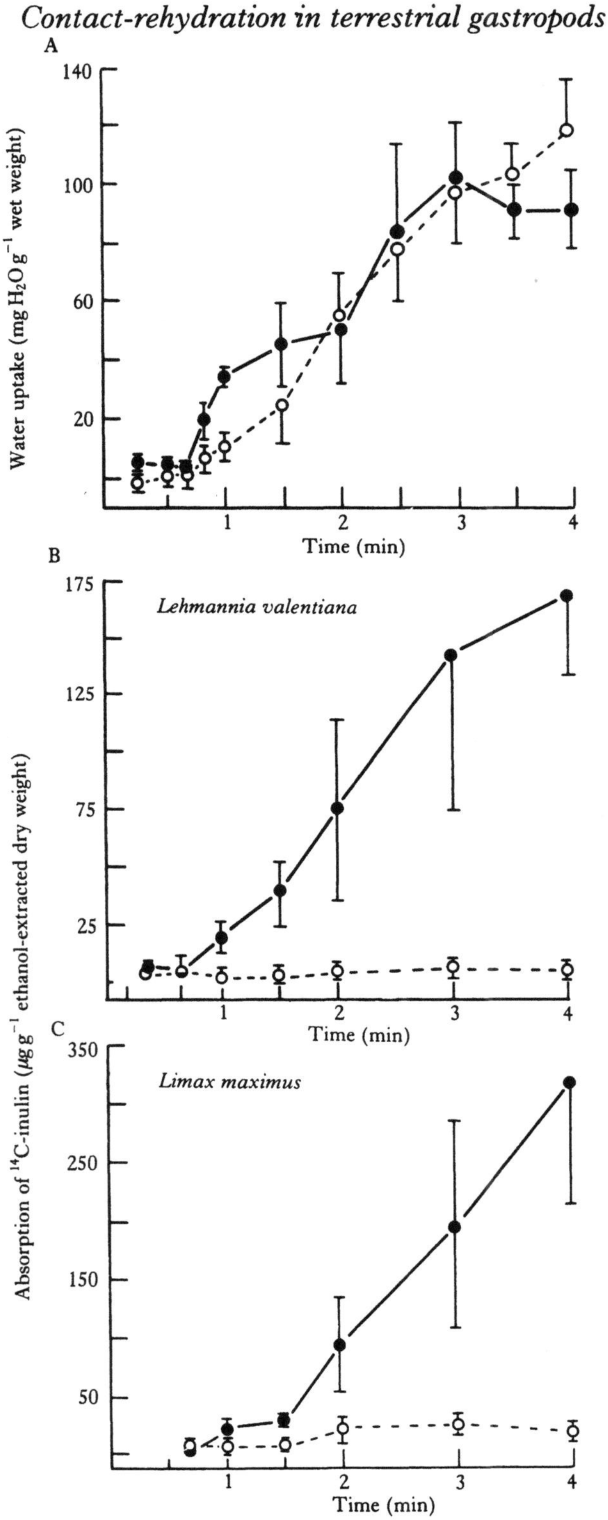

Fig. 4A–C. The time course of water uptake by two species of slugs (*Limax maximus* ● – ● and *Lehmannia valentiana* ○ – ○) during contact-rehydration. Each point is the mean of five to nine measurements, with only one measurement from each slug. The data on *Limax* are redrawn from Prior (1984). The rates of absorption of [^{14}C]inulin by dehydrated *Lehmannia (B)* and *Limax (C)* parallel that of water absorption during the 4-min test period (compare *A* with *B* and *C*). There was negligible [^{14}C]inulin uptake by hydrated control slugs of either species (○ – ○ in *B* and *C*). Each point in *B* and *C* is the mean ±SE of three to five replicates (Prior and Uglem 1984)

themselves to 93.5% ± 12.2% IBW. Once rehydrated, they terminate contact-rehydration by moving off the pad and coming to rest on the dry surface. Thus, there is a dehydration threshold (i.e., 65.1% IBW) for initiation of the behavior and a rehydration set-point (i.e., 93.5% IBW) at which the behavior is terminated.

Specific changes in hemolymph osmolality which accompany the changes in body hydration are involved in the control of contact-rehaydration. Dehydration to 65% IBW results in an increase in hemolymph osmolality from the normal (i.e., 100% IBW) of 140 mOsm kg^{-1} H_2O- up to 200–220 mOsm kg^{-1} H_2O (Prior et al. 1983). When fully hydrated slugs are injected with hyperosmotic mannitol to raise their hemolymph osmolality to 200 mOsm kg^{-1} H_2O, contact-rehydration is initiated. It has also been shown that experimental dilution of the hemolymph results in termination of contact-rehydration (Prior 1984).

The high rate of water absorption during contact-rehydration (i.e., 20–50 μl min^{-1}; see Fig. 4) occurs only when the hemolymph osmolality is increased to about 200 mOsm kg^{-1} H_2O. As seen in Fig. 4 the time course of ^{14}C-inulin absorption closely resembles that of water absorption. Also, subsequent experiments have shown comparable rates of absorption of ^3H_2O and ^{14}C-mannitol. However, larger markers appear to be excluded according to their molecular sizes (Prior and Uglem 1984; Uglem et al. 1985).

Recent results show that contact-rehydration and water uptake in *Limax maximus* can be initiated by injections of the octapeptide hormone angiotensin II (5×10^{-9} M) or arginine vasotocin (5×10^{-7} M; Makra and Prior 1985; Table 1). Although genuine AII has not yet been demonstrated to exist in mollusks, AVT-like immunoreactivity has recently been found in the central nervous systems of *Limax* (Sawyer et al. 1984) and *Aplysia* (Moore et al. 1981). Furthermore, injection of AVT has been shown to cause fluid exudation across the integument of the slug *Ariolimax columbianus* (Deyrup-Olsen and Martin 1982). Thus, the increase in epithelial permeability that occurs during contact-rehydration in *Limax* is dependent upon changes in hemolymph osmolality and appears to be neurally modulated, possibly via centrally released neuropeptides.

This is certainly comparable to the scheme proposed for modulation of integumental permeability in the freshwater snail *Lymnaea stagnalis*. There is considerable literature documenting the role of the integument in water balance in *Lymnaea* (see Greenaway 1970). The high affinity sodium uptake mechanism of the integument in *Lymnaea* is modifiable, increasing net Na^+ uptake four- to five-fold depending upon external Na^+ concentration (Greenaway 1970; Schlichter 1981). The potential role of CNS neurosecretory neurons in the control of integumental permeability has been suggested by Wendelar Bonga (1970a, b, 1971, 1972) who demonstrated the existence of neurosecretory dark green cells (DGC) in the pleural ganglion with axonal projections to the integument of foot, head, and mantle (Swindale and Benjamin 1976). Changes in the osmolality of the external environment result in variations in the neurosecretory content of the DGCs. Thus, it appears possible that neurosecretory neurons in the pleural ganglia may have a direct effect on the changes in integumental permeability that occur in response to osmotic stress.

More recently Grimm-Jorgensen (1978, 1979, 1980, 1983) has shown that injection of thyroid releasing hormone (TRH) into *Lymnaea stagnalis* causes a

Table 1. Effects of angiotensin II and subthreshold dehydration on contact-rehydration in *Limax maximus*[a]

Experimental conditions	Number of slugs that initiated the behavioral posture	Mean rate of water absorption ($\mu l\ cm^{-2}\ min^{-1}$ $\pm$SEM)	Mean rehydration setpoint (% IBW $\pm$ SEM)
Normal air-dehydration to 65% IBW (Prior 1984)	36/39	7.8 ± 0.3 ($N = 119$)[b]	93.6 ± 1.1 ($N = 119$)
AII (5×10^{-9} mol l^{-1}) injected into slugs at 80% IBW	29/29	7.2 ± 0.9 ($N = 18$)[c]	92.8 ± 1.4 ($N = 29$)
AII (5×10^{-9} mol l^{-1}) and saralasin (5×10^{-7} mol l^{-1}) injected into slugs at 80% IBW	11/29	7.0 ± 1.1 ($N = 7$)[c]	95.2 ± 3.2 ($N = 11$)
$1.0 \times$ Saline injected into slugs at 80% IBW	7/29	2.2 ± 0.5 ($N = 5$)[c]	88.1 ± 2.6 ($N = 7$)

[a] Calculated hemolymph concentrations of AII and saralasin are given in parentheses and each dehydration level is $\pm 2\%$ IBW.
[b] This includes additional trials in which the complete behavior was observed.
[c] Only those slugs which were completely on the moist pads were included in the calculation of the rates of water absorption.
The only rate of water absorption that was significantly different from that in the normal air-dehydration trial was that of the $1.0 \times$ saline injection trial ($P < 0.05$ by least square means). None of the rehydration set points observed in the experimental groups differed significantly (by any test) from that observed during normal dehydration induced contact-rehydration (Makra and Prior 1985).

marked loss of body water. Furthermore, an immunoreactive TRH-like substance has been demonstrated in the CNS, hemolymph, and integument of the foot and mantle of *Lymnaea* (see Grimm-Jorgensen et al. 1984). These observations have led to the hypothesis that the pleural ganglia contain a TRH-like material that serves as an osmoregulatory neurohormone affecting integumental transport.

Thus, in *Lymnaea* as in terrestrial forms, there is evidence for central neuronal modulation of epithelial permeability in response to osmotic stress. There does not appear to be a behavioral response that would correspond to the posturing that occurs during contact-rehydration in *Limax*. There may, however, be alterations in the responsiveness of certain aspects of the animal's behavior that facilitate the overall osmoregularity response pattern. Such responses in other species are discussed in the following sections.

3.2 Permeability of the Renocardiac Complex

3.2.1 Renal Function

Because the production of dilute urine could be a significant osmoregulatory mechanism in gastropods, numerous studies have involved analyses of the mechanisms controlling the rate of urine production. A number of anatomical examinations of the kidney sac have been made. This is the site of primary urine formation. In *Achatina achatina*, for example, nephrocytes are bathed apically by primary urine and basally by an ultrafiltrate of hemolymph (Skelding 1973a, b). Pro-urine within the kidney sac is from the pericardial sac, which is an ultrafiltrate of the blood produced by the contraction of the heart (see Robertson 1964; Harrison and Martin 1966; Machin 1975).

A number of microscopic studies have indicated that the intercellular contacts between kidney sac cells may be a site for regulation of urine production (e.g., Newell and Skelding 1973). Recently, Khan and Saleuddin (1979a, b, 1981) have demonstrated that the kidney sac epithelium of the freshwater snail *Helisoma trivolvis* is highly responsive to changes in the osmolality of the environment. Expanded intercellular spaces were observed in snails kept in hyposmotic medium. Similar observations have been made with *Helix aspersa* kept under high humidity conditions (Saleuddin et al. 1983). In both cases the changes in hemolymph osmolality appear to initiate the morphological responses. These workers have further related the ultrastructural changes seen in *Helisoma* in response to osmotic stress to release of a neurosecretory diuretic material from the visceral ganglion (Kahn and Saleuddin 1979a, b, 1981). When *Helisoma* kidney is treated in vitro with homogenates of visceral ganglion from snails exposed to hyposmotic medium, the intercellular spaces of the kidney sac epithelium are widened and the septate junctions appear to be reduced. These effects are not seen with cerebral or visceral ganglion homogenates from isosmotically acclimated snails (Kahn and Saleuddin 1981). Thus, it is proposed that a decrease in hemolymph osmolality initiates release of a diuretic neurosecretory material which increases epithelial permeability resulting in an increase in the production of hyposmotic urine.

3.2.2 Cardiac Function

Analysis of the neural control of cardiac function in gastropods is highlighted by the studies on the opistobranch *Aplysia californica*. It has been shown that the neuronal network involved in the regulation of cardiac function is in the abdominal ganglion and consists of an array of motoneurons, interneurons, and neurosecretory neurons that can cause depression or excitation of cardiac activity (e.g., Koester et al. 1973, 1974, 1979; Mayeri et al. 1974; Sawada et al. 1981a, b; Rittenhouse and Price 1986).

It is known that various physiological or environmental stimuli can modify cardiac activity in gastropods. For example, in both *Aplysia* (Koch et al. 1984) and *Limax* (Grega and Prior 1985) feeding results in an increase in heart rate. In *Limax* the increase in cardiac activity is initiated by proprioceptive afferent input to the CNS resulting from stretch of the crop. It has also been reported that in *Helix* and *Achatina* afferent activity from the renocardiac system is directed to the visceral ganglion (Salanki and S.-Rozsa 1973; S.-Rozsa 1979; S.-Rozsa et al. 1980). With such afferent systems, it is reasonable to expect modulation of the cardiac system by a variety of physiological variables such as hemolymph PO_2, pH, osmolality, and perfusion pressure.

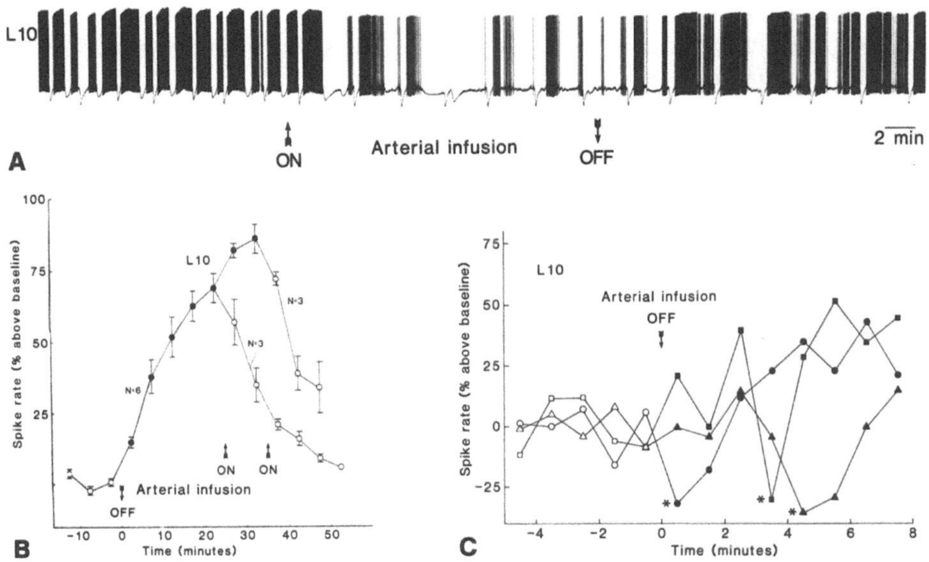

Fig. 5A–C. Effects of ganglionic circulation on impulse activity of the heart exciter interneuron, L10. *A* Infusion of oxygenated saline into the caudal artery reversibly inhibited L10 spike activity. *Arrows* indicate the beginning (on: 2.5 µl min^{-1}) and end (off: 0 µ/min^{-1}) of arterial infusion. *B* Measurement of L10 responses in 6 isolated ganglia. Average (±SEM) spike rate gradually increased during the period when arterial infusion was off and returned to baselin levels when flow resumed. The period and extent of L10's response depended on the period (25 or 35 min) of reduced circulation. *C* Three examples showing short-term effects of decreased ganglionic circulation of L10 activity. During the first 1 to 4 min of response, L10 was briefly inhibited (*) by synaptic input from the respiratory command interneuron complex (Int II). In this and subsequent figures, *open symbols* indicate arterial infusion on, *closed symbols* indicate infusion off. Baseline spike rate is the average of measurements shown preceding the stimulus (Bablanian and Treistman 1985)

In an elegant study by Furgal and Brownell (1987) the basic neurophysiological approach in the cardiac system of *Aplysia* has been extended to an integrative analysis of circulatory and respiratory responses to variables such as hemolymph PO_2 and cardiovascular perfusion pressure. An isolated preparation of the abdominal ganglion was used in which the major artery was cannulated to allow manipulation of perfusion pressure. The experimental design allowed simultaneous intracellular recordings of activity in identified heart regulatory neurons and measurement of the frequency and force of ventilatory gill movements. Figure 5 illustrates the inhibitory effect of arterial perfusion of the abdominal ganglion on the activity of the heart exciter interneurons L10. Within several minutes the strongly rhythmic bursting pattern of L10 was significantly reduced, as was that in several other neurons as well (Fig. 6). In the case of the neurosecretory neurons R11, R12, R9, and R14, which innervate arteries and pericardial structures, reduction of arterial perfusion again resulted in immediate excitation (Fig. 7).

The results from isolated ganglion preparations are well correlated with those from preparations that included the innervated heart and gill. In such preparations, reduced perfusion of the abdominal ganglion resulted in increased ven-

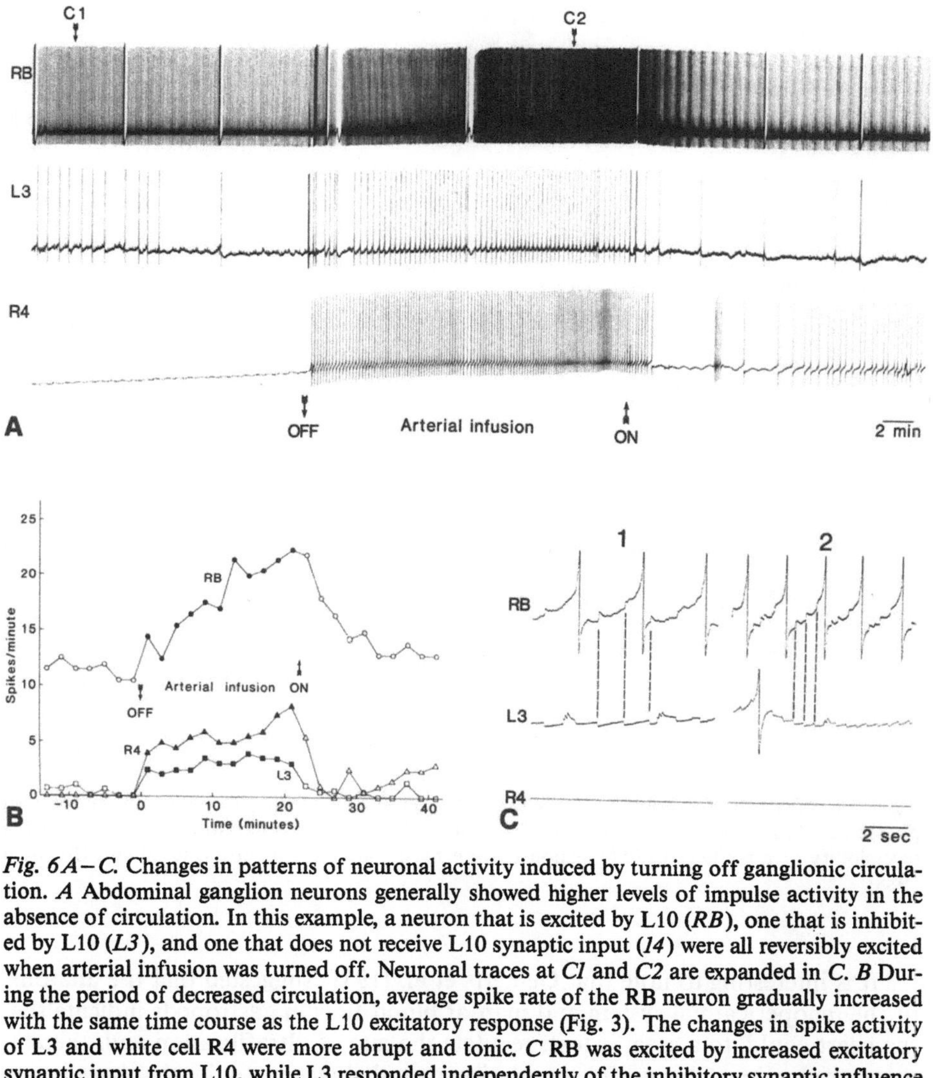

Fig. 6A–C. Changes in patterns of neuronal activity induced by turning off ganglionic circulation. *A* Abdominal ganglion neurons generally showed higher levels of impulse activity in the absence of circulation. In this example, a neuron that is excited by L10 (*RB*), one that is inhibited by L10 (*L3*), and one that does not receive L10 synaptic input (*14*) were all reversibly excited when arterial infusion was turned off. Neuronal traces at *Cl* and *C2* are expanded in *C*. *B* During the period of decreased circulation, average spike rate of the RB neuron gradually increased with the same time course as the L10 excitatory response (Fig. 3). The changes in spike activity of L3 and white cell R4 were more abrupt and tonic. *C* RB was excited by increased excitatory synaptic input from L10, while L3 responded independently of the inhibitory synaptic influence from L10 (indicated by co-occurrence of EPSPs in RB and IPSPs in L3, *dotted lines*) which increased while arterial infusion was off (Furgal and Brownell 1987)

tilatory gill pumping and sustained increases in heart activity. Thus, in this preparation, compensatory changes in cardiovascular and respiratory function in response to arterial pressure are neuronally modulated. Conceivably, such modulation could mediate osmoregulatory responses to changes in hemolymph volume that occur during osmotic stress.

In that pro-urine is produced by way of ultrafiltration of hemolymph in the pericardial sac, or in some species the kidney sac, the effective perfusion pressure generated by the heart could control urine output and thereby serve as a major osmoregulatory mechanism.

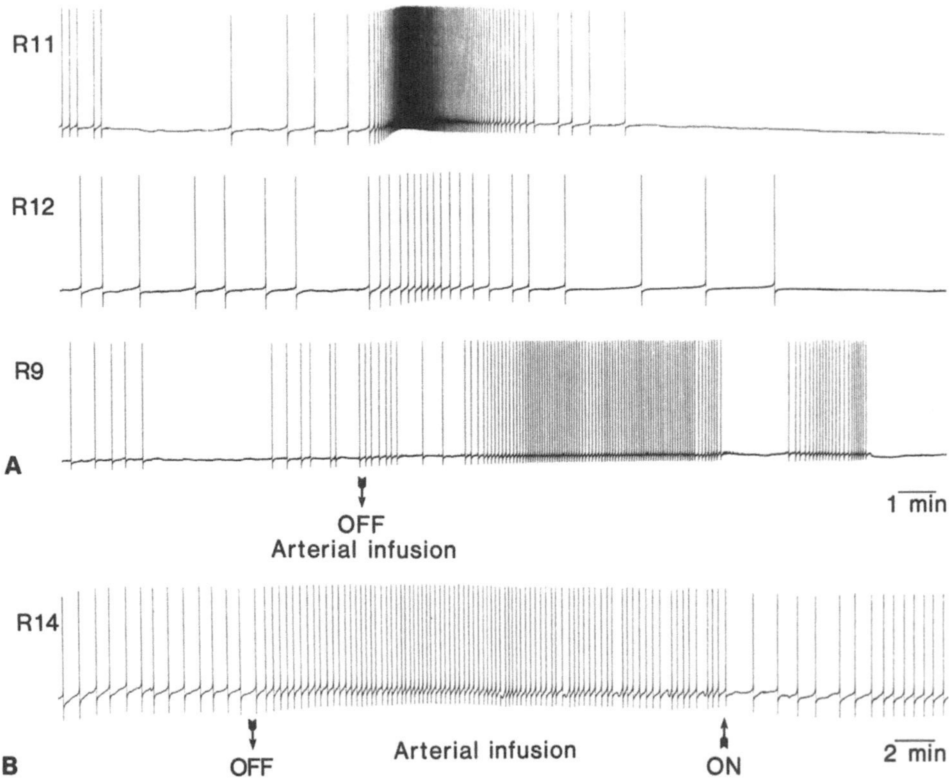

Fig. 7. A Influence of ganglionic circulation on spontaneous activity of with cells shows both phasic and tonic responses to decreased arterial infusion. *B* A separate experiment showing reversibility of a white cell (R14) response to decreased arterial infusion (Furgal and Brownell 1988)

It is interesting to note that Geraerts et al. (1981) suggested that the anatomy of the renopericardial system and plan of blood flow in gastropods indicate that the heart could, to a great extent, regulate the rate of ultrafiltration and, thereby, urine production. This useful speculation preceded recent studies that have begun to elucidate the interactions between the neural control of heart function and water balance.

In *Lymnaea* there are numerous reports of interactions between renopericardial function and various physiological conditions such as feeding and ion balance (e.g., De With 1977, 1978, 1980; Geraerts et al. 1981). It has been shown that in fasted *Lymnaea* heart rate is depressed (from 38 to 22 contractures min^{-1}). Furthermore, there is a concomitant reduction in urine output, which De With (1980) attributed to the reduction in cardiac activity. Similarly, in short-term observations, the heart rate of *Limax* was found to increase 16.7% immediately following a meal (Grega and Prior 1985).

In addition, in a recent study it was shown that in *Limax*, progressive dehydration of intact animals to 70% of initial body wight resulted in as much as an 18%

Table 2. Effects of increased hemolymph osmolality on heart rate in intact *Limax*

Solution injected	*N*	Calculated hemolymph osmolality (mOsm kg^{-1} H_2O)	Simulated level of dehydration (% IBW)	Increase in heart rate (%)
Mannitol	7	196.1 ± 14.4	71.2 ± 7.2	9.6 ± 4.5
NaCl	5	203.8 ± 8.3	67.5 ± 3.9	15.6 ± 2.5
Saline	4	140 ± 0.0	100.0 ± 0.0	0.8 ± 1.7

Hemolymph osmolality of fully hydrated slugs was increased by injections of 1.0 molal mannitol or 0.5 molal NaCl. Data are expressed as mean ± SD (or SEM for the percent increase in heart rate). The simulated level of corresponding dehydration was determined from Fig. 5 of Prior et al. (1983). The increase in heart rate was statistically significant for both mannitol ($p < 0.05$) and NaCl ($p < 0.1$; Spearman rank coefficient).

Effects of hyperosmotic saline perfusion of heart or CNS on heart rate in a heart-CNS preparation in *Limax*

Organ perfused	Type of hyperosmotic saline	Percentage change in heart contraction		
		Rate	Force	*N*
Heart	Ionic	12.1 ± 3.1 decrease	15.4 ± 3.3 decrease	4
	Mannitol	12.0 ± 2.3 decrease	14.9 ± 1.1 decrease	4
Brain	Ionic	5.2 ± 1.6 increase	8.8 ± 3.8 increase	10
	Mannitol	5.1 ± 0.4 increase	9.7 ± 3.1 increase	4

The increases in heart contraction rate and force between ionically and osmotically (with mannitol) hyperosmotic salines are not significantly different from each other, but are significantly different from control values (Mann-Whitney; $p < 0.05$) (Grega and Prior 1986).

increase in heart rate (Grega and Prior 1986). Using an isolated CNS-heart preparation in which the CNS or heart could be separately exposed to hyperosmotic saline it was determined that the CNS is primarily responsible for the increase in heart rate and contraction force (Table 2).

Quantitative electron microscopy and immunocytochemical studies have demonstrated that axonal projections of centrally located neurosecretory cells heavily innervate the kidney sac, pericardial sac, and heart. In *Lymnaea* the yellow-green and yellow cells of the visceral and parietal ganglia are known to project to renocardiac structures and have been shown to have greater synthesis and release of stores of neurosecretory granules following exposure of the animal to a hyperosmotic environment (Wendelaar Bonga 1971, 1972; Soffe et al. 1979). Earlier work has documented that lesioning of the pleural ganglia of *Lymnaea* leads to water uptake while injection of pleural ganglia homogenates results in decreased body weight (Lever and Joosse 1961).

Thus, there is considerable evidence for neuromodulation of the renocardiac system in gastropods.

3.3 Osmotically Sensitive Osphradium and Neuron R15 in *Aplysia*

For many years the marine opistobranch, *Aplysia*, has been a primary model system for the study of neuronal mechanisms underlying behavior (see Kandel 1976). The responses of *Aplysia* to osmotic stress have been examined by numerous workers in studies ranging from analysis of osmotic tolerance and volume regulation (e.g., Van Weel 1957) to electrophysiological analyses of the osmotically sensitive osphradium and abdominal ganglion neuron R15 (e.g., Stinnakre and Tauc 1966; Bablanian and Treistman 1983, 1985).

Early electrophysiological studies indicated that neuron R15 became hyperpolarized following application of hyposmotic sea water to a specialized external epithelial sensory structure known as the osphradium (Stinnakre and Tauc 1966). The osphradium was therefore identified as an osmoreceptor structure with at least one identified central nervous system projection. In that *Aplysia* is known to occur in intertidal areas, it was proposed that the osphradium-R15 interaction could be involved in either avoidance responses or in water and ion regulation (Prosser 1973).

Kupfermann and Weiss (1976) reported that injection of an homogenate of neuron R15 into intact *Aplysia* resulted in a rapid increase in body weight of 3% to 10%. Using the weight increase as a bioassay to test R15 extracts, it was concluded that R15 contains one or more stable low molecular weight polypeptides that are involved in the response to hyposmotic stress. Thus, the putative R15 hormone was considered to be an antidiuretic factor.

More recently Bablanian and Treistman (1983, 1985) have made intracellular recordings from R15 in an intact "Willows-type" preparation of *Aplysia*. The

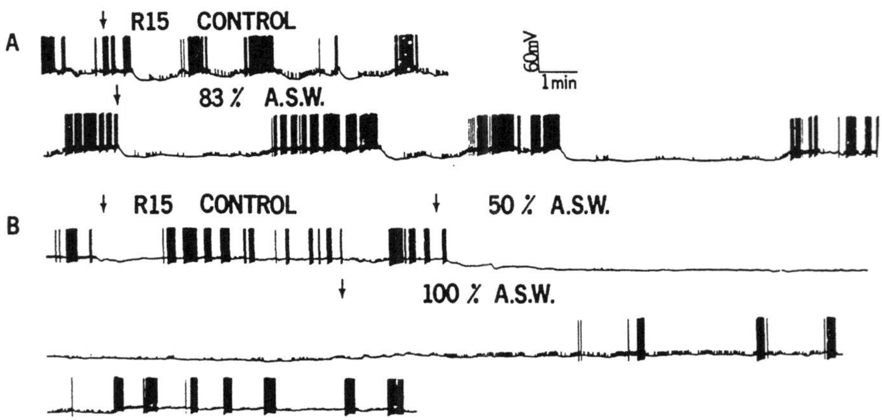

Fig. 8A, B. The effect of hypotonic sea water on activity of cell R15 in the intact preparation is illustrated. *A* Exposure of the animal to 83% sea water results in the appearance of long interburst intervals in cell R15. *Arrows* indicate the beginning of a control water bath change and a bath change into 83% A.S.W. Spikes have been clipped by the limited time response to the penwriter. *B* Exposure of the animal to 50% sea water resulted in complete suppression of BPA. The *first arrow* indicates the start of a control bath change, and the *second arrow* indicates introduction of 50% sea water. After the return of the animal to 100% ASW, BPA was resumed (Bablanian and Treistman 1985)

Table 3. Changes in the interburst interval (*IBI*) of cell R15 upon exposure of the intact preparation to hypo- and hypertonic sea water compared with exposure to 100% sea water (control)

Sea water (%)	Average increase of IBI[a] (%)	*P* Value (Student's *t*-test)
115	28.5±12.5 (n = 2)	Not significant
93	125.0±22.8 (n = 9)	$P<0.005$
83	140.0±31.8 (n = 3)	$P<0.025$
50	Complete inhibition (n = 2)	

[a] Computed for the 10 min after each stimulus presentation. These values are obtained by computing

$$\frac{IBI_{(\text{experimental presentation})} - IBI_{(\text{control presentation})}}{IBI_{(\text{control presentation})}} \times 100$$

(Bablanian and Treistman 1985).

spontaneous burst pattern of R15 is inhibited when the exterior of the animal is exposed to dilute artificial sea water (Fig. 8, Table 3). Although this is a confirmation of earlier findings, in this particular preparation it was possible to measure hemolymph composition while R15 was active or quiescent. When R15 was silenced by injection of hyperpolarizing current through the recording microelectrode, there was a 13% increase in hemolymph K^+ concentration and a 70% increase in ammonia concentration but no change in Na^+. Similarly, they reported a severalfold increase in the concentrations of free amino acids (especially taurine, glycine, glutamic acid, and histidine). Thus, the correlation between activity in R15 and the concentrations of hemolymph K^+, ammonia, and certain free amino acids is consistent with the hypothesis that R15 is involved in ion and water balance in *Aplysia*. To date, however, possible mechanisms of action of the R15 peptide have not been examined. Determination of targets for the peptide (e.g., neural networks, kidney, heart, integument) will be necessary in order to develop a scheme of the control system in this weakly osmoregulatory organism.

3.4 Osmotic Modulation of Feeding Behavior and Central Pattern Generators

Many experimental approaches, ranging from behavioral observations to molecular analyses of hormone function have been combined to study the neuronal control of feeding behavior in gastropods.

In several gastropod species feeding responsiveness has been shown to be correlated with the state of body hydration (see Prior 1985 for a review). In the terrestrial slugs *Limax maximus* and *Limax pseudoflavus*, dehydration to 65%–70% of initial body weight results in a dramatic reduction in feeding responsiveness (Prior 1983). Slugs orient to food, evert their lips, and make contact with the food but do not initiate the prolonged pattern of cyclic radular movements involved in the ingestive phase of feeding. Thus, the appetitive phase of the behavior is apparently unaffected, while initiation of the consumatory phase is blocked.

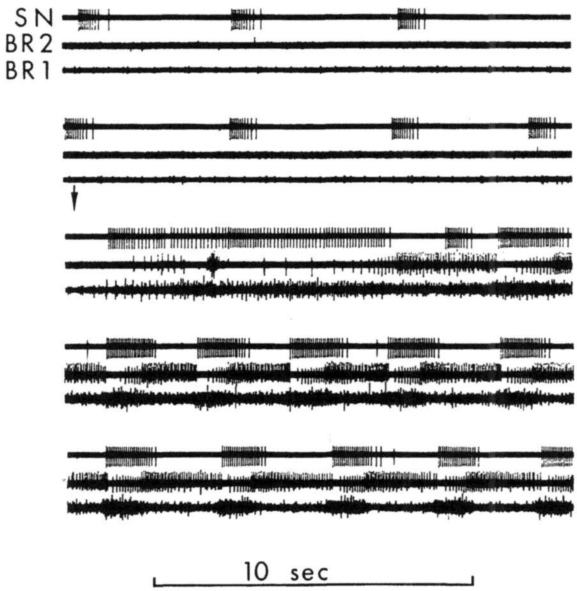

Fig. 9. Activation of the feeding motor program in vitro by chemical stimulation of the lips. Preparations consisted of lips, lip nerves, cerebral ganglion, and buccal ganglia. Suction electrodes monitored the salivary nerve and two buccal motor roots before and after application of potato extract selectively to the lip region (Prior and Gelperin 1977)

A neuronal correlate of this dehydration-induced reduction in feeding responsiveness has been recently described in *Limax* (Phifer and Prior 1985). In this study, isolated CNS preparations were used to examine the effect of saline osmolality on expression of the feeding motor program. The feeding motor program (FMP) in *Limax* is a discrete pattern of cyclic motor activity recorded from the buccal ganglion nerves that underlies the alternation of protraction and retraction of the radula during feeding (Fig. 9, Gelperin et al. 1978). The FMP, which can be initiated by chemical stimulation of the lips or electrical stimulation of the lip nerves, often continues for 30–40 bite cycles.

It has been shown that in several species of terrestrial slug feeding behavior is suppressed by dehydration to 65%–70% IBW. Furthermore, this level of dehydration is known to result in an increase of hemolymph osmolality from 140 mOsm kg^{-1} H_2O (100% IBW) to approximately 200–220 mOsm kg^{-1} H_2O. Based upon these observations, the effect of increased hemolymph osmolality could be simulated by exposing a CNS preparation, in which the FMP could be repeatedly stimulated, to 1.0 and 1.5×saline, which correspond to the hemolymph of slugs at 100 and 70% IBW, respectively. The results of a series of such experiments are shown in Fig. 10. In each case the number of bite cycles/FMP was significantly reduced when the preparation was perfused with 1.5×saline.

A comparison of the details of the initial bursts (e.g., interburst interval) in the full length FMPs in 1.0×saline with those of the truncated FMPs in 1.5×saline revealed that the initial phase of the FMP is unaffected (Fig. 11). Thus, it appears that the effect is one of shortening the number of cycles in a feeding bout, not disruption of the central pattern. This correlates well with the behavioral observation that dehydrated animals occasionally made several radular movements but did not initiate prolonged bouts of feeding. Thus, in *Limax* the dehydra-

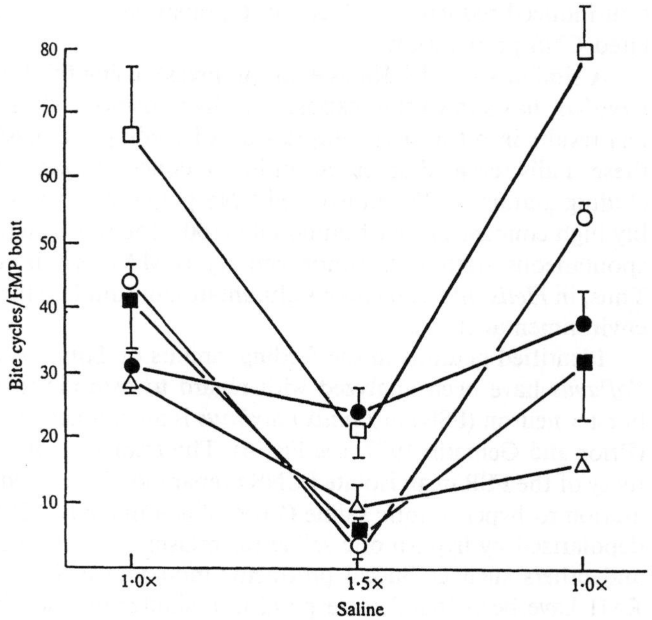

Fig. 10. The effect of saline concentration changes on the number of bite cycles per FMP bout. This figure shows the results of five experiments (five different preparations) in which FMP bouts were elicited by identical electrical stimulation in 1.0×, 1.5×, and 1.0× salines. FMP bout lengths for each experiment are plotted as the mean±SEM number of bite cycles per FMP bout for two to six stimulations in each saline (Phifer and Prior 1985)

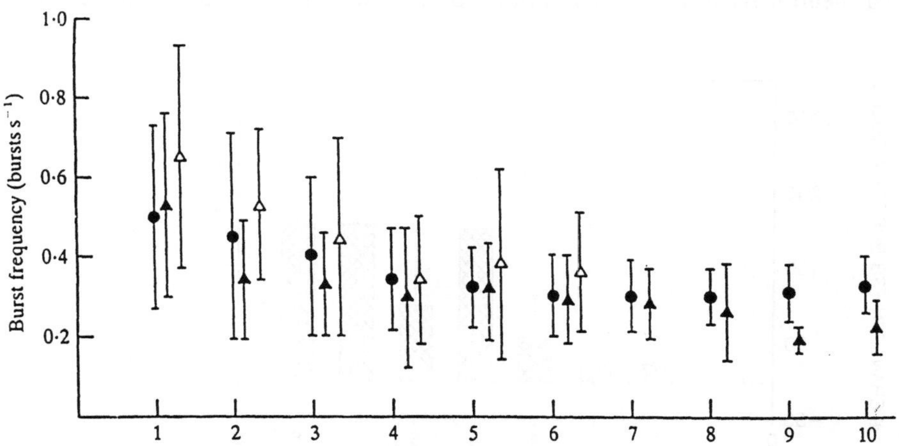

Fig. 11. Burst frequencies for the salivary burster neuron (SB) in salines of different concentration or osmolality. These burst frequencies are plotted against the burst number beginning with the first burst in an FMP bout. Each *point* represents the $\bar{x}\pm$SEM for: 16–29 bursts from eight preparations in 1.0×saline (●), 3–9 bursts from four preparations in 1.5× (▲) saline, or 3–6 bursts from three preparations in 1.0×+mannitol (△) saline (Phifer and Prior 1985)

tion-induced reduction in feeding responsiveness can be demonstrated in an isolated CNS preparation.

A similar study by Jones et al. (in press), using the freshwater snail, *Helisoma trivolvis,* has shown that exposure to hyperosmotic stress (20% artificial sea water) results in a transient suppression of feeding. Analysis of the hemolymph of these snails revealed an elevation in the concentrations of acidic amino acids including glutamate. When isolated CNS preparations were exposed to a comparably high concentration of glutamate (100–150 mM) there was suppression of the spontaneous patterned motor activity (PMA) that is associated with feeding. Thus, in *Helisoma* exogeneous glutamate can simulate the effects of hyperosmotic environmental stress.

Identified neurons in the feeding circuits of *Limax maximus* and *Limax pseudoflavus* have been analyzed with regard to osmotic tolerance. The fast salivary burster neuron (FSB) in *Limax maximus* is an autoactive salivary duct motoneuron (Prior and Gelperin 1977; see Fig. 6). The frequency of the spontaneous burst activity of the FSB in an isolated CNS preparation is reduced by exposure of the preparation to hyperosmotic saline (1.0×saline+mannitol, 225 mOsm kg^{-1} H_2O) and depolarized by hyposmotic saline (85 mOsm kg^{-1} H_2O; Prior 1981). This neuron and others such as buccal protractor motoneuron B7 and retractor motoneuron RM1 have been found to respond in a similar manner, becoming hyperpolarized in response to hyperosmotic (i.e., dehydration) stress. These in vitro responses correlate well with the general depression of feeding behavior described above.

3.5 Speculation: A Multifunctional Modulatory Neuron in *Limax maximus*

It has recently been shown that activation of the multifunctional modulatory buccal neuron B1 in *Limax* can modify both feeding neurons and cardiac function

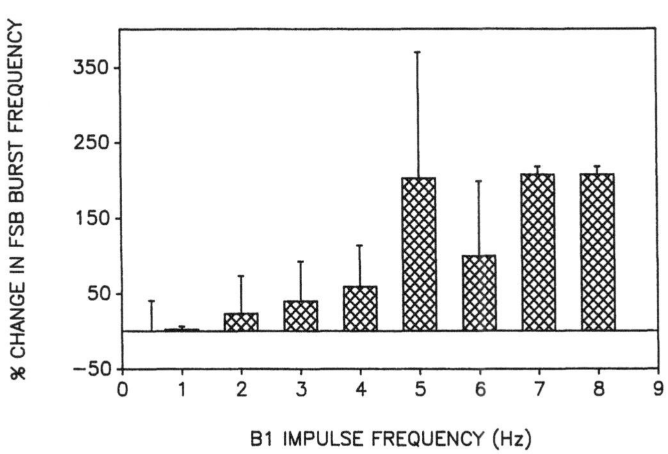

Fig. 12. The burst frequency of the fast salivary burster (*FSB*) in an isolated buccal ganglion-brain preparation of *Limax maximus* is plotted as a function of impulse frequency of a single buccal neuron, *B1*. *B1* was stimulated by pulses of injected current so that ten impulses at the various frequencies were elicited

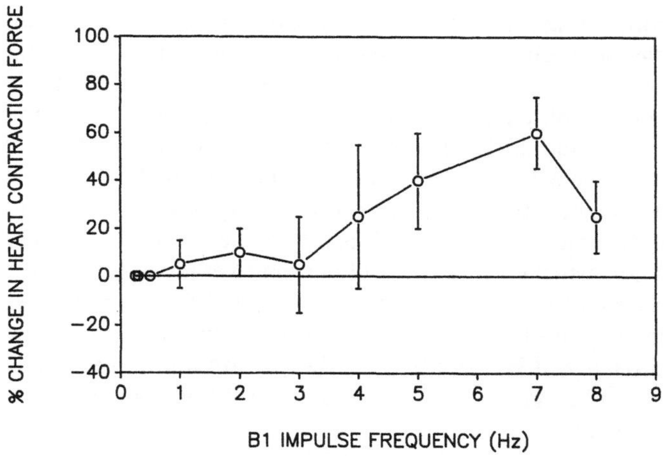

Fig. 13. Cumulative frequency response curve for the effect of B1 stimulation on heart contractile force in *Limax maximus.* Data are from nine heart/CNS preparations. Single B1 neurons were driven at each instantaneous spike frequency in each preparation. The number of action potentials was held constant at ten for each frequency. Changes in heart contractile force are expressed as a percentage of the prestimulation value within each preparation

(Prior and Delaney 1986; Welsford and Prior 1987). In isolated CNS and CNS-heart preparations it has been possible to demonstrate that activation of the right or left buccal neuron B1 by current injection results in increased burst frequency in the fast salivary burster (Fig. 12) and an increase in the force of contraction in the heart (Fig. 13). In both cases, current injection into B1 was adjusted so that the number of action potentials was maintained constant (ten) regardless of the impulse frequency. It is clear from the data in Figs. 12 and 13 that impulse frequencies of 5 to 7 Hz are most effective in activating both targets, the FSB and the heart.

Recent observations on the feeding motor program have revealed that the small cardioactive peptide, SCP_B, is effective in increasing the responsiveness of the feeding motor program to chemosensory stimuli (Prior and Watson 1988; Prior 1988). In this study it was noted that the large buccal neuron, B1, contained an SCP_B-like immunoreactive substance. It was known that the B1 had a major axonal projection that extended from the buccal ganglion through the cerebral ganglion and visceral ganglion nerve 11, which innervates the renocardiac region (Prior, unpublished observation). It had also been found that B1 becomes hyperpolarized in response to hyperosmotic stress (Phifer 1983). In this study it was found that the impulse frequency of B1 showed an average decline of 18% in response to hyperosmotic saline. These observations, together with those from studies of other multifunction, multitransmitter neurons (e.g., Rittenhouse and Price 1986; Lloyd et al. 1988; Willows et al. 1988; Lloyd and Willows 1988), suggest that the B1-feeding-cardiovascular system might serve as a useful model for the study of integration of the multiple osmoregulatory responses observed in gastropods.

In conclusion, our understanding of the effects of dehydration on hemolymph composition in gastropods far exceeds that which is known about the mechanisms controlling the responses to osmotic stress. Unquestionably, water regulatory behaviors such as contact-rehydration in terrestrial forms involve neuromodulatory systems. In addition, it would appear that the rate of urine formation can likewise be controllted by neurally-modulated changes in epithelial permeability of the kidney sac and pericardial sac and in heart contractility. These observations, together with the possibility that integumental permeability is neurally modulated, suggest that target sites for an osmoregulatory hormone in gastropods may well include both central neuronal networks and diverse epithelial structures. Thus, neuron to neuron and neuron to epithelial cell interactions may, in concert, represent the array of responses utilized for the maintenance of body hydration in slugs and snails. It would appear that an environmental neurobiological approach will be essential for the elucidation of the coordinated interactions between "behavioral" and "physiological" osmoregulatory mechanisms in the Gastropoda.

References

Adolph EF (1933) Exchanges of water in the frog. Biol Rev 8:224–240
Bablanian GM, Treistman SN (1983) Seawater osmolarity influences bursting pacemaker activity in intact *Aplysia californica*. Brain Res 271:342–345
Bablanian GM, Treistman SN (1985) The effect of hyperpolarization of cell R15 on the hemolymph composition of intact *Aplysia*. J Comp Physiol 155:297–303
Benson JA, Treherne JE (1978a) Axonal adaptation to osmotic and ionic stress in an invertebrate osmoconformer (*Mercierella enigmatica* Fauvel). II. Effects of ionic dilution on the resting and action potentials. J Exp Biol 76:205–219
Benson JA, Treherne JE (1978b) Axonal adaptation to osmotic and ionic stress in an invertebrate osmoconformer (*Mercierella eniqmatica* Fauvel) III. Adaptations to hypoosmotic dilution J Exp Biol 76:221–235
Bentley PJ (1971) Endocrines and osmoregulation. Springer, Berlin Heidelberg New York
Beres LS, Pierce SK (1981) The effects of salinity stress on the electrophysiological properties of *Mya arenaria* neurons. J Comp Physiol 144:165–173
Brown D, Grosso A, DeSousa RC (1980) Isoproterenol-induced intramembrane particle aggregation and water flux in toad epidermis. Biochem Biophys Acta 596:158–164
Brunn F (1921) Beitrag zur Kenntnis der Wirkung von Hypophysenextrakt auf den Wasserhaushalt des Frosches. Z Ges Exp Med 25:170–175
Burton RF (1964) Variations in the volume and concentration of the blood of the snail, *Helix pomatia* L., in relation to the water content of the body. Can J Zool 42:1085–1097
Burton RF (1965a) Sodium, potassium and magnesium in the snail, *Helix pomatia* L. Physiol Zool 38:335–342
Burton RF (1965b) Variations in the water and mineral contents of some organs of the snail, *Helix pomatia* L. Can J Zool 43:771–779
Burton RF (1965c) Relationships between the cation contents of slime and blood in the snail, *Helix pomatia* L. Comp Biochem Physiol 15:339–345
Burton RF (1966) Aspects of ionic regulation in certain terrestrial pulmonata. Comp Biochem Physiol 17:1007–1018
Burton RF (1968a) Ionic regulation in the snail, *Helix aspersa*. Comp Biochem Physiol 25:501–508
Burton RF (1968b) Ionic balance in the blood of pulmonata. Comp Biochem Physiol 25:509–516
Burton RF (1971a) Natural variations in cation levels in the blood of three species of land snail (Pulmonata: Helicidae). Comp Biochem Physiol A 39:267–275

Burton RF (1971b) Concentrations of cations in the blood of some terrestrial snails. Comp Biochem Physiol A 39:875–878

Burton RF (1983) Ionic regulation and water balance. In: Saleuddin ASM, Wilbur KM (eds) The mollusca, vol 5: Physiology, pt 2. Academic Press, New York London, pp 292–352

Camhi JM (1983) Neuroethology. Sinauer, Sunderland, MA

Carlson AD, Treherne JE (1969) The ionic basis of the fast action potentials in the isolated cerebro-visceral connective of *Anodonta cygnea*. J Exp Biol 51:297–318

Carlson AD, Treherne JE (1977) Ionic basis of axonal excitability in an extreme euryhaline osmoconformer, the serpulid worm, *Mercierella enigmatica* (Fauvel). J Exp Biol 67:205–215

Dainton BH (1954a) The activity of slugs. I. The induction of activity by changing temperatures. J Exp Biol 31:165–187

Dainton BH (1954b) The activity of slugs. II. The effect of light and air currents. J Exp Biol 31:188–197

DeSousa RC, Grosso A (1982) Osmotic water flow across the abdominal skin of the toad *Bufo marinus*: effect of vasopressin and isoprenaline. J Physiol 329:281–296

De With ND (1977) Evidence for the independent regulation of specific ions in the haemolymph of *Lymnaea stagnalis* (L). Proc Konigl Nederl Akad Wetensch Ser C80:144–157

De With ND (1978) The effects of starvation and feeding on the haemolymph in the freshwater snail *Lymnaea stagnalis*. Proc Konigl Nederl Akad Wetensch Ser C81:241–248

De With ND (1980) Water turn-over, ultrafiltration, renal water reabsorption and renal circulation in fed and starved specimens of *Lymnaea stagnalis* adapted to different external osmolarities. Proc Konigl Nederl Akad Wetensch Ser C 83:109–120

Deyrup-Olsen I, Martin AW (1982) Surface exudation in terrestrial slugs. Comp Biochem Physiol 72c:45–51

DiBona DR (1983) Cytoplasmic involvement in ADH-mediated osmosis across toad urinary bladder. Am J Physiol 245 (Cell Physiol 14):C297–C307

Fromter E, Diamond J (1972) Route of passive ion permeation in epithelia. Nature (London) 235:9–13

Furgal SM, Brownell PH (1987) Ganglionic circulation and its effects on neurons controlling cardiovascular functions in *Aplysia californica*. J Exp Zool 244:347–364

Gelperin A, Chang JJ, Reingold SC (1978) Feeding motor program in *Limax*. I. Neuromuscular correlates and control by chemosensory input. J Neurobiol 9:285–300

Geraerts WPM, De With ND, Roubos EW, Joosse J (1981) "Endocrine aspects of hydromineral regulation in molluscs". In: Farner S, Lederis K (eds) *Neurosecretion, molecules, cells, systems*. Plenum, New York, pp 337–347

Gilles R (ed) (1979) Intracellular organic osmotic effectors. In: Mechanisms of osmoregulation in animals (Maintenance of cell volume). John Wiley & sons, New York, pp 111–154

Greenaway P (1970) Sodium regulation in the freshwater mollusc *Limnaea stagnalis* (L.) (Gastropoda: Pulmonata). J Exp Biol 53:147–163

Grega DS, Prior DJ (1985) The effects of feeding on heart activity in the terrestrial slug, *Limax maximus*: central and peripheral control. J Comp Physiol A 156:539–545

Grega DS, Prior DJ (1986) Modification of cardiac activity in response to dehydration in the terrestrial slug, *Limax maximus*. J Exp Zool 237185–190

Grimm-Jørgensen Y (1978) Immunoreactive thyrotopin releasing factor in a gastropod: distribution in the central nervous system and hemolymph of *Lymnaea stagnalis*. Gen Comp Endocrinol 35:387–390

Grimm-Jørgensen Y (1979) Effect of thyrotropin releasing factor on body weight of the pond snail *Lymnaea stagnalis*. J Exp Zool 208:169–175

Grimm-Jørgensen Y (1980) Effect of thyrotropin-releasing hormone and ^{22}Na uptake by the pond snail, *Helisoma carabaceum*. J Exp Zool 212:471–473

Grimm-Jørgensen Y (1983) Possible Physiological Roles of Thyrotropin Releasing Hormone and a Seomatostatin-like Peptide in Gastropods. In: Lever J, Boer HH (eds) Molluscan Neuroendocrinology. Mon Royal Neth Acad Arts and Sci. North Holland Publishing Co. Amsterdam, Oxford, New York, pp 21–27

Grimm-Jørgensen Y, Connolly SM, Visser TJ (1984) Effect of thyrotropin-releasing hormone and its metabolites on the secretion of sulfated polysaccharides by foot integument of a pond snail. Gen Comp Endocrinol 55:410–417

Harrison FM, Martin AW (1965) Excretion in the cephalopod, *Octopus dofleini*. J Exp Biol 42:71–98
Hasegawa K, Takeda N (1986) Responses of giant neurons to osmotic stress in the giant african snail *Achatina fulica*. Comp Biochem Physiol 84A (2):289–294
Hays RM (1983) Alteration of luminal membrane structure by antidiuretic hormone. Am J Physiol 245 (Cell Physiol 14):C289–C296
Heller H (1941) Differentiation of an (amphibian) water balanace principle from the antidiuretic principle of the posterior pituitary gland. J Physiol 100:125–141
Hughes GM, Kerkut GA (1956) Electrical activity in a slug ganglion in relation to the concentration of Locke solution. J Exp Biol 33:282–294
Jones PG, Rosser SJ, Bulloch AGM (1988) Glutamate suppression of feeding and the underlying output of effector neurons in *Helisoma*. (in press)
Kandel ER (1976) Cellular basis of behavior. Freeman, San Francisco
Kerkut GA, Taylor BJR (1956) The sensitivity of the pedal ganglion of the slug to osmotic pressure changes. J Exp Biol 33:493–501
Khan HR, Saleuddin ASM (1979a) Effects of osmotic changes and neurosecretory extracts on kidney ultrastructure in the freshwater pulmonate *Helisoma*. Can J Zool 57:1256–1270
Khan HR, Saleuddin ASM (1979b) Osmotic regulation and osmotically induced changes in the neurosecretory cells of the pulmonate snaul *Helisoma*. Can J Zool 57:1371–1383
Khan HR, Saleuddin ASM (1981) Cell contacts in the kidney epithelium of *Helisoma* (Mollusca: gastropoda) – effects of osmotic pressure and brain extracts: a freeze-fracture study. J Ultrastruct Res 75:23–40
Koch VT, Koester J, Weiss KR (1984) Neuronal mediation of cardiovascular effects of food arousal in *Aplysia*. J Neurophysiol 51:126–135
Koester J, Mayeri E, Liebeswar G, Kandel ER (1973) Cellular regulation of homeostasis: neuronal control of circulation in *Aplysia*. Fed Proc 32:2179–2187
Koester J, Mayeri E, Liebeswar G, Kandel ER (1974) Neural control of circulation in *Aplysia*. II. Interneurons. J Neurophysiol 37:476–496
Koester J, Dieringer N, Mandelbaum DE (1979) Cellular neuronal control of molluscan heart. Am Zool 19:103–116
Kupfermann I, Weiss KR (1976) Water regulation by a presumptive hormone contained in identified neurosecretory cell R15 of *Aplysia*. J Gen Physiol 67:113–123
Lever J, Joosse J (1961) On the influence of the salt content of the medium on some special neurosecretory cells in the lateral lobes of the cerebral ganglia of *Lymnaea stagnalis*. Konigl Nederl Akad Wetensch Ser C64:631–639
Lewis S (1983) Control of Na^+ and water absorption across vertebrate 'tight' epithelia by ADH and aldosterone. J Exp Biol 106:9–24
Lloyd PE, Willows AOD (1988) Identified multiple transmitter neurons in *Tritonia*. II. Control of gut motility. J Neurobiol (in press)
Lloyd PE, Masinovsky BP, Willows AOD (1988) Multiple transmitter neurons in *Tritonia*: I. Biochemical studies. J Neurobiol (in press)
Machin J (1964a) The evaporation of water from *Helix aspersa* I. The nature of the evaporating surface. J Exp Biol 41:759–769
Machin J (1964b) The evaporation of water from *Helix aspersa*. II. Measurement of air flow and the diffusion of water vapour. J Exp Biol 41:771–781
Machin J (1964c) The evaporation of water from *Helix aspersa*. III. The application of evaporation formulae. J Exp Biol 41:783–792
Machin J (1966) The evaporation of water from *Helix aspersa*. IV. Loss from the mantle of the inactive snail. J Exp Biol 45:269–278
Machin J (1975) In: Fretter V, Peake J (eds) "Water relationships"; in *Pulmonates*, vol 1. Academic Press, New York London, pp 105–163
Makra ME, Prior DJ (1985) Angiotensin II can initiate contact-rehydration in terrestrial slugs. J Exp Biol 119:385–388
Mayeri E, Koester J, Kupfermann I, Liebeswar G, Kandel ER (1974) Neural control of circulation in *Aplysia*. I. Motoneurons. J Neurophysiol 37:458–475

Moore GJ, Thornhill JA, Gill V, Lederis K, Lukowiak K (1981) An arginine vasotocin-like neuropeptide is present in the nervous system of the marine mollusc *Aplysia californica*. Brain Res 206:213–218
Newell PF, Skelding JM (1973) Structure and permeability of the septate junction in the kidney sac of *Helix pomatia* L. Z Zellforsch 147:31–39
Parker HT, Pierce SK (1985) Comparative electrical properties of identified neurons in *Elysia chlorotica* before and after low salinity acclimation. Comp Biochem Physiol 82A:367–372
Phifer CB (1983) Dehydration-induced and osmotically-mediated modification of feeding behavior and its neural correlate in the slug *Limax maximus*. Ph D Diss, Univ Kentucky
Phifer CB, Prior DJ (1985) Body hydration and haemolymph osmolality affect feeding and its neural correlate in the terrestrial gastropod, *Limax maximus*. J Exp Biol 118:405–421
Phillips J, Lewis S (1983) Epithelia and cellular mechanisms in osmoregulation. J Exp Biol 106 (Rev vol)
Pickering BT, Heller H (1959) Chromatographic and biological characteristics of fish and frog neurohypophysial extracts. Nature (London) 184:1463–1464
Prior DJ (1981) Hydration-related behaviour and the effects of osmotic stress of motor function in the slugs. *Limax maximus* and *Limax pseudoflavus*. In: Salanki J (ed) Advances in physiological sciences, vol 23: *Neurobiology of invertebrates*. Pergamon, Oxford, pp 131–145
Prior DJ (1982) Osmotic control of drinking behavior in terrestrial slugs. Am Zool 22 (4):978
Prior DJ (1983) Hydration-induced modulation of feeding responsiveness in terrestrial slugs. J Exp Zool 227:15–22
Prior DJ (1984) Analysis of contact-rehydration in terrestrial gastropods: osmotic control of drinking behaviour. J Exp Biol 111:63–73
Prior DJ (1985) Water-regulatory behaviour in terrestrial gastropods. Biol Rev 60:403–424
Prior DJ (1988) Central actions of SCP_B in the gastropod mollusc, *Limax maximus*. Am Zool (in press)
Prior DJ, Delaney K (1986) Activation of buccal neuron B1 in the edible slug, *Limax maximus*, mimics the actions of exogenous SCP_B. Am Zool 26:126A
Prior DJ, Gelperin A (1977) Autoactive molluscan neuron: reflex function and synaptic modulation during feeding in the terrestrial slug, *Limax maximus*. J Comp Physiol 114:217–232
Prior DJ, Pierce SK (1981) Adaption and tolerance of invertebrate nervous system to osmotic stress. J Exp Zool 215:237–245
Prior DJ, Uglem GL (1984) Analysis of contact-rehydration on terrestrial gastropods. Absorption of ^{14}C-inulin through the epithelium of the foot. J Exp Biol II:75–80
Prior DJ, Watson WH (1988) The molluscan neuropeptide, SCP_B, increases the responsiveness of the feeding motor program of *Limax maximus*. J Neurobiol 19 (1):87–105
Prior DJ, Hume M, Varga D, Hess SD (1983) Physiological and behavioural aspects of water balance and respiratory function in the terrestrial slug, *Limax maximus*. J Exp Biol 104:111–127
Prosser CL (1973) Comparative animal physiology. Saunders, Philadelphia
Quinn RH, Pierce SK (1987) The relationship of solute efflux to hyposmotic depolarization in neurons from *Elysia chlorotica*. Am Zool 27 (4):131 A
Riddle WA (1983) Physiological ecology of land snails and slugs. In: Russell-Hunter WD (ed) The mollusca, vol 6. Ecology, pp 431–461 Academic Press, New York
Rittenhouse AR, Price CH (1986) Electrophysiological and anatomical identification of the peripheral axons and target tissues of *Aplysia* neurons R3–14 and their status as multifunctional, multimessenger neurons. J Neurosci 6:2071–2084
Robertson JD (1964) Osmotic and ionic regulation. In: Wilbur KM, Yonge CM (eds) Physiology of mollusca, vol 1, pp 283–311. Academic Press, New York
S.-Rozsa K (1979) Heart regulatory neural network in the central nervous system of *Achatina fulica* Ferussac (Gastropoda, Pulmonata). Comp Biochem Physiol 63A:435–445
S.-Rozsa K, Salanki J, Vero M, Koeacevic N, Konjevic D (1980) Neural network regulating heart activity in *Aplysia depilans* and its comparison with other gastropod species. Comp Biochem Physiol 65A (1):61–68
Salanki J, S.-Rozsa K (1973) Single neuron responses to tactile stimulation of the heart in the snail *Helix pomatia* L. J Comp Physiol 84:267–279

Saleuddin ASM, Farrell CL, Gomot L, Khan HR (1983) Relative humidity affects the intercellular spaces and cell contacts of the kidney epithelium of the terrestrial snail, *Helix aspersa* Muller. J Morphol 178:313–322
Sawada M, Blankenship JE, McAdoo DJ (1981a) Neural control of a molluscan blood vessel, anterior aorta of *Aplysia*. J Neurophysiol 46:967–985
Sawada M, McAdoo DJ, Blankenship JE, Price CH (1981b) Modulation of arterial muscle contraction in *Aplysia* by glycine and neuron R14. Brain Res 207:486–490
Sawyer WH (1951) Effect of posterior pituitary extract on permeability of frog skin to water. Am J Physiol 164:44–48
Sawyer W, Deyrup-Olsen I, Martin AW (1984) Immunological and biological characteristics of the vasotocin-like activity in the head ganglia of gastropod molluscs. Gen Comp Endocrinol 54:97–108
Schlichter LC (1981) Ion relations of haemolymph, palial fluid, and mucus of *Lymnaea stagnalis*. Can J Zool 59:605–613
Skelding JM (1973a) Studies on the renal physiology of *Achatina achatina* (L.). Malacologia 14:93–96
Skelding JM (1973b) The fine structure of the kidney of *Achatina achatina* (L.). Z Zellforsch 147:1–29
Soffe SR, Slade CT, Benjamin PR (1979) Environmental osmolarity and neurosecretory neurons in *Lymnaea stagnalis* (L.). Malacologia 18:583–586
Stinnakre J, Tauc L (1966) Effects de l'activation osmotique de l'osphradium sur les neurones du système nerveux central de l'Aplysie. J Physiol (Paris) 58:266–267
Swindale NV, Benjamin PR (1976) The anatomy of neurosecory neurons in the pond snail *Lymnaea stagnalis* L. Philos Trans R Soc London Ser B 274:169–202
Taylor A, Palmer LG (1982) Hormonal regulation of sodium chloride and water transport in epithelia. In: Goldberger RF, Yamamoto KR (eds) Biological regulation and development, vol 3A. Plenum, New York, pp 253–298
Treherne JE (1980) Neuronal adaptations to ionic and osmotic stress. Comp Biochem Physiol 67B:455–463
Treherne J, Pichon Y (1978) Long-term adaptation of *Sabella* giant axons to hyposmotic stress. J Exp Biol 75:252–263
Uglem GL, Prior DJ, Hess SD (1985) Analysis of contact-rehydration in terrestrial gastropods: estimation of the pore size and molecular sieving of the integumental paracellular pathway. J Comp Physiol 156:285–289
Van Weel PB (1957) Observations on the osmoregulation in *Aplysia juliana* Pease (Aplysiidae, Mollusca). Z Vergl Physiol 39:492–506
Wells GP (1944) The water relations of snails and slugs. III. Factors determining activity in *Helix pomatia L.* J Exp Biol 20:79–87
Welsford IG, Prior DJ (1987) The effect of SCP_B application and buccal neuron, B1, stimulation on heart activity in the slug, *Limax maximus*. Am Zool 27 (4):138A
Wendelaar Bonga SE (1970a) Investigations on neurosecretion in the central and peripheral nervous system of the pulmonate snail *Lymnaea stagnalis*. In: Bargmann W, Scharrer B (eds) Aspects of neuroendocrinology. Springer, Berlin Heidelberg New York, pp 43–46
Wendelaar Bonga SE (1970b) Ultrastructure and histochemistry of neurosecretory cells and neurohaemal areas in the pond snail *Lymnaea stagnalis*. Z Zellforsch 108:190–224
Wendelaar Bonga SE (1971) Osmotically induced changes in the activity of neurosecretory cells located in the pleural ganglia of the freshwater snail *Lymnaea stagnalis* (L.). Studied by quantitative electron microscopy. Neth J Zool 21:127–158
Wendelaar Bonga SE (1972) Neuroendocrine involvement in osmoregulation in a freshwater mollusc, *Lymnaea stagnalis*. Gen Comp Endocrinol [Suppl] 3:308–316
Willmer PG (1978a) Volume regulation and solute balance in the nervous tissue of an osmoconforming bivalve (*Mytilus edulis*). J Exp Biol 77:157–179
Willmer PG (1978b) Electrophysiological correlates of ionic and osmotic stress in an osmoconforming bivalve (*Mytilus edulis*). J Exp Biol 77:181–205
Willmer PG (1978c) Sodium fluxes and exchange pumps: further correlates of osmotic conformity in the nerves of an estuarine bivalve (*Mytilus edulis*). J Exp Biol 77:207–223
Willows AOD, Lloyd PE, Masinovsky BP (1988) Multiple transmitter neurons in *Tritonia*. III. Modulation of a central pattern generator controlling feeding. J Neurobiol (in press)

Chapter 2

Chemosensory Systems in Lower Organisms: Correlations with Internal Receptor Systems for Neurotransmitters and Hormones

William E. S. Carr, Richard A. Gleeson, and *Henry G. Trapido-Rosenthal*[1]

Contents

1 Introduction

The presence of specific chemical substances in the environment evokes behavioral responses in many lower eukaryotic organisms such as protozoans and invertebrates (for reviews, see Mackie and Grant 1974, Gleeson 1978, Atema 1985, Carr 1987). Beginning with Haldane (1954), investigators have noted that several of the chemical substances evoking these responses in lower life forms include some of the very same compounds that function internally as neurotransmitters, modulators, or hormones (see Table 1). Hence, substances such as acetylcholine and the adenine nucleotides may evoke behavioral responses when present in an external aquatic environment just as they evoke physiological responses when present in a synaptic cleft.

Multicellular organisms possess many different neurotransmitters and hormones that serve as chemical signals coordinating the activities of various cell

[1] The Whitney Laboratory and Department of Zoology, University of Florida, 9505 Ocean Shore Boulevard, St. Augustine, FL 32086, USA

Advances in Comparative and Environmental Physiology, Vol. 5

Table 1. Neuroactive substances evoking behavioral responses in lower life forms

Neuroactive agent	Molecular structure	Organism responding	Behavior affected	References[a]
Acetylcholine	$CH_3\overset{O}{\overset{\|}{C}}-OCH_2CH_2N(CH_3)_3$	Dinoflagellate	Light sensitivity	1
		Ciliate protozoan	Swimming	2
Adenosine	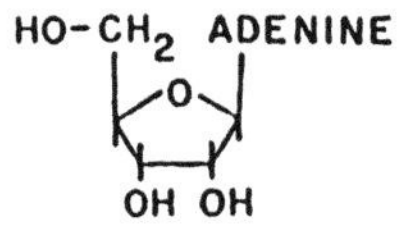	Army worm	Feeding	3
AMP[b]	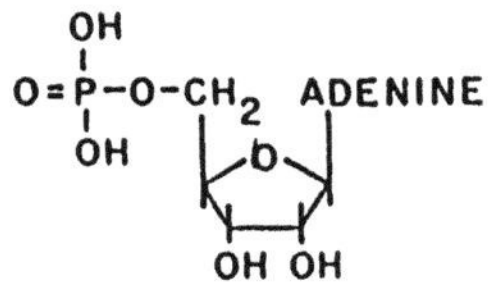	Shrimp	Attraction	4
ATP[b]	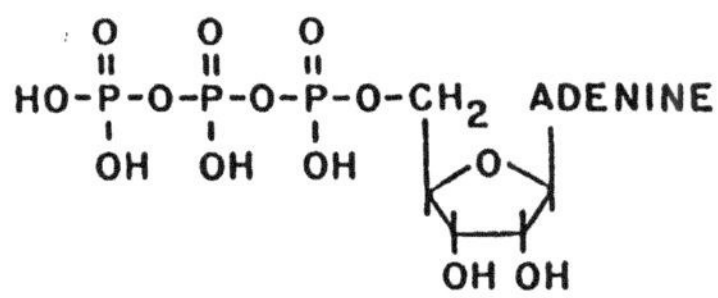	Assassin bug	Liquid ingestion	5
DOPA[c]	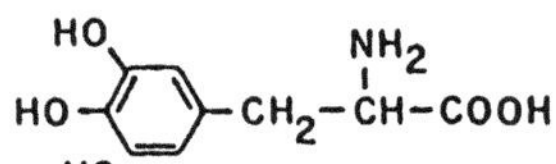	Dinoflagellate	Light sensitivity	1
		Oyster larva	Settlement and metamorphosis	6
Epinephrine	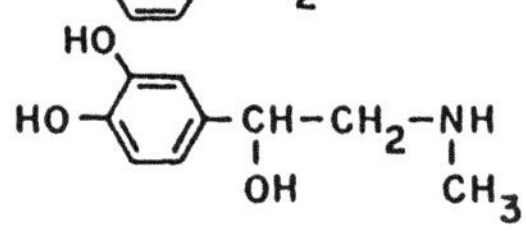	Ciliate protozoan	Glucose uptake	7
GABA[c]	$H_2N-CH_2-CH_2-CH_2-COOH$	Abalone larva	Settlement and metamorphosis	8
		Beetle larva	Feeding	9
Glutamate	$HOOC-CH_2-CH_2-\underset{NH_2}{\underset{\mid}{C}}H-COOH$	Colonial alga	Induces sexual stage	10
		Barnacle	Feeding	11
Glycine	H_2N-CH_2-COOH	Mud snail	Proboscis extension	12
Taurine	$H_2N-CH_2-CH_2-SO_3H$	Spiny lobster	Attraction	13
Thyroxine	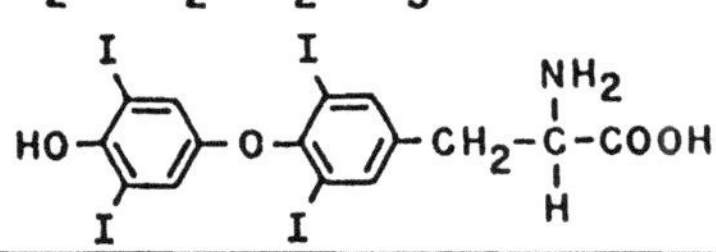	Ciliate protozoan	Phagocytosis	14

[a] References: *1* Forward 1977; *2* Doughty 1978; *3* Ma 1977; *4* Carr and Thompson 1983; *5* Friend and Smith 1982; *6* Coon et al. 1985; *7* Csaba 1980; *8* Morse et al. 1979; *9* Mitchell 1985; *10* Starr et al. 1980; *11* Allison and Dorsett 1977; *12* Carr 1967; *13* Fine-Levy et al. 1987; *14* Csaba et al. 1977.

[b] Additional examples of organisms with chemoreceptors for purine nucleotides are given in Table 2 (Sect. 4).

[c] DOPA = dihydroxyphenylalanine; GABA = γ-aminobutyric acid.

types. With most of these signal molecules, interaction with the target cell occurs without the molecule itself actually entering the interior of the cell (see elegant description of chemical signaling by Berridge 1985). Instead, most signal substances bind to receptor molecules present on the outer surface of the cell membrane. This binding event then activates a transduction process that provides the link between the chemical signal and the initiation of an appropriate cellular response.

Figure 1 depicts two generalized schemes whereby information about a signal molecule is transduced across a cell membrane. Whereas the first step in each scheme involves the binding of the signal molecule to a receptor binding site on the outer membrane surface, the transduction processes of the two systems are very different. In one scheme, the receptor has a transmembrane organization in which an ion channel is an integral part of the receptor molecule (Fig. 1 A). In this system, known for acetylcholine (nicotinic), $GABA_A$, and glutamate receptors, the binding of the signal molecule to the receptor directly influences the electrical properties of the membrane by affecting the movement of ions through a "ligand-gated" ion channel (e.g., McCarthy et al. 1986; Barnard et al. 1987; Changeux and Rivah 1987). In the second scheme (Fig. 1 B), known for epinephrine, acetylcholine (muscarinic), serotonin, and several other receptor systems (e.g., Berridge 1985; Levitzki 1986), an ion channel is not an integral part of the receptor molecule. In this case, the binding of the signal molecule to the receptor triggers a cascade of transduction events that begin with structural changes in a family of membrane proteins, called guanine nucleotide-binding proteins (G-proteins). Conformational changes in the G-proteins induce changes in the structure of an enzyme that enhances or inhibits the production of one of a small family of "second messenger" substances such as cyclic AMP, cyclic GMP, inositol trisphosphate, or diacylglycerol. The second messenger may directly affect a cellular response (e.g., Nakamura and Gold 1987) or, more commonly, it may activate an additional enzyme such as a protein kinase. The activated kinase utilizes ATP to phosphorylate a cellular component, for example, an ion channel, which in turn may contribute to ion movements leading to an overt response (e.g., depolarization, hyperpolarization, secretion, or contraction). Although there are some notable exceptions, an important revelation that has stemmed from the characterization of the mechanism of action of many transmitters and hormones is an awareness that many (most?) of the receptors for these signal molecules share major elements of one or the other of the transduction schemes depicted in Fig. 1 (for recent reviews see Levitzki 1986, Barnard et al. 1987, Gilman 1987, Dolphin 1987).

The procedures of biochemistry, physiology, and molecular pharmacology, long used in studies with internal signal molecules, are now being applied in studies of external chemosensory systems. Consequently it has become apparent that the analogies between the substances serving as behavioral stimulants in lower organisms, and also as transmitters or hormones in various organisms (see Table 1), are more than just anecdotal. Indeed, recent studies have revealed that external chemosensory systems of lower organisms are very much like the internal signaling systems present at synapses and other internal interfaces where specific chemicals regulate the activities of neighboring cells. This review will show that the par-

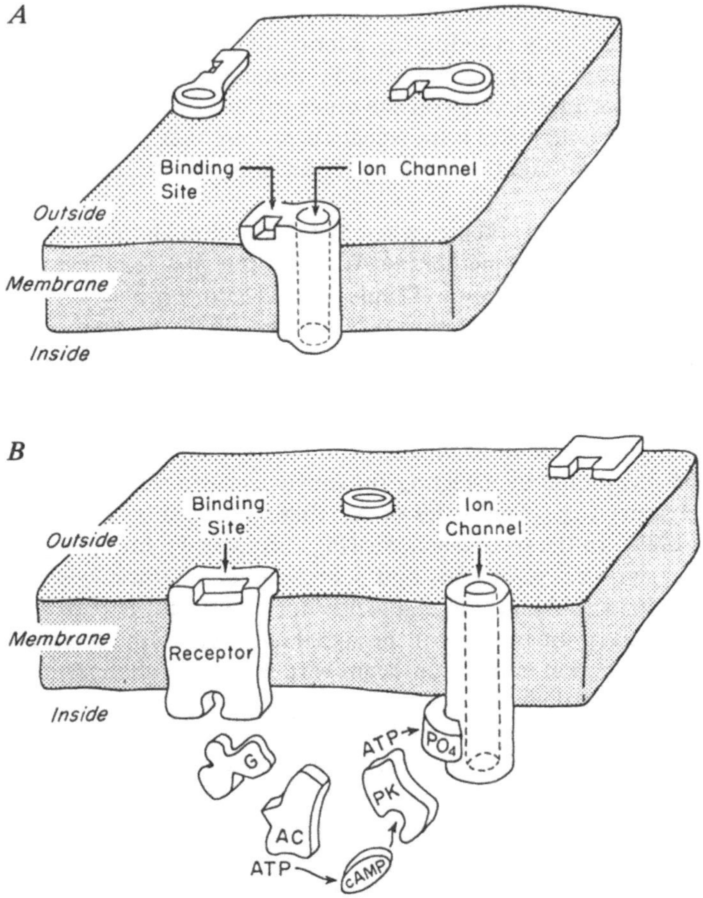

Fig. 1A, B. Generalized schemes for the binding and transduction of extracellular chemical signals. *A* Receptor with a transmembrane organization that includes an ion channel as an integral part of the receptor molecule. *B* Receptor molecule that is coupled to other membrane proteins functioning as transduction components. The components depicted in this scheme are: G-proteins (*G*), the enzyme, adenylate cyclase (*AC*), a second messenger, cyclic adenosine 3′ : 5′-monophosphate (*cAMP*) and the enzyme protein kinase (*PK*) which phosphorylates (PO_4) an ion channel or another cellular protein. Alternative transduction components are known to involve a phospholipase enzyme that hydrolyzes a membrane phospholipid to yield the second messengers inositol trisphosphate and diacylglycerol. Further details and references are given in the text

allels between external and internal "chemosensory" systems span the following tiers of complexity:

1. Receptors may be activated by identical molecules;
2. Receptors exhibit similar structure-activity relationships;
3. Similar molecules and mechanisms contribute to transduction:
   A) Involve G-proteins;
   B) Employ second messengers;
   C) Control ion channels;

4. Similar processes are employed to remove stimulatory molecules from the receptor environment;
5. Receptor activity can be modulated;
6. Some signal molecules are synthesized and stored in an inactive form;
7. Amino acid sequences of receptor proteins indicate evolutionary relationships.

This review does not use an exhaustive phyletic approach. Rather, we have selected a series of in-depth studies of chemosensory systems in lower eukaryotic organisms that permit us to make clear comparisons between aspects of these systems and those of analogous internal signaling systems. General reviews on chemoreception in lower organisms are available as follows: bacteria (Ordal 1985; Kleene 1986), algae (Maier and Muller 1986), slime molds and other protozoa (van Houten and Preston 1987), coelenterates (Lenhoff 1981), nematodes (Zukkerman and Jansson 1984), mollusks (Croll 1983), crustaceans (Ache 1982) and insects (Kaissling 1987).

2 Chemosensory Systems of Slime Molds, Protozoa, and Yeast

Features of the chemosensory systems of certain unicellular organisms prompted van Houten and Preston (1987) to note that "unicells might be considered as prototype receptor cells whose basic sensory mechanisms have been conserved through evolution" (see also Hinrichsen and Schultz 1988). As shown below, this statement applies not only to the receptor types and transmembrane signaling mechanisms employed by unicellular organisms, but also to the manner in which certain signal molecules are synthesized.

2.1 Cellular Slime Molds: Chemosensory System for Cyclic AMP

2.1.1 Background

Several species of cellular slime molds live in fertile soil where they have a life cycle consisting of both unicellular and multicellular stages (e.g., Bonner 1983). The unicellular stage is represented by free-living amoebae that feed on bacteria and multiply by mitosis. When the bacterial population is depleted, starved amoebae are stimulated to aggregate into a multicellular, sluglike mass which eventually differentiates into a stalk and a spore-producing body. In *Dictyostelium discoideum*, the most thoroughly studied species, it is now well known that the aggregation of individual amoebae into a multicellular mass is induced by a chemotactic response to cyclic adenosine 3′:5′-monophosphate (cAMP) that is released into the external environment by neighboring amoebae (ibid). Two major points will be made in the following description of the signaling system mediating this response:

(1) The amoebae have a chemosensory system exhibiting major similarities to the signaling systems present in the internal tissues of higher organisms, and (2)

the life history of the slime mold exhibits a succession of events wherein the external signaling system utilized by the unicellular amoebae becomes an internal communication system for the multicellular aggregate. For reviews of slime mold chemotaxis, see Frazier et al. (1984), McRobbie (1986), Gerisch (1987).

2.1.2 The cAMP Receptor

Slime mold amoebae detect cAMP with receptors on the cell surface since the signal molecule itself is unable to pass through the cell membrane (Moens and Konijn 1974). The cAMP receptor is highly specific and most changes in the adenine, ribose, or phosphate moieties result in marked decreases in chemotactic activity (Mato et al. 1978). The results of receptor binding studies with an array of analogs led to the development of a model depicting how the cAMP molecule is bound to the receptor by two hydrogen bonds and a hydrophobic interaction with the adenine moiety (Fig. 2) (van Haastert and Kien 1983). Binding studies also provided evidence that the surface receptor for cAMP is distinct from both a surface phosphodiesterase and an intracellular cAMP-dependent phosphokinase (ibid; Klein et al. 1985).

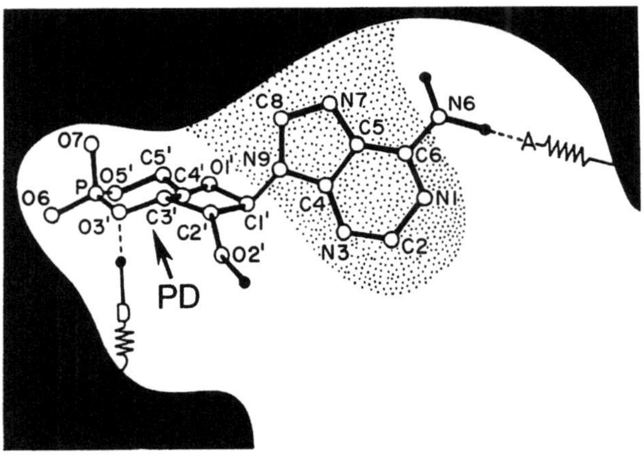

Fig. 2. Hypothetical model of the binding of cAMP to a surface receptor of *D. discoideum*. It is proposed that the receptor protein binds the cAMP molecule via hydrogen bonds with the oxygen (O-3′) of the ribose diester moiety and the hydrogen at N-6 of the adenine moiety. The adenine moiety is also bound in a hydrophobic cleft of the receptor (shown by *stippling*). *Arrow* indicates the site at which the phosphodiesterase (*PD*) hydrolyzes the bond between C-3′ and O-3′ (Receptor model from van Haastert and Kien 1983)

2.1.3 Transduction of the cAMP Signal

Activation of the cAMP receptors on the surface of the amoebae results in the immediate interaction of the receptors with a G-protein (= N-protein) present in the cell membrane (van Haastert 1984). Receptor activation with cAMP in intact amoebae rapidly converts the cAMP receptor from a high affinity state to a low affinity state. In isolated membrane preparations, this change in receptor affinity can be induced by incubations with cAMP plus GTP. Van Haastert (1984) has noted that the interactions of cAMP, GTP, and the cAMP receptor are quite similar to the interactions of adrenergic agonists with the β-adrenergic receptor, GTP, and the membrane-coupled G-protein in mammalian cells (e.g., see Levitzki 1986).

It has been known for several years that stimulation of slime mold amoebae by cAMP results in a rapid increase in the intracellular concentration of cGMP (e.g., Mato et al. 1977; Wurster et al. 1977). This increase in cGMP is thought to be due to the activation of guanylate cyclase by the receptor-activated G-protein (van Haastert et al. 1986). Evidence for the involvement of cGMP as a second messenger in transduction stems primarily from the close correlation existing between the binding affinity of cAMP analogs to the membrane receptor, and the potency of the same analogs as inducers of cGMP production (van Haastert and Kien 1983). Intracellular proteins that bind cGMP with high affinity have been identified, but their role in transduction has not as yet been elucidated (van Haastert et al. 1982). Models summarizing the marked analogies between receptor-mediated cyclase activation in both mammals and slime molds are provided by van Haastert et al. (1986).

2.1.4 Inactivation of the cAMP Signal

The residence time of cAMP molecules in the extracellular space around slime mold amoebae is governed by phosphodiesterase (PD) enzymes. These enzymes inactivate the signal molecule by hydrolysis at the 3′-position of the ribose diester moiety to yield the inactive product, 5′-AMP (see Fig. 2) (e.g., Gerisch et al. 1974). Two major types of PD occur; one type is bound to the outer membrane surface, and the other is released to the external environment as a soluble enzyme. The activity of the soluble PD is regulated by a protein inhibitor (Franke and Kessin 1981). Studies have shown that both types of PD enzymes make important contributions to the chemotactic process wherein oscillating cAMP signals are utilized to direct aggregation of the amoebae. By inactivating signal molecules, the enzymes function to steepen the cAMP gradient at the immediate cell surface and to decrease the signal to noise ratio in the extracellular environment (Nanjundiah and Malchow 1976; Darmon et al. 1978).

2.1.5 Persistence of the cAMP Signaling System in the Multicellular Stage

In the life cycle of *D. discoideum* the signaling system for cAMP does more than just mediate the aggregation of unicellular amoebae to form the multicellular slug. The same system also persists in the multicellular stage in which cAMP is secreted internally and functions, together with other cellular secretions, to induce cell movements and various cell-cell interactions. These interactions lead to the organization and differentiation of the specific stalk and spore cell types that appear during this stage (e.g., Kay 1983; Schaap and Wang 1986; Gerisch 1987). Hence, the complete cAMP signaling system, including specific receptors on the membrane surface, the transduction components within the membrane, and the inactivating phosphodiesterase enzymes, occurs not only as an external chemosensory system in the unicellular stage of *D. discoideum*, but also in the subsequent developmental stage, the multicellular stage, as a primitive internal "transmitter" system.

2.2 Protozoa and Yeast: Chemosensory Systems for Acetylcholine, Catecholamines, and Peptides

2.2.1 Background

Although not as intensively studied as the slime molds, the chemosensory systems of other unicellular organisms also involve receptors and membrane transduction processes exhibiting marked parallels to the internal signaling systems of higher organisms. For example, it is known that the membrane of the ciliated protozoan *Paramecium* displays an electrical excitability quite like that of a nerve cell (Eckert and Brehm 1979; Naitoh 1982; Hinrichsen and Schultz 1988). When exposed to certain chemotactic agents which are behavioral attractants, *Paramecium* shows a membrane hyperpolarization; conversely, a membrane depolarization is evoked by agents that are behavioral repellents (van Houten 1979; van Houten and Preston 1987). These opposite effects on membrane potential are due to the selective opening or closing of specific ion channels and appear to be analogous to the hyperpolarizing or depolarizing effects of inhibitory and excitatory transmitters, respectively, upon postsynaptic membranes.

In this section, the focus will be upon two topics: (1) evidence for the occurrence of receptor systems for acetylcholine and catecholamines on the membranes of unicellular flagellates and ciliates, and (2) peptide mating factors in yeast that, like many bioactive peptides in internal tissues, are produced by the hydrolysis of an erstwhile inactive protein.

2.2.2 Protozoan Receptors for Acetylcholine and Catecholamines

Two major types of receptors have been described for the neurotransmitter acetylcholine (ACh). These cholinergic receptors are classified as muscarinic and nicotinic receptors because of their selective stimulation by the plant alkaloids muscarine and nicotine (Weiner and Taylor 1985). An enzyme, acetylcholinesterase, serves to inactivate ACh by hydrolyzing the ester bond. Two major types of receptors also occur for the catecholamines. These receptors are classified as α- and β-adrenergic receptors on the basis of the relative effectiveness of epinephrine, norepinephrine, and isoproterenol (ibid). For vertebrate organisms, a series of selective agonists and antagonists have been identified for the cholinergic and adrenergic receptor types. Some of these chemical agents have been employed in studies with protozoans to show that particular behavioral responses are affected by surface chemoreceptors with marked similarities to cholinergic or adrenergic receptor types.

In the ciliate *Paramecium aurelia* a reversal of ciliary motion and a subsequent reversed swimming direction is evoked when defined concentrations of K^+ are applied externally (Doughty and Dodd 1978; Doughty 1978). Aspects of this ciliary reversal response are potentiated by ACh and nicotine, but not by muscarine, thereby indicating the response may be mediated by a surface nicotinic receptor. The acetylcholinesterase inhibitor neostigmine, like ACh itself, potentiates the ciliary reversal response evoked by K^+. In the absence of K^+, defined concentrations of either ACh, nicotine, neostigmine, or the ACh analog butylcholine, also evoke behavioral responses leading to the attraction and aggregation of these unicells (Doughty 1979). The aggregation response is inhibited by d-tubocurarine, a specific nicotinic receptor antagonist. It is uncertain whether the membrane cholinergic system of *Paramecium* is adapted to detect ACh secreted externally by the individual itself, or ACh released by conspecifics or other species (Doughty 1978).

In the dinoflagellate *Gymnodinium splendens*, the light sensitivity of a behavioral phototactic response is affected in an antagonistic manner by externally applied catecholamines and cholinergic agents (Forward 1977). The catecholamines 3,4-dihydroxyphenylalanine (L-DOPA) and 3,4-dihydroxyphenylethylamine (dopamine) cause a *decrease* in light sensitivity, whereas the opposite effect is caused by the catecholamine blocking agents dichloroisoproterenol, propranolol, and dibenzyline. Conversely, the light sensitivity of *G. splendens* is *increased* by ACh and by the acetylcholinesterase inhibitor eserine; light sensitivity is decreased by atropine, a blocking agent for muscarinic cholinergic receptors. Although considerable evidence supports the occurrence of functional catecholamine and cholinergic systems in dinoflagellates, the biological significance of these antagonistic systems is not clear. However, they may function to mediate intra- or interspecific cues affecting light and dark adaptation (ibid).

Growth of the parasitic protozoan *Trypanosoma cruzi* is affected by the β-adrenergic agents epinephrine and isoproterenol (Oliveira et al. 1984). The receptors mediating growth are surface receptors whose stimulation results in the activation of adenylate cyclase and a subsequent increase in the cAMP content of the cells. This effect is reversed by the β-adrenergic blocker propranolol, but not by the α-adrenergic blocker phentolamine. Recent radioligand binding studies with the β-adrenergic antagonist dihydroalprenolol (DHA) provide further evidence for the existence of functional β-adrenergic receptors in *T. cruzi* (De Castro and Oliveira 1987). Bound ^{3}H-DHA is displaced by the β-adrenergic ligands propranolol, alprenolol, and epinephrine but not by the α-adrenergic ligands phentolamine or catechol. Since *T. cruzi* is an obligatory parasite that lives its entire life in close contact with host cells and their products, it is proposed that the β-adrenergic system may function in providing the parasite with information about either the host itself, or the population of conspecifics living within the host environment (ibid).

Additional reports of externally applied neuroactive substances affecting behavioral responses of protozoans include those of Hauser et al. (1975), Csaba (1980), van Houten et al. (1981), Brown and Kerkut (1981), Levandowsky et al. (1984), Leick and Hellung-Larsen (1985), and Hellung-Larsen et al. (1986).

2.2.3 Peptide Mating Factors in Yeast

In mammals, most biologically active peptide hormones such as oxytocin, vasopressin, ACTH, and many others are initially synthesized as larger inactive protein molecules (proproteins) that are only later cleaved and otherwise modified at particular loci to yield the active peptide products (see Gainer et al. 1985). The production of biologically active peptides from inactive proproteins is not, however, unique to higher organisms. As described below, an identical process occurs in the production of peptide mating factors by yeast (see Sprague et al. 1983 for review).

The unicellular yeast *Saccharomyces cerevisiae* produces specific peptides that are released externally and function to attract and induce fusion of opposite mating types to yield the diploid stage of the life cycle. The active yeast mating factors, termed α-factor and a-factor, are peptides of 13 and 12 amino acids, respectively (Kurjan and Herskowitz 1982). Multiple copies of the α-factor are synthesized as part of a putative proprotein consisting of 165 amino acid residues. Each proprotein molecule is later cleaved to provide four copies of the α-factor (ibid) which is then released externally to function as a mating pheromone. Though studies of synthesis of the a-factor are less complete, it is also presumed to be produced from a proprotein.

In addition to the marked similarities between the manner of synthesis of external mating factors in yeast and bioactive peptides in higher organisms, other major analogies exist in the signaling systems. The two identified mating factors of yeast (α-factor, a-factor) activate distinctly different surface receptors that are, however, coupled to functionally identical and interchangeable membrane transduction systems (Bender and Sprague 1986; Herskowitz and Marsh 1987). Recent studies indicate that these transduction systems involve a G-protein that may be analogous to G-proteins associated with transduction systems of higher organisms (Herskowitz and Marsh 1987; Dietzel and Kurjan 1987).

The production of a developmental pheromone by the colonial alga *Volvox capensis* occurs by a process somewhat similar to that used to produce the peptide mating factors in yeast. Each *Volvox* colony consists of both somatic and reproductive cells that are held together by a gelatinous matrix composed of glycoprotein. Reproductive cells within the matrix may divide to produce either asexual or sexual daughter colonies. Sexual daughter colonies are produced preferentially when the amino acid glutamate appears at a sufficient concentration in the external environment. Glutamate molecules serving this signal function are released (= "activated") when the glycoprotein in the matrix is hydrolyzed by enzymes associated with daughter colonies escaping from the parental colony (Starr et al. 1980; Maier and Muller 1986). This process whereby active signal molecules are cleaved from an inactive "proprotein" is hence very similar to the process used to produce bioactive peptides in forms as disparate as yeast and mammals.

3 Marine Larvae: Factors Affecting Settlement and Metamorphosis

For the free-swimming larvae of many marine invertebrates, specific environmental signals are used to activate a sequence of developmental changes culminating in settlement and metamorphosis to a benthic form. These specific signals are frequently chemical in nature (Crisp 1974; Chia 1978; Hadfield 1978; Burke 1983; Morse 1985). Further, the metamorphic events induced by these exogenous chemicals are often mediated by the larval nervous system. The ability of specific chemicals to initiate metamorphosis suggests that the interface between the exogenous signals and the larval nervous system is provided by chemoreceptors (Chia and Koss 1982; Burke 1983). Moreover, it has been proposed that these chemoreceptors bear an ancestral relationship to some of the internal receptors for neurotransmitters (Lenhoff and Heagy 1977; Morse et al. 1979; Morse 1985). The interaction of chemical inducers with the chemosensory systems of larval organisms thus provides another interesting example of analogy between internal and external chemical signaling systems.

3.1 Natural Inducers of Larval Metamorphosis

In cases where the exogenous chemicals inducing metamorphosis are known, it has been shown that these chemicals bear some logical relationship to the environment of the larval or adult form of the organism. For example, larvae of the sand dollar *Dendraster excentricus* metamorphose in response to a peptide of about 1000 daltons that is released into sand beds by adults of this species (Burke 1984). Larvae of the gregarious, reef-building tube worm *Phragmatopoma californica* metamorphose in response to compounds present in the cementing material used by conspecific adults to build their tubes (Jensen and Morse 1984; Pawlik 1986). Larvae of the nudibranch *Phestilla sibogae* are induced to undergo metamorphosis by a low molecular weight (300 to 500 daltons), water-soluble material released from the coral *Porites compressa* that is fed upon by the adult nudibranch (Hadfield 1977, 1978; Hadfield and Scheuer 1985). In the case of larvae of the red abalone *Haliotis rufescens* the natural inducer of metamorphosis is normally encountered at the surface of a crustose red alga such as *Lithothamnium californicum*, which provides both food and camouflage for the juvenile stage of this gastropod mollusk (Morse et al. 1979; Morse and Morse 1984). For abalone larvae, the inducing compound is a peptide (600 to 1000 daltons MW) found in association with the algal phycobiliproteins (Morse and Morse 1984).

3.2 The Involvement of the Larval Nervous System in Metamorphosis

As described below, several lines of evidence indicate that the larval nervous system is directly involved in the process of metamorphosis. (1) Neuroactive compounds can induce all or part of the metamorphic programs in larvae of several species. For example, larvae of the polychaete worm *P. californica* can be induced to settle by DOPA (Jensen and Morse 1984), as can those of the bivalve mollusk *Crassostrea gigas* (Coon et al. 1985). Larvae of the nudibranch *P. sibogae* can be induced to undergo metamorphosis by choline, a compound that, in addition to being a precursor of the neurotransmitter acetylcholine, also leads to an increase in the synthesis and release of catecholamines (Hirata and Hadfield 1986). Furthermore, larvae of the abalone *H. rufescens* and several other molluscan species can be induced to metamorphose into their benthic form by the neurotransmitter γ-aminobutyric acid (GABA), or by GABA analogs (Morse et al. 1979, 1980a, b; Morse 1985; Trapido-Rosenthal and Morse 1986a, b). (2) The response of larvae to inducing compounds can be modulated by mechanisms very similar to those that modulate the responsiveness of neurons to neurotransmitters. For example, the response of abalone larvae to inducing GABA-mimetic compounds can be facilitated by diamino acids (Trapido-Rosenthal and Morse 1985, 1986a; Baxter and Morse 1987) in a manner that is similar to the way in which neurally mediated responses to GABA can be facilitated by compounds that are not themselves GABA analogs (Haefely et al. 1979; Choi et al. 1981). Also, larvae can be habituated or desensitized to inducing compounds (Hadfield and Scheuer 1985; Trapido-Rosenthal and Morse 1986a, b), just as a variety of neurons can be habituated to their stimuli (Posner et al. 1982; Borden et al. 1984). (3) Transduction of the GABA-triggered signal in abalone larvae appears to involve both a depolarizing transmembrane ion flux (Baloun and Morse 1984; Morse 1985) and an intracellular second messenger (Trapido-Rosenthal and Morse 1986b; Baxter and Morse 1987; see Sect. 3.3). Both of these phenomena are also known to be involved in signal transduction in neurons (Berridge 1985; Barnard et al. 1987). (4) Modifications of the ionic environment that cause depolarization of neurons and other cells (Hodgkin and Horowicz 1959; Hagiwara et al. 1961) also initiate the metamorphosis of both molluscan and polychaete larvae (Baloun and Morse 1984; Yool et al. 1986; see Sect. 3.3). (5) Finally, the direct electrical stimulation of parts of the nervous system of larval sand dollars induces metamorphosis (Burke 1983).

3.3 Larval Chemoreceptors with Properties of GABA Receptors

In mammals, two classes of GABA receptors, termed $GABA_A$ and $GABA_B$, have been described. $GABA_A$ receptors have ligand-gated channels for chloride ions (Olsen et al. 1984; Barnard et al. 1987), whereas $GABA_B$ receptors are coupled

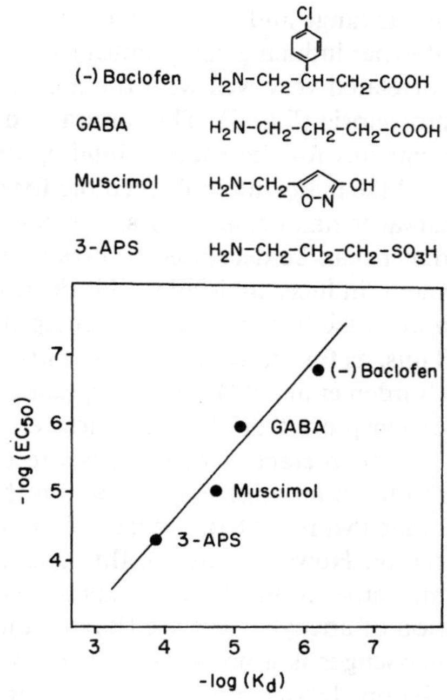

Fig. 3. Correlation of biological activity with receptor affinity. The biological activity of the indicated metamorphosis-inducing molecules correlates well (correlation coefficient = 0.97) with the affinity of the receptor for these molecules. Structures of the molecules (−)baclofen, GABA, muscimol, and 3-aminopropane sulfonic acid (*APS*) are shown at the *top* of the figure. The dissociation constants (K_d) were determined by using these compounds to compete for the specific binding of ^{3}H(−)baclofen. Molar concentrations of these compounds capable of inducing a metamorphic response in 50% of the larvae in a bioassay are defined as EC_{50} values (Trapido-Rosenthal and Morse 1986b)

to a transduction system involving G-proteins and a second messenger (Hill 1985; Watling and Bristow 1986; Dolphin and Scott 1987). Other distinguishing features of these receptor classes are cited below in conjunction with a description of the receptors mediating the metamorphosis of abalone larvae.

The characterization of receptors for compounds inducing larval metamorphosis is most advanced for receptors of the red abalone *H. rufescens*. In this organism metamorphosis is induced by the interaction of molecules with receptors that exhibit a high degree of stereochemical specificity. GABA, a simple four-carbon compound with a carboxylic acid group at one end and an amino group at the other, is capable of inducing the complete metamorphic response. However, any change in the number of carbons separating the two functional groups or any deletions or alterations of these functional groups decreases the activity of the molecule (Morse et al. 1980b). The high degree of stereospecificity is perhaps best exemplified by the response to the GABA analog baclofen (*β*-para-chlorophenyl-GABA). This synthetic compound has a bulky side chain on the *β*-carbon atom (see Fig. 3) that restricts the alternative positions available to the amino and carboxylic ends of the molecule. The (−) isomer of baclofen is an agonist at the $GABA_B$ class of GABA receptors in the mammalian CNS (Bowery et al. 1980, 1983). Most importantly, (−)baclofen is an even more potent inducer of larval metamorphosis than GABA itself, whereas the stereoisomer (+)baclofen is several orders of magnitude less potent than GABA (Trapido-Rosenthal and Morse 1986b).

The interaction of tritiated (−)baclofen with competent larvae of *H. rufescens* was characterized by receptor binding studies that revealed a K_D in the micromolar range and a B_{max} of 15 fmol per larva (ibid). More importantly, the ability of other inducing compounds to compete for the specific binding of (−)baclofen correlated very well with the ability of these compounds to induce larval metamorphosis (Fig. 3). The natural inducer, the GABA-mimetic algal peptide, also competes for the specific binding of (−)baclofen.

Like many internal receptors for neurotransmitters, the receptor that mediates larval metamorphosis is also subject to modulation. The larval response to inducers can be decreased or down-regulated by premature exposure of young larvae to inducer molecules; this decrease in larval responsiveness is correlated with a decrease in the number of receptors (Trapido-Rosenthal and Morse 1986a, b). Thus, as for many internal neurotransmitter receptors, including GABA receptors (Borden et al. 1984), down-regulation appears to be accomplished by the removal of receptors from the cell surface.

The interaction of the larval receptor with GABA or a GABA-mimetic ligand results in an apparent increase in the flux of ions across a chemosensory membrane (Morse 1985); the latter is a property held in common with the $GABA_A$ receptor. However, this ion flux does not appear to be directly gated by the ligand but rather to involve the second messenger cAMP that is generated by the activation of adenylate cyclase (Baxter and Morse 1987). This involvement of a second messenger is a property of the $GABA_B$ receptor. Hence, the chemoreceptors of abalone larvae appear to share properties with both the $GABA_A$ and $GABA_B$ receptors present in other organisms (Trapido-Rosenthal and Morse 1986a). Whether the larvae have both receptor types or a single type having both $GABA_A$ and $GABA_B$ characteristics remains to be determined. In either case the larval receptors are postulated to have an ancestral relationship with internal GABA receptors (Morse 1985).

4 Crustaceans: Chemosensory Systems for Adenine Nucleotides

Several classes of chemosensory cells exhibiting sensitivity to particular neuroactive compounds have been characterized in crustaceans (reviewed by Carr et al. 1987). These include cells responsive to *taurine* (Shepheard 1974; Fuzessery et al. 1978; Derby and Atema 1982b; Johnson and Atema 1983; Johnson et al. 1984; Gleeson and Ache 1985; Gleeson et al. 1987), *glutamate* (Case and Gwilliam 1961; Shepheard 1974; Johnson and Ache 1978; Derby and Atema 1982a, b; Derby and Ache 1984), *glycine* (Bauer et al. 1981; Derby and Atema 1982a; Derby and Ache 1984), *ecdysones* (Spencer and Case 1984), and *pyridines* (Bauer and Hatt 1980; Hatt and Bauer 1982; Hatt and Schmiedel-Jakob 1984, 1985). Physiological studies of structure activity relationships for cells selectively activated by taurine (Fuzessery et al. 1978) and glutamate (Derby and Atema 1982b) have yielded some important insights on similarities and differences with internal receptor systems for these agents. In this section, however, we will focus on the purinergic chemoreceptor systems of crustaceans which have been investigated utilizing a combination of behavioral, physiological, and biochemical approaches.

Table 2. Organisms with chemoreceptors sensitive to purine nucleotides and related compounds

Organism		Compound	Reference
Slime mold		cAMP	Konijn et al. (1967)
Crustacean	Shrimp	AMP	Carr and Thompson (1983)
	Florida spiny lobster	AMP	Derby et al. (1984)
	Prawn	AMP	Harpaz et al. (1987)
		ATP	Carr et al. (1986)
	California spiny lobster	ATP	Zimmer-Faust (1987)
Insect	Army worm	Adenosine	Ma (1977)
	Assassin bug	ATP	Friend and Smith (1982)
	Black fly	ADP, ATP	Sutcliffe and McIver (1979)
	Horsefly	ADP, ATP	Friend and Stoffolano (1983)
	Mosquito (larvae)	AMP	Barber et al. (1982)
	Mosquito (adult)	ATP	Galun et al. (1985)
	Tsetse fly	ATP	Mitchell (1976)
Mollusk	Octopus	AMP	Chase and Wells (1986)
Fish	Turbot	IMP	Mackie and Adron (1978)
	Puffer	ADP, IMP	Kiyohara et al. (1975)
Mammal	Cat	ATP, ADP	Boudreau et al. (1975)
	Dog, rat	ATP, IMP	Boudreau (1987)
	Cow	GMP, IMP	Torii and Cagan (1980)
	Human	IMP, GMP	Yamaguchi (1979)
		GMP	Rifkin and Bartoshuk (1980)
		Adenosine(?)	Schiffman et al. (1985)

Within the past 20 years adenosine and its noncyclic nucleotides have been implicated as important neurotransmitter/modulator substances in mammalian species. Receptors for these compounds (purinergic receptors) have been identified in a variety of tissues including central and peripheral neurons (Burnstock 1980; Phillis and Wu 1981), smooth muscle cells of the viscera (Brown and Burnstock 1981), and components of the cardiovascular system (Olsson et al. 1979). Burnstock (1978) distinguished two major types of purinergic receptors based on ligand potency, which he designated P_1 and P_2. For P_1 receptors the order of ligand potency is adenosine ≥ AMP > ADP ≥ ATP, whereas for P_2 receptors this sequence is reversed (i.e., ATP ≥ ADP > AMP ≥ adenosine).

In addition to the slime mold example described earlier, external chemoreceptors sensitive to adenine nucleotides and related compounds also occur in a number of other organisms, especially insects, crustaceans, and mammals (Table 2). Of particular note are the studies in marine crustaceans demonstrating the presence of P_1- and P_2-like chemoreceptors that mediate detection of AMP and ATP, respectively, in seawater. The behavioral, physiological, and biochemical findings of these studies, which are summarized below, reveal striking parallels between these external purinergic systems and those described for the internal tissues of mammals.

4.1 P_1-Like Chemoreceptors

In experiments to assess the chemical nature of chemoattractants present in foods consumed by the marine shrimp *Palaemonetes pugio* a behavioral bioassay was developed for quantifying the stimulatory capacity of extract fractions and pure substances. These experiments showed that of the eight compounds found to be chemoattractants, five were purine nucleotides and AMP was clearly the most potent of all the substances tested (Carr and Derby 1986). Behavioral assays of AMP analogs with *P. pugio* revealed structure activity relationships (SAR) which suggested that the attraction induced by AMP is mediated by chemoreceptors akin to internal P_1-type purinergic receptors (Carr and Thompson 1983).

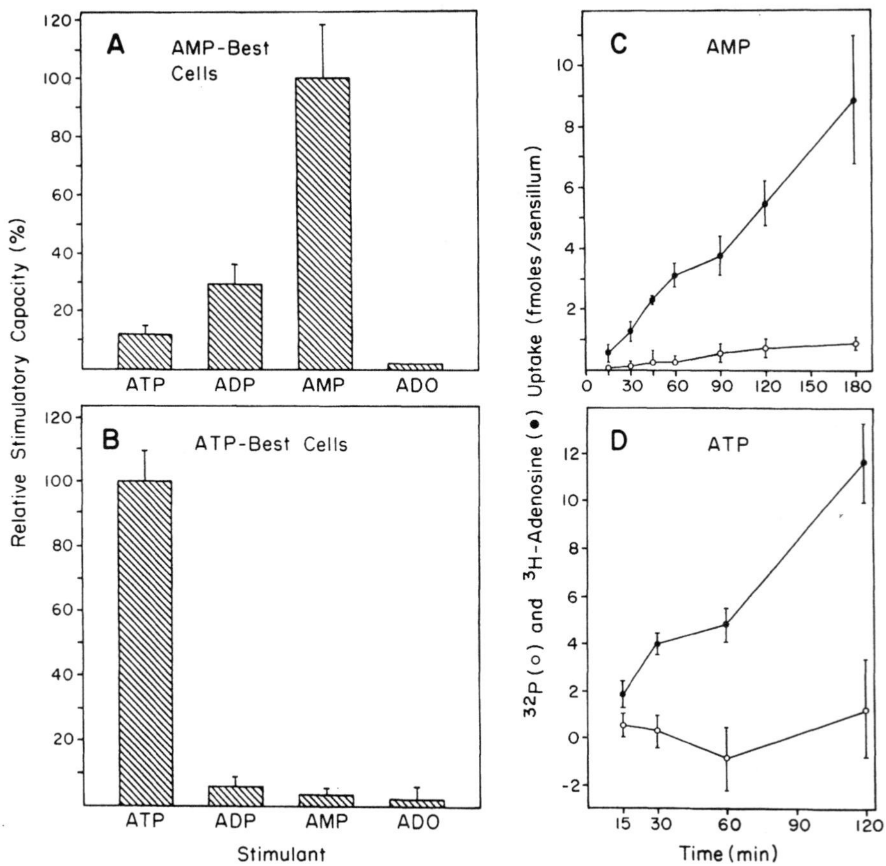

Fig. 4A–D. Comparisons of the physiological and biochemical properties of the olfactory purinergic systems of *P. argus*. Physiological response profiles to adenine nucleotides and adenosine (100 μM) for "AMP-best" cells (*A*) and "ATP-best" cells (*B*) (data from Derby et al. 1984 and Zimmer-Faust et al. 1988). *C* and *D* compare sensillar uptake of ^{3}H- versus ^{32}P-label from AMP and ATP. ^{3}H-labeling of AMP and ATP was at the 2 and 8 positions of the adenine ring; ^{32}P-labeling was at the α-phosphate group and γ-phosphate group of AMP and ATP, respectively (After Trapido-Rosenthal et al. 1987)

Physiological evidence for the presence of external chemoreceptors sensitive to AMP in seawater was first obtained from the olfactory organ (antennule) of the Florida spiny lobster *Panulirus argus*. Utilizing a perfused antennule preparation, extracellular recordings from single olfactory cells are obtained with suction electrodes applied to the axons of the sensory neurons (see Gleeson and Ache 1985). The procedure allows monitoring the responses of cells to chemical stimuli introduced into artificial seawater that continuously flushes over the chemosensory sensilla.

Using the above technique in the spiny lobster, Derby et al. (1984) identified a population of chemosensory cells showing marked similarities to P_1-type purinergic receptors. These cells, referred to as "AMP-best" cells (Fig. 4A), exhibited activity profiles to AMP analogs which were remarkably similar to the activity profiles shown behaviorally by the shrimp *P. pugio*. Namely, AMP is the most active stimulant with the potency sequence for the analogs being AMP > ADP > ATP or adenosine; structural modifications in the ribose phosphate moiety result in a greater loss in activity than do changes in the purine base; the *anti*-conformation of the purine moiety is more effective than the *syn*-conformation; and theophylline is an antagonist of the AMP response. As with the behavioral evidence for *P. pugio*, these response properties for the AMP-best chemoreceptors of *P. argus* closely align with the characteristics of internal P_1-type purinergic receptors. The major difference is that activation of the crustacean P_1-like chemoreceptor requires a phosphate group at the 5′-position of the adenosine moiety, whereas the internal P_1 receptor is equally activated by AMP and adenosine.

4.2 P_2-Like Chemoreceptors

In contrast to the behavioral results from the shrimp *P. pugio*, ATP was found to be more potent than AMP as a stimulant of locomotory behavior associated with food acquisition in the Pacific spiny lobster *Panulirus interruptus* (Zimmer-Faust 1987). Indeed, in a study to further define the SAR for this behavior, the results suggested activation of chemoreceptors similar to P_2- rather than P_1-type purinergic receptors (Zimmer-Faust et al. 1988).

In physiological studies with *P. argus*, ATP-best chemoreceptor cells (Fig. 4B) have been identified in the antennule (Carr et al. 1986). These cells share many of the features characteristic of P_2-type purinergic receptors (Burnstock 1978; Burnstock and Kennedy 1985) including: (1) a potency sequence of ATP > ADP > AMP or adenosine, (2) activation by the slowly hydrolyzable ATP analogs β,γ-imido ATP, α,β-methylene ATP, and β,γ-methylene ATP, (3) rapid desensitization, and (4) activation by a range of nucleotide triphosphates with modifications in both the adenine and ribose moieties. Interestingly, a population of ATP-best cells with virtually identical response characteristics to those for *P. argus* has recently been identified in the antennule of *P. interruptus* and may mediate the behavioral response of this species to ATP (Zimmer-Faust et al. 1988). These results, therefore, demonstrate that P_2-like as well as P_1-like chemoreceptors are present in the olfactory system of spiny lobsters.

4.3 Biochemical Processes Associated with the Olfactory Purinergic System

Additional parallels with the internal purinergic systems of mammals have been found in the biochemical processing of external adenine nucleotides by the antennule of the lobster. In mammals extracellular adenine nucleotides are removed from the receptor environment in a two-step process. First, they are dephosphorylated to adenosine by ectonucleotidases located on the external membrane surfaces of neurons, glia, and other types of cells (Kreutzberg et al. 1978; Cusack et al. 1983; Pearson 1985); secondly, the adenosine so generated is then internalized by a specific uptake system involving facilitated diffusion (Hertz 1978; Bender et al. 1981; Kreutzberg et al. 1983).

A similar two-step scheme for the processing of adenine nucleotides has been revealed in the olfactory sensilla of the lobster *P. argus* (Trapido-Rosenthal et al. 1987). The incubation of sensilla with ^{3}H-adenosine demonstrated the presence of a saturable, sodium-dependent uptake system for adenosine which exhibits a K_m of 7.1 μM and Vmax of 5.2 fmol/sensillum/min. Evidence that extracellular dephosphorylation of adenine nucleotides to adenosine precedes uptake was obtained in double-label experiments in which it was shown that ^{3}H from adenine-labeled AMP and ATP is internalized, whereas ^{32}P from the phosphate-labeled nucleotides is not (Fig. 4C, D). Since adenosine is not a stimulus for either the AMP- or ATP-best cells in the lobster's antennule (see Fig. 4A, B), the dephosphorylation of excitatory nucleotides functions as a means of inactivating these substances in the receptor environment. This function is analogous to the inactivation of acetylcholine by acetylcholinesterases at synapses.

An additional biochemical parallel between internal purinergic systems and those of the lobster's antennule was the finding that the extracellular dephosphorylation of AMP by enzymes in the olfactory sensilla is competitively inhibited by ADP and the ADP analog, *α, β*-methylene ADP (Trapido-Rosenthal et al. 1987). A comparable inhibition by both of these compounds has been described for 5′-ectonucleotidases associated with smooth muscle, endothelium, and heart tissues in mammals (Burger and Lowenstein 1970; Pearson et al. 1980; Lowenstein et al. 1983; Gordon et al. 1986).

These biochemical results, therefore, complement the behavioral and physiological evidence of similarities between the purinergic chemoreceptors of crustaceans and the internal purinergic receptors of mammals, and accordingly lend further support to the notion of a possible common ancestry for these receptor systems.

5 Conclusions

Major parallels exist between the external chemosensory systems of lower organisms and the internal "chemosensory" systems that detect transmitters and hormones. In this regard the chemotactic response of the slime mold to cAMP is particularly noteworthy in that the same mechanisms for the detection and transduction of a specific signal molecule persist in both the uni- and multicellular stages of the life cycle (see Sect. 2.1.5). The ontogenetic progression shown by the slime mold, wherein an external chemosensory system for cAMP becomes internalized to function as a "transmitter" system, is quite analogous to a similar progression that occurs in vertebrate development. In the latter, receptors for the transmitters acetylcholine, epinephrine, and GABA are present on the surface membrane of oocytes from various amphibians and mammals (Kusano et al. 1982; Eusebi et al. 1984; Dolci et al. 1985). Furthermore, the acetylcholine response mediated by muscarinic receptors on frog oocytes exhibits a sensitivity to the second messenger, inositol trisphosphate, and to Ca^{2+}, which suggests that these receptors employ the same transduction components as present in synapses of the fully differentiated adult (Gillo et al. 1987). In short, unicellular stages of both the frog and slime mold possess functional membrane transduction systems that are initially positioned to monitor particular chemical substances occurring in the external environment. With the ontogeny of multicellularity in each organism, it appears that the same systems, complete with receptors and transduction components, become internalized within the organism to mediate chemical communication between neighboring cells.

G-proteins and second messenger cyclic nucleotides are ubiquitous constituents of the transduction systems employed with several transmitters and hormones as well as with the cAMP chemosensory system of the slime mold (Sects. 1 and 2.1.3). Although G-proteins and second messengers are not yet well characterized in chemoreceptors of invertebrates, it is likely that these transduction components are present since they are associated with internal receptor systems throughout the animal kingdom (Venter et al. 1988). Moreover, several recent studies of vertebrate chemoreceptors indicate that both G-proteins (Pace et al. 1985; Pace and Lancet 1986; Huque and Bruch 1986; Weinstock et al. 1986) and cyclic AMP (Heldman and Lancet 1986; Sklar et al. 1986; Shirley et al. 1986; Bruch and Kalinoski 1987; Nakamura and Gold 1987) contribute to transduction processes in these organisms.

Additional parallels between external and internal signaling systems are evident when the chemotactic behavior of the mammalian leukocyte is considered. Leukocytes are blood cells that move about in an aqueous extracellular milieu in much the same manner as protozoans move about in a pond or culture medium. Also, like protozoans (see Sect. 2.2), leukocytes show receptor-mediated behavioral responses to external chemical signals. For leukocytes, these chemicals are specific peptides and other chemotactic agents released into the body fluids by bacteria or by immunological processes occurring at sites of inflammation (Schiffmann 1982; Snyderman and Pike 1984). Most importantly, G-proteins and second messengers are constituents of the signal transduction process in leukocytes (for reviews see Smith et al. 1987; van Houten and Preston 1987). Hence, these migratory blood cells, like certain free-living unicellular organisms, possess a chemosensory system consisting of the same basic components as employed for the detection and transduction of chemical signals in synapses.

With the advent of modern techniques for gene cloning and DNA sequencing, the deduced amino acid sequences of an ever-increasing number of proteins are becoming known. As a consequence of this new technology, the extent of homology can be determined between the amino acid sequences of different proteins such as receptor proteins that are activated by specifically different ligands. A truly astonishing finding is that a high degree of homology exists between several "different" receptor proteins. The extent of this homology indicates that some of these receptors have very close evolutionary relationships (see Venter et al. 1988 for review). One family of receptors exhibiting an apparent common ancestry includes the muscarinic acetylcholine, β-adrenergic and yeast α-mating factor receptors (Gocayne et al. 1987; Hall 1987). All of these receptors exhibit a highly conserved protein structure and all utilize G-proteins in the transduction process. A second family of receptors having apparent close evolutionary ties includes the $GABA_A$, glycine, and nicotinic acetylcholine receptors (Barnard et al. 1987). The receptors within this family also exhibit a highly conserved protein structure and all have ligand-gated channels as an integral part of the receptor molecule.

When Haldane (1954) first suggested that some internal receptor types may have evolved from external chemoreceptors of primitive organisms, he possessed only a rudimentary knowledge of receptors and had no knowledge of either transduction events or homologies between disparate receptor proteins. Haldane's hypothesis was based on the simple observations that neuroactive substances such as acetylcholine and epinephrine influenced the swimming patterns of certain unicellular protozoans. He reasoned that the processes whereby particular chemical substances affect the activity of nerve and muscle cells might be similar to the processes whereby these same substances affect protozoans. Today it is known that the similarities between external and internal "chemoreceptors" include not only the signal molecules and receptor types, but also the mechanisms and molecular components that contribute to transduction, membrane excitability, signal inactivation, receptor regulation, and even the manufacture of certain types of signal substances. Collectively these similarities provide very strong arguments in support of Haldane's original hypothesis. However, the most compelling argument for Haldane's proposal is probably the recent evidence that an external chemoreceptor for a mating factor in yeast, and the internal muscarinic and β-adrenergic receptors of mammals, share a common ancestral origin (Gocayne et al. 1987). These discoveries of apparent evolutionary relationships between external and internal receptor types will certainly encourage the sequencing of additional genes for chemoreceptors of lower organisms. If the amino acid sequences coded by these genes reveal further homologies with internal receptor types, then it would seem that Haldane and later others (see Lentz 1968; Kittredge et al. 1974; Lenhoff and Heagy 1977; Carr et al. 1987) were indeed correct in proposing that internal chemical signaling systems evolved from the external chemosensory systems of primitive organisms.

Acknowledgments. This work was supported by grant no. BNS-8607513 from the National Science Foundation. We are grateful to Dr. Charles D. Derby for providing constructive criticism of the manuscript. Ms. Marsha Lynn Milstead prepared the illustrations.

References

Ache BW (1982) Chemoreception and thermoreception. In: Atwood HL, Sandeman DC (eds) The biology of Crustacea. Academic Press, New York London, pp 369–398

Allison P, Dorsett D (1977) Behavioral studies on chemoreception in *Balanus hameri.* Mar Behav Physiol 4:205–217

Atema J (1985) Chemoreception in the sea: adaptations of chemoreceptors and behavior to aquatic stimulus conditions. Soc Exp Mar Biol Symp 39:387–423

Baloun AJ, Morse DE (1984) Ionic control of settlement and metamorphosis in larval *Haliotis rufescens* (Gastropoda). Biol Bull 167:124–138

Barber JT, Ellgard EG, Herskowitz K (1982) The attraction of larvae of *Culex pipiens quinquefasciatus* Say to ribonucleic acids and nucleotides. J Insect Physiol 18:585–588

Barnard EA, Darlison MG, Seeburg P (1987) Molecular biology of the $GABA_A$ receptor: the receptor/channel superfamily. Trends Neurosci 10:502–509

Bauer U, Hatt H (1980) Demonstration of three different types of chemosensitive units in the crayfish claw using a computerized evaluation. Neurosci Lett 17:209–214

Bauer U, Dudel J, Hatt H (1981) Characteristics of single chemoreceptive units sensitive to amino acids and related substances in the crayfish leg. J Comp Physiol A 144:67–74

Baxter G, Morse DE (1987) G protein and diacylglycerol regulate metamorphosis of planktonic molluscan larvae. Proc Natl Acad Sci USA 84:1867–1870

Bender A, Sprague GF Jr (1986) Yeast peptide pheromones, a-factor and α-factor, activate a common response mechanism in their target cells. Cell 47:929–937

Bender AS, Wu PH, Phillis JW (1981) The rapid uptake and release of [^{3}H]adenosine by rat cerebral cortical synaptosomes. J Neurochem 36:651–660

Berridge MJ (1985) The molecular basis of communication within the cell. Sci Am 253:142–152

Bonner JT (1983) Chemical signals in social amoebae. Sci Am 248:114–120

Borden LA, Czajkowski C, Chan CY, Farb DH (1984) Benzodiazepine receptor synthesis and degradation by neurons in culture. Science 226:857–860

Boudreau JC (1987) Mammalian neural taste responses to amino acids and nucleotides. In: Kawamura Y, Kare MR (eds) Umami: a basic taste. Dekker, New York, pp 201–217

Boudreau JC, Anderson W, Oravec J (1975) Chemical stimulus determinants of cat geniculate ganglion chemoresponsive group II unit discharge. Chem Senses Flav 1:495–517

Bowery NG, Hill DR, Hudson AL, Doble A, Middlemiss DN, Shaw J, Turnbull M (1980) (−)Baclofen decreases neurotransmitter release in the mammalian CNS by an action at a novel GABA receptor. Nature (London) 283:92–94

Bowery NG, Hill DR, Hudson AL (1983) Characteristics of GABA receptor binding sites on rat whole brain synaptic membranes. Br J Pharmacol 78:191–206

Brown CM, Burnstock G (1981) Evidence in support of the P_1/P_2 purinoceptor hypothesis in the guinea-pig taenia coli. Br J Pharmacol 73:617–624

Brown ID, Kerkut GA (1981) A study of the chemokinetic effects of various pharmacological agents upon *Tetrahymena vorax.* Comp Biochem Physiol 69C:275–280

Bruch RC, Kalinoski DL (1987) Interaction of GTP-binding proteins with chemosensory receptors. J Biol Chem 262:2401–2404

Burger RM, Lowenstein JM (1970) Preparation and properties of 5′-nucleotidase from smooth muscle of small intestine. J Biol Chem 245:6274–6280

Burke RD (1983) The induction of marine invertebrate larvae: stimulus and response. Can J Zool 61:1701–1719

Burke RD (1984) Pheromonal control of metamorphosis in the Pacific sand dollar *Dendraster excentricus.* Science 225:442–443

Burnstock G (1978) A basis for distinguishing two types of purinergic receptors. In: Straub RW, Bolis L (eds) Cell membrane receptors for drugs and hormones. Raven, New York, pp 107–118

Burnstock G (1980) Purinergic nerves and receptors. Prog Biochem Pharmacol 16:141–154

Burnstock G, Kennedy C (1985) Is there a basis for distinguishing two types of P_2-purinoceptors? Gen Pharmacol 16:433–440

Carr WES (1967) Chemoreception in the mud snail *Nassarius obsoletus*. II. Identification of stimulatory substances. Biol Bull 13:106–127
Carr WES (1987) The molecular nature of chemical stimuli in the aquatic environment. In: Atema J, Fay RR, Popper AN, Tavolga WN (eds) Sensory biology of aquatic animals. Springer, Berlin Heidelberg New York Tokyo, pp 3–27
Carr WES, Derby CD (1986) Behavioral chemoattractants for the shrimp, *Palaemonetes pugio*: identification of active components in food extracts and evidence of synergistic mixture interactions. Chem Senses 11:49–64
Carr WES, Thompson HW (1983) Adenosine 5′-monophosphate, an internal regulatory agent, is a potent chemoattractant for a marine shrimp. J Comp Physiol A 153:47–53
Carr WES, Gleeson RA, Ache BW, Milstead ML (1986) Olfactory receptors of the spiny lobster: ATP-sensitive cells with similarities to P_2-type purinoceptors of vertebrates. J Comp Physiol A 158:331–338
Carr WES, Ache BW, Gleeson RA (1987) Chemoreceptors of crustaceans: similarities to receptors for neuroactive substances in internal tissues. Environ Health Perspect 71:31–46
Case J, Gwilliam GF (1961) Amino acid sensitivity of the dactyl chemoreceptors of *Carcinides maenas*. Biol Bull 121:449–455
Changeux J-P, Revah F (1987) The acetylcholine receptor molecule: allosteric sites and the ion channel. Trends Neurosci 10:245–250
Chase R, Wells MJ (1986) Chemotactic behaviour in the octopus. J Comp Physiol 158:375–381
Chia F-S (1978) Perspective: settlement and metamorphosis of marine invertebrate larvae. In: Chia F-S, Rice ME (eds) Settlement and metamorphosis of marine invertebrate larvae. Elsevier/North Holland Biomedical Press, Amsterdam New York, pp 283–285
Chia F-S, Koss R (1982) Fine structure of the larval rhinopores of the nudibranch, *Rostanga pulchra*, with an emphasis on the sensory receptor cells. Cell Tissue Res 25:235–248
Choi DW, Farb DH, Fischbach GD (1981) Chlordiazepoxide selectively potentiates GABA conductance of spinal cord and sensory neurons in cell culture. J Neurophysiol 45:621–631
Coon SL, Bonar DB, Weiner RM (1985) Induction of settlement and metamorphosis of the Pacific oyster, *Crassostrea gigas* (Thunberg), by L-DOPA and catecholamines. J Exp Mar Biol Ecol 94:211–221
Crisp DJ (1974) Factors influencing settlement of marine invertebrate larvae. In: Grant PT, Mackie AM (eds) Chemoreception in marine organisms. Academic Press, New York London, pp 177–265
Croll RG (1983) Gastropod chemoreception. Biol Rev 58:293–319
Csaba G (1980) Phylogeny and ontogeny of hormone receptors: the selection theory of receptor formation and hormonal imprinting. Biol Rev Cambridge Philos Soc 55:47–63
Csaba G, Sudar F, Nagy SU, Dobozy O (1977) Localization of hormone receptors in *Tetrahymena*. Protoplasmia 91:179–181
Cusack NJ, Pearson JD, Gordon JL (1983) Stereoselectivity of ectonucleotidases on vascular endothelial cells. Biochem J 214:975–981
Darmon M, Barra J, Brachet P (1978) The role of phosphodiesterase in aggregation of *Dictyostelium discoideum*. J Cell Sci 31:233–243
de Castro SL, Oliveira MM (1987) Radioligand binding characterization of beta-adrenergic receptors in the protozoa *Trypanosoma cruzi*. Comp Biochem Physiol 87C:5–8
Derby CD, Ache BW (1984) Quality coding of a complex odorant in an invertebrate. J Neurophysiol 51:906–924
Derby CD, Atema J (1982a) Chemosensitivity of walking legs of the lobster *Homarus americanus*: neurophysiological response spectrum thresholds. J Exp Biol 98:303–315
Derby CD, Atema J (1982b) Narrow-spectrum chemoreceptor cells in the walking legs of the lobster *Homarus americanus*: taste specialists. J Comp Physiol A 146:181–189
Derby CD, Carr WES, Ache BW (1984) Purinergic olfactory cells of crustaceans: response characteristics and similarities to internal purinergic cells of vertebrates. J Comp Physiol A 155:341–349
Dietzel C, Kurjan J (1987) The yeast SCG1 gene: a G α-like protein implicated in the a- and α-factor response pathway. Cell 50:1001–1010
Dolci S, Eusebi F, Siracusa G (1985) γ-amino butyric-n-acid sensitivity of mouse and human oocytes. Dev Biol 109:242–246

Dolphin AC (1987) Nucleotide binding proteins in signal transduction and disease. Trends Neurosci 10:53–57
Dolphin AC, Scott RH (1987) Calcium channel currents and their inhibition by (–)baclofen in rat sensory neurones: modulation by guanine nucleotides. J Physiol (London) 386:1–17
Doughty MJ (1978) Control of ciliary activity in *Paramecium* – II. Modification of K^+-induced ciliary reversal by cholinergic ligands and quaternary ammonium compounds. Comp Biochem Physiol 61C:375–384
Doughty MJ (1979) Control of ciliary activity in *Paramecium*. – III. Evidence for specific membrane binding sites for ions and cholinergic ligands. Comp Biochem Physiol 63C:183–197
Doughty MJ, Dodd GH (1978) Chemical modification of the excitable membrane of *Paramecium*. Effect of a cross-linking agent. Comp Biochem Physiol 59C:21–31
Eckert R, Brehm P (1979) Ionic mechanisms of excitation in *Paramecium*. Annu Rev Biophys Bioeng 8:353–383
Eusebi F, Pasetto N, Siracusa G (1984) Acetylcholine receptors in human oocytes. J Physiol (London) 346:321–330
Fine-Levy JB, Derby CD, Daniel PC (1987) Chemosensory discrimination: behavioral abilities of the spiny lobster. Ann N Y Acad Sci 510:280–283
Forward RB Jr (1977) Effects of neurochemicals upon a dinoflagellate photoresponse. J Protozool 24:401–405
Franke J, Kessin RH (1981) The cyclic nucleotide phosphodiesterase inhibitory protein of *Dictyostelium discoideum*: purification and characterization. J Biol Chem 256:7628–7637
Frazier WA, Meyers-Hutchins BL, Jamieson GA, Galvin NJ (1984) Chemotactic transduction in the cellular slime molds. In: Elson E, Frazier W, Glaser L (eds) Cell membranes: methods and reviews. Plenum, New York, pp 1–41
Friend MG, Smith JJB (1982) ATP analogs and other phosphate compounds as gorging stimulants for *Rhodnius prolixus*. J Insect Physiol 28:371–376
Friend WG, Stoffolano JG (1983) Feeding responses of the horsefly, *Tabanus nigrovittatus*, to phagostimulants. Physiol Entomol 8:377–383
Fuzessery ZM, Carr WES, Ache BW (1978) Antennular chemosensitivity in the spiny lobster, *Panulirus argus*: studies of taurine sensitive receptors. Biol Bull 154:226–240
Gainer H, Russell JT, Loh YP (1985) The enzymology and intracellular organization of peptide precursor processing: the secretory vesicle hypothesis. Prog Neuroendocrinol 40:171–184
Galun R, Koontz LC, Gwadz RW, Ribeiro JMC (1985) Effect of ATP analogues on the gorging response of *Aedes aegypti*. Physiol Entomol 10:275–281
Gerisch G (1987) Cyclic AMP and other signals controlling cell development and differentiation in *Dictyostelium*. Annu Rev Biochem 56:853–879
Gerisch G, Malchow D, Hess B (1974) Cell communication and cyclic-AMP regulation during aggregation of the slime mold, *Dictyostelium discoideum*. In: Jaenicke L (ed) Biochemistry of sensory functions. Springer, Berlin Heidelberg New York, pp 279–298
Gillo B, Lass Y, Nadler E, Oron Y (1987) The involvement of inositol 1,4,5-trisphosphate and calcium in the two-component response to acetylcholine in *Xenopus* oocytes. J Physiol (London) 392:349–361
Gilman AG (1987) G proteins: transducers of receptor-generated signals. Annu Rev Biochem 56:615–649
Gleeson RA (1978) Functional adaptations in chemosensory systems. In: Ali MA (ed) Sensory ecology: review and perspectives. Plenum, New York, pp 291–317
Gleeson RA, Ache BW (1985) Amino acid suppression of taurine-sensitive chemosensory neurons. Brain Res 335:99–107
Gleeson RA, Trapido-Rosenthal HG, Carr WES (1987) A taurine receptor model: taurine-sensitive olfactory cells in the lobster. In: Huxtable RJ, Franconi F, Giotti A (eds) The biology of taurine. Plenum, New York, pp 253–263
Gocayne J, Robinson DA, FitzGerald MG, Chung F-Z, Kerlavage AR, Lentes K-U, Lai J, Wang C-D, Fraser CM, Venter JG (1987) Primary structure of rat cardiac β-adrenergic and muscarinic cholinergic receptors obtained by automated DNA sequence analysis: further evidence for a multigene family. Proc Natl Acad Sci USA 84:8296–8300

Gordon EL, Pearson JD, Slakey LL (1986) The hydrolysis of extracellular adenine nucleotides by cultured endothelial cells from pig aorta. J Biol Chem 261:15496–15504
Hadfield MG (1977) Chemical interactions in larval settling of a marine gastropod. In Faulkner DJ, Fenical WH (eds) Marine natural products chemistry. Plenum, New York, pp 403–413
Hadfield MG (1978) Metamorphosis in marine molluscan larvae: an analysis of stimulus and response. In: Chia F-S, Rice ME (eds) Settlement and metamorphosis of marine invertebrate larvae. Elsevier/North Holland Biomedical Press, Amsterdam New York, pp 165–175
Hadfield MG, Scheuer D (1985) Evidence for a soluble metamorphic inducer in *Phestilla*: ecological, chemical and biological data. Bull Mar Sci 37:556–566
Haefely W, Polc P, Schaffner R, Keller H, Pieri L, Mohler H (1979) Facilitation of GABA-ergic transmission by drugs. In: Krogsgaard-Larsen P, Scheel-Kruger J, Kofod H (eds) GABA neurotransmitters: pharmacochemical, biochemical, and pharmacological aspects. Academic Press, New York, London, pp 357–375
Hagiwara SK, Kusano K, Saito N (1961) Membrane changes of *Onchidium* nerve cells in potassium-rich media. J Physiol (London) 155:470–489
Haldane JBS (1954) La signalisation animale. Ann Biol 58:89–98
Hall ZW (1987) Three of a kind: the β-adrenergic receptor, the muscarinic acetylcholine receptor, and rhodopsin. Trends Neurosci 10:99–101
Harpaz S, Kahan D, Galun R, Moore I (1987) Responses of freshwater prawn, *Macrobrachium rosenbergii*, to chemical attractants. J Chem Ecol 13:1957–1965
Hatt H, Bauer U (1982) Electrophysiological properties of pyridine receptors in the crayfish walking leg. J Comp Physiol A 148:221–224
Hatt H, Schmiedel-Jakob I (1984) Electrophysiological studies of pyridine-sensitive units on the crayfish walking leg. I. Characteristics of stimulatory molecules. J Comp Physiol A 154:855–863
Hatt H, Schmiedel-Jakob I (1985) Specific antagonists at the pyridine receptor: evidence from electrophysiological studies with acetylpyridines. Chem Senses 10:317–323
Hauser DCR, Levandowski, Glassgold JM (1975) Ultrasensitive chemosensory responses by a protozoan to epinephrine and other neurochemicals. Science 190:285–286
Heldman J, Lancet D (1986) Cyclic AMP-dependent protein phosphorylation in chemosensory neurons: identification of cyclic nucleotide-regulated phosphoproteins in olfactory cilia. J Neurochem 47:1527–1533
Hellung-Larsen P, Leick V, Tommerup N (1986) Chemoattraction in *Tetrahymena*: on the role of chemokinesis. Biol Bull 170:357–367
Herskowitz I, Marsh L (1987) Conservation of a receptor/signal transduction system. Cell 50:995–996
Hertz L (1978) Kinetics of adenosine uptake into astrocytes. J Neurochem 31:55–62
Hill DR (1985) $GABA_B$ receptor modulation of adenylate cyclase activity in rat brain slices. Br J Pharmacol 84:249–257
Hinrichsen RD, Schultz JE (1988) *Paramecium*: a model system for the study of excitable cells. Trends Neurosci 11:27–32
Hirata KY, Hadfield MG (1986) The role of choline in metamorphic induction of *Phestilla* (Gastropoda, Nudibranchia). Comp Biochem Physiol 84C:15–21
Hodgkin AL, Horowicz P (1959) The influence of potassium and chloride ions on the membrane potential of single muscle fibres. J Physiol (London) 148:127–160
Huque T, Bruch RC (1986) Odorant- and guanine nucleotide-stimulated phosphoinositide turnover in olfactory cilia. Biochem Biophys Res Commun 137:36–42
Jensen RA, Morse DE (1984) Intraspecific facilitation of larval recruitment: gregarious settlement of the polychaete *Phragmatopoma californica* (Fewkes). J Exp Mar Biol Ecol 83:107–126
Johnson BR, Ache BW (1978) Antennular chemosensitivity in the spiny lobster, *Panulirus argus*: amino acids as feeding stimuli. Mar Behav Physiol 5:145–157
Johnson BR, Atema J (1983) Narrow-spectrum chemoreceptor cells in the antennules of the American lobster. Neurosci Lett 41:145–150
Johnson BR, Voigt R, Borroni PF, Atema J (1984) Response properties of lobster chemoreceptors: tuning of primary taste neurons in walking legs. J Comp Physiol A 155:593–604

Kaissling K-E (1987) Wright RH Lectures on insect olfaction. In: Colbow C (ed) Simon Fraser Univ, Burnaby, British Columbia, Can
Kay R (1983) Cyclic AMP and development in the slime mould. Nature (London) 301:659
Kittredge JS, Takahashi FT, Lindsey J, Lasker R (1974) Chemical signals in the sea: marine allelochemics and evolution. Fish Bull 72:1–11
Kiyohara S, Hidaka I, Tamura T (1975) Gustatory response in the puffer. II. Single fiber analyses. Bull Jpn Soc Sci Fish 41:383–391
Kleene SJ (1986) Bacterial chemotaxis and vertebrate olfaction. Experientia 42:241–250
Klein P, Theibert A, Fontana D, Devreotes PN (1985) Identification and cyclic AMP-induced modification of the cyclic AMP receptor in *Dictyostelium discoideum*. J Biol Chem 260:1757–1764
Konijn TM, van de Meene JGC, Bonner JT, Barkley DS (1967) The acrasin activity of adenosine-3′,5′-cyclic phosphate. Proc Natl Acad Sci USA 58:1152–1154
Kreutzberg GW, Barron KD, Schubert P (1978) Cytochemical localization of 5′-nucleotidase in glial plasma membranes. Brain Res 158:247–257
Kreutzberg GW, Reddington M, Lee KW, Schubert P (1983) Adenosine: transport, function and interactions with receptors in the CNS. J Neur Transmiss Suppl 18:112–119
Kurjan J, Herskowitz I (1982) Structure of a yeast pheromone gene (MF): a putative α-factor precursor contains four tandem copies of mature α-factor. Cell 30:933–943
Kusano K, Miledi R, Stinnakre J (1982) Cholinergic and catecholaminergic receptors in the *Xenopus* oocyte membrane. J Physiol (London) 328:143–170
Leick V, Hellung-Larsen P (1985) Chemosensory responses in *Tetrahymena*: the involvement of peptides and other signal substances. J Protozool 32:550–553
Lenhoff HM (1981) Biology and physical chemistry of feeding response of hydra. In: Cagan RH, Kare MR (eds) Biochemistry of taste and olfaction. Academic Press, New York London, pp 475–497
Lenhoff HM, Heagy W (1977) Aquatic invertebrates: model systems for the study of receptor activation and evolution of receptor proteins. Annu Rev Pharmacol Toxicol 17:243–258
Lentz TL (1968) Primitive nervous systems. Yale Univ Press, New Haven
Levandowsky M, Cheng T, Kehr A, Kim J, Gardner L, Silvern L, Tsang L, Lai G, Chung C, Prakash E (1984) Chemosensory responses to amino acids and certain amines by the ciliate *Tetrahymena*: a flat capillary assay. Biol Bull 167:322–330
Levitzki A (1986) β-Adrenergic receptors and their mode of coupling to adenylate cyclase. Physiol Rev 66:819–854
Lowenstein JM, Yu MK, Naito Y (1983) Regulation of adenosine metabolism by 5′-nucleotidases. In: Berne RM, Rall TW, Rubio R (eds) Regulatory functions of adenosine. Nijhoff, Boston, pp 117–131
Ma WC (1977) Electrophysiological evidence for chemosensitivity to adenosine, adenine and sugars in *Spodoptera exempta* and related species. Experientia 33:356–358
Mackie AM, Adron JW (1978) Identification of inosine and inosine 5′-monophosphate as the gustatory feeding stimulants for the turbot *Scophthalamus maximus*. Comp Biochem Physiol 60A:79–83
Mackie AM, Grant PT (1974) Interspecies and intraspecies communication by marine invertebrates. In: Grant PT, Mackie AM (eds) Chemoreception in marine organisms. Academic Press, New York London, pp 105–141
Maier I, Muller DG (1986) Sexual pheromones in algae. Biol Bull 170:145–175
Mato JM, Krens FA, van Haastert JM, Konijn TM (1977) 3′:5′-Cyclic AMP-dependent 3′:5′-cyclic GMP accumulation in *Dictyostelium discoideum*. Proc Natl Acad Sci USA 74:2348–2351
Mato JM, Jastorff B, Morr M, Konijn TM (1978) A model for cyclic AMP-chemoreceptor interaction in *Dictyostelium discoideum*. Biochim Biophys Acta 544:309–314
McCarthy MP, Earnest JP, Young EF, Choe S, Stroud RM (1986) The molecular neurobiology of the acetylcholine receptor. Annu Rev Neurosci 9:383–413
McRobbie SJ (1986) Chemotaxis and cell mobility in the cellular slime molds. CRC Crit Rev Microbiol 13:335–375
Mitchell BK (1976) Physiology of an ATP receptor in labellar sensilla of the tsetse fly *Glossina morsitans morsitans* Westw. (Diptera: Glossinidae). J Exp Biol 65:259–271

Mitchell BK (1985) Specificity of an amino acid-sensitive cell in the adult Colorado Beetle, *Leptinotarsa decemlineata*. Physiol Entomol 10:421–429

Moens PB, Konijn TM (1974) Cyclic AMP as a cell surface activating agent in *Dictyostelium discoideum*. FEBS Lett 45:44–46

Morse ANC, Morse DE (1984) Recruitment and metamorphosis of *Haliotis* larvae induced by molecules uniquely available at the surfaces of crustose red algae. J Exp Mar Biol Ecol 75:191–215

Morse DE (1985) Neurotransmitter-mimetic inducers of larval settlement and metamorphosis. Bull Mar Sci 37:697–706

Morse DE, Hooker N, Duncan H, Jensen L (1979) γ-Aminobutyric acid, a neurotransmitter, induces planktonic abalone larvae to settle and begin metamorphosis. Science 204:407–410

Morse DE, Duncan H, Hooker N, Baloun A, Young G (1980a) GABA induces metamorphosis in planktonic molluscan larvae. Fed Proc 39:3237–3241

Morse DE, Hooker N, Duncan H (1980b) GABA induces metamorphosis in *Haliotis*. V: stereochemical specificity. Brain Res Bull 5:381–387

Naitoh Y (1982) Protozoa. In: Shelton GAB (ed) Electrical conduction and behavior in "simple" invertebrates. Clarendon, Oxford, pp 1–48

Nakamura T, Gold GH (1987) A cyclic nucleotide-gated conductance in olfactory receptor cilia. Nature (London) 325:442–444

Najundiah V, Malchow D (1976) A theoretical study of the effects of cyclic AMP phosphodiesterases during aggregation in *Dictyostelium*. J Cell Sci 22:49–58

Oliveira MM, Antunes A, De Mello FG (1984) Growth of *Trypanosoma cruzi* epimastigotes controlled by shifts in cyclic AMP mediated by adrenergic ligands. Mol Biochem Parasitol 11:283–292

Olsen RW, Wong EHF, Stauber GB, King RG (1984) Biochemical pharmacology of the γ-aminobutyric acid receptor/ionophore protein. Fed Proc 43:2773–2778

Olsson RA, Khouri EM, Bedznek JL Jr, McLean J (1979) Coronary vasoactivity of adenosine in the conscious dog. Circ Res 45:468–478

Ordal GW (1985) Bacterial chemotaxis: biochemistry of behavior in a single cell. CRC Crit Rev Microbiol 12:95–130

Pace U, Lancet D (1986) Olfactory GTP-binding protein: signal-transducing polypeptide of vertebrate chemosensory neurons. Proc Natl Acad Sci USA 83:4947–4951

Pace U, Hanski E, Salomon Y, Lancet D (1985) Odorant-sensitive adenylate cyclase may mediate olfactory reception. Nature (London) 316:255–258

Pawlik JR (1986) Chemical induction of larval settlement and metamorphosis in the reef-building tube-worm *Phragmatopoma californica* (Sabellariidae, Polychaeta). Mar Biol 91:59–68

Pearson JD (1985) Ectonucleotidases. In: Paton DM (ed) Methods in pharmacology, vol 6: Methods in adenosine research. Plenum, New York, pp 83–107

Pearson JD, Carleton JS, Gordon JL (1980) Metabolism of adenine nucleotides by ectoenzymes of vascular endothelial and smooth-muscle cells in culture. Biochem J 190:421–429

Phillis JW, Wu PH (1981) The role of adenosine and its nucleotides in central synaptic transmission. Prog Neurobiol 16:187–239

Posner BI, Khan MN, Bergeron JJM (1982) Endocytosis of peptide hormones and other ligands. Endocrinol Rev 3:280–298

Rifkin B, Bartoshuk LM (1980) Taste synergism between monosodium glutamate and disodium 5′-guanylate. Physiol Behav 24:1169–1172

Schaap P, Wang M (1986) Interactions between adenosine and oscillatory cAMP signaling regulate size and pattern in *Dictyostelium*. Cell 45:137–144

Schiffmann E (1982) Leukocyte chemotaxis. Annu Rev Physiol 44:553–568

Schiffman SS, Gill JM, Diaz C (1985) Methyl xanthines enhance taste: evidence for modulation of taste by adenosine receptor. Pharmacol Biochem Behav 22:195–203

Shepheard P (1974) Chemoreception in the antennule of the lobster, *Homarus americanus*. Mar Behav Physiol 2:261–273

Shirley SG, Robinson CJ, Dickinson K, Aujla R, Dodd GH (1986) Olfactory adenylate cyclase of the rat. Biochem J 240:605–607

Sklar PB, Anholt RRH, Snyder SH (1986) The odorant-sensitive adenylate cyclase of olfactory receptor cells. J Biol Chem 261:15538–15543

Smith CD, Verghese MW, Snyderman R (1987) Regulation of leukocyte responses to chemoattractants: guanine nucleotide regulatory (N) proteins and phospholipase C. In: Konijn TM, van Haastert PJM, van der Starre H, van der Wel H, Houslay MD (eds) Molecular mechanisms of desensitization to signal molecules. Springer, Berlin Heidelberg New York Tokyo, pp 277–289

Snyderman R, Pike MC (1984) Chemoattractant receptors on phagocytic cells. Annu Rev Immunol 2:257–281

Spencer M, Case JF (1984) Exogenous ecdysteroids elicit low threshold sensory responses in spiny lobsters. J Exp Zool 229:163–166

Sprague GF Jr, Blair LC, Thorner J (1983) Cell interactions and regulation of cell type in the yeast *Saccharomyces cerevisiae*. Annu Rev Microbiol 37:623–660

Starr RC, O'Neil RM, Miller CE III (1980) L-glutamic acid as mediator of sexual morphogenesis in *Volvox capensis*. Proc Natl Acad Sci USA 77:1025–1028

Sutcliffe JF, McIver SB (1979) Experiments on biting and gorging behavior in the black fly, *Simulium venustum*. Physiol Entomol 4:393–400

Torii K, Cagan RH (1980) Biochemical studies of taste sensation. IX. Enhancement of L-[^{3}H] glutamate binding to bovine taste papillae by 5′-ribonucleotides. Biochim Biophys Acta 627:313–323

Trapido-Rosenthal HG, Morse DE (1985) L-*α*, *ω*-diamino acids facilitate GABA induction of larval metamorphosis in a gastropod mollusc (*Haliotis rufescens*). J Comp Physiol B 155:403–414

Trapido-Rosenthal HG, Morse DE (1986a) Regulation of receptor-mediated settlement and metamorphosis in larvae of a gastropod mollusc (*Haliotis rufescens*). Bull Mar Sci 39:383–392

Trapido-Rosenthal HG, Morse DE (1986b) Availability of chemosensory receptors is down-regulated by habituation of larvae to a morphogenetic signal. Proc Natl Acad Sci USA 83:7658–7662

Trapido-Rosenthal HG, Carr WES, Gleeson RA (1987) Biochemistry of an olfactory purinergic system: dephosphorylation of excitatory nucleotides and uptake of adenosine. J Neurochem 49:1174–1182

van Haastert PJM (1984) Guanine nucleotides modulate cell surface cAMP-binding sites in membranes of *Dictyostelium discoideum*. Biochem Biophys Res Commun 124:597–604

van Haastert PJM, Kien E (1983) Binding of cAMP derivatives to *Dictyostelium discoideum* cells: activation mechanism of the cell surface cAMP receptor. J Biol Chem 258:9636–9642

van Haastert PJM, van Walsum H, Pasveer FJ (1982) Nonequilibrium kinetics of a cyclic GMP-binding protein in *Dictyostelium discoideum*. J Cell Biol 94:271–278

van Haastert PJM, DeWit RJW, Janssens PMW, Kesbeke F, DeGoede J (1986) G-protein-mediated interconversions of cell-surface cAMP receptors and their involvement in excitation and desensitization of guanylate cyclase in *Dictyostelium discoideum*. J Biol Chem 261:6904–6911

van Houten J (1979) Membrane potential changes during chemokinesis in *Paramecium*. Science 240:1100–1103

van Houten J, Preston RR (1987) Chemoreception in single-celled organisms. In: Finger TE (ed) Neurobiology of taste and smell. John Wiley & Sons, New York, pp 11–38

van Houten J, Hauser DCR, Levandowsky M (1981) Chemosensory behavior in protozoa. In: Levandowsky M, Hutner SH (eds) Biochemistry and physiology of protozoa, vol 4. Academic Press, New York London, pp 67–124

Venter JC, DiPorzio U, Robinson DA, Shreeve SM, Lai J, Kerlavage AR, Fracek SP, Lentes K-U, Fraser CM (1988) Evolution of neurotransmitter receptor systems. Progr Neurobiol 30:105–169

Watling KJ, Bristow DR (1986) $GABA_B$ receptor-mediated enhancement of vasoactive intestinal peptide-stimulated cyclic AMP production in slices of rat cerebral cortex. J Neurochem 46:1755–1761

Weiner N, Taylor P (1985) Drugs acting at synaptic and neuroeffector junctional sites. In: Gilman AG, Goodman LS, Rall TW, Murad F. The pharmacological basis of therapeutics. Macmillan, New York, pp 66–99

Weinstock RS, Wright HN, Spiegel AM, Levine MA, Moses AM (1986) Olfactory dysfunction in humans with deficient guanine nucleotide-binding protein. Nature (London) 322:635–636

Wurster B, Schubiger K, Wick U, Gerisch G (1977) Cyclic GMP in *Dictyostelium discoideum*: oscillations and pulses in response to folic acid and cyclic AMP signals. FEBS Lett 76:141–144

Yamaguchi S (1979) The umami taste. In: Boudreau JC (ed) Food taste chemistry. Am Chem Soc, Washington, DC, pp 33–51

Yool AJ, Grau SM, Hadfield MG, Jensen RA, Markell DA, Morse DE (1986) Excess potassium induces larval metamorphosis in four marine invertebrate species. Biol Bull 170:255–266

Zimmer-Faust RK (1987) Crustacean chemical perception: towards a theory of optimal chemoreception. Biol Bull 172:10–29

Zimmer-Faust RK, Gleeson RA, Carr WES (1988) The behavioral response of spiny lobsters to ATP: evidence for mediation by P_2-like chemosensory receptors. Biol Bull 175:167–174

Zuckerman BM, Jansson HB (1984) Nematode chemotaxis and possible mechanisms of host/prey recognition. Annu Rev Phytopathol 22:95–113

Chapter 3

Metallothionein and Metallothionein-Like Proteins: Physiological Importance[1]

David W. Engel[2] and *Marius Brouwer*[3]

Contents

1 Introduction

Since the initial demonstration of a low-molecular weight cadmium-binding protein in mammals, the research on the function of these proteins, which also can bind Cu, Zn, and Hg, has been focused primarily on their detoxifying properties. Such a direction is understandable, since the synthesis of the proteins can be induced by these potentially toxic metal ions. These proteins, isolated from mammals, have structural characteristics that are very different from those of most proteins, such as having 30% of the amino acid residues as cysteine and having the capacity to bind 7 mol of cadmium, zinc, or mercury or 12 mol copper per mole of protein. The binding of metal to these proteins is through thiol bonding,

[1] The US Government has the right to retain a nonexclusive royalty-free licence in and to any copyright covering this paper

[2] National Marine Fisheries Service, NOAA, Southeast Fisheries Center, Beaufort Laboratory, Beaufort, North Carolina 28516, USA
[3] Duke University Marine Laboratory, Marine Biomedical Center, Beaufort, North Carolina 28516, USA

and therefore, the name metallothionein was proposed and accepted (throughout the text metallothionein will be denoted as MT).

The majority of research on MT has been devoted to the characterization of the molecular structure of the protein, and its importance in metal toxicology. It is thus not surprising that other than making generalized statements concerning the "natural" function of MT, little progress has been made toward demonstrating linkages between specific physiological processes and the functional properties of MT in either mammals or other animal groups. Since the authors of this review are interested in functional/regulatory physiology and biochemistry, the emphasis will be in that direction. To provide some background information, however, it will be necessary to discuss both the history of MT research and the structural and chemical characteristics of the molecule before dealing with the functional and comparative aspects of the protein's involvement in normal physiological processes.

2 History

Early information on cadmium concentrations in the tissues of man and other mammals led to investigations to identify the protein carrier of cadmium (Vallee 1979). A low-molecular weight cadmium-binding protein was first identified in the cortex of the horse kidney by Margoshes and Vallee (1957). Later it was purified, characterized, and designated metallothionein (Kagi and Vallee 1960, 1961), because of the high cadmium, zinc, and sulfur concentrations. The characteristics of this cadmium-binding protein include (1) a molecular weight of 6000–7000 daltons, (2) 30% of the amino acid residues as cysteine, (3) no aromatic amino acids, (4) absorption maximum at 250 nm with little absorbance at 280 nm, and (5) metal binding through thiolate bonds. Later Kojima et al. (1976) determined the amino acid sequence of the protein and Otvos and Ar-

Table 1. Summary of the characteristics of metallothionein and metallothionein-like proteins in unicellular, invertebrate, and vertebrate organisms (Information used in this table taken from Kagi and Nordberg 1979, Cherian and Goyer 1978, Engel and Roesijadi 1987)

1. Soluble intracellular protein
2. Low molecular weight of 6000 to 10000
3. Binds transition metals strongly in vivo Hg>Cu>Cd>Zn (4 to 12 atoms mol^{-1}) and will bind Ni and Co in vitro
4. High concentration of cysteine (20% to 33%)
5. Heat stable
6. Few or no aromatic or hydrophobic amino acid residues
7. Absorption spectra characteristic of metal-thiolate complexes (Cd 250 nm; Cu 270 nm; Zn 220 nm)
8. Unique amino acid sequence including a fixed distribution of cysteinyl residues
9. Structural and functional homologies with horse kidney metallothionein

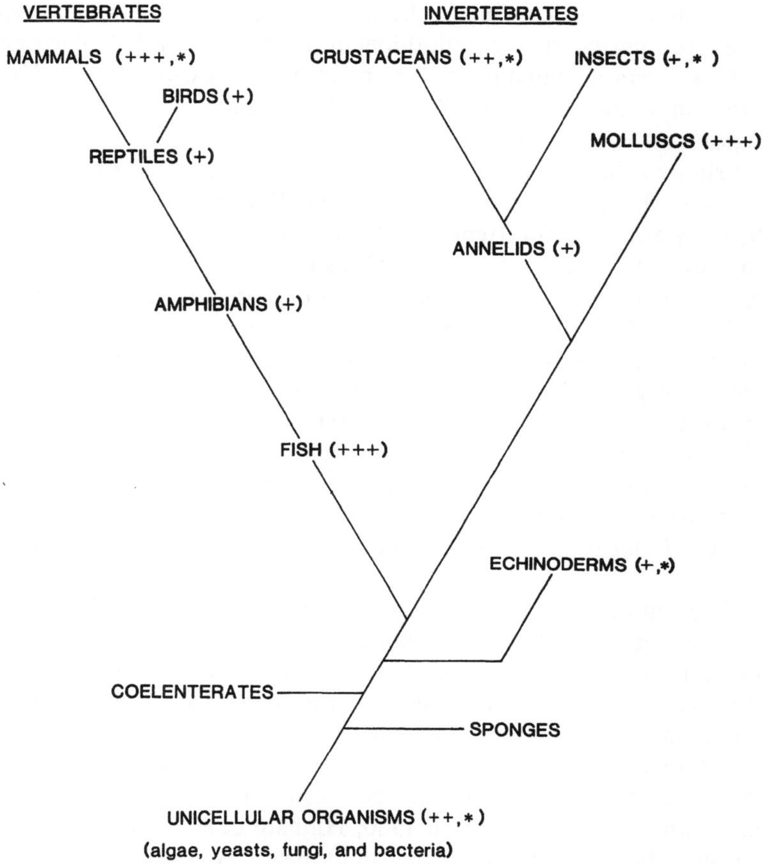

Fig. 1. A schematic phylogenetic tree showing the phyla in which metallothioneins have been identified, characterized, and sequenced. The numbers of +'s and *'s indicate the number of different species examined. Observations indicated in this diagram are referred to in the text (After L. Hyman in Barnes 1968). + = Metallothionein identified and amino acid analysis determined; * = Metallothionein amino acid sequence determined

mitage (1980) determined the molecular configuration. It was agreed at the First International Meeting on Metallothionein and other low-molecular weight metal-binding proteins in 1978 that for a protein to be considered an MT, it must conform to the characteristics of horse kidney MT (Kagi and Nordberg 1979). During the Second International Meeting (1985) the nomenclature was changed so that all MTs that resemble equine renal metallothionein in several of their features are defined as "metallothionein" (Fowler et al. 1987). A more complete listing of the characteristics of MT are given in Table 1. Since then MTs have been identified in the tissues of other mammals and also in other vertebrates and invertebrates (Fig. 1; Kagi and Nordberg 1979; Roesijadi 1981; Hamer 1986; Kagi and Kojima 1987).

Many of the proteins isolated from nonmammalian vertebrates and invertebrates are different from the typical horse kidney MT, and have been referred to as MT-like proteins. Due to the changes in nomenclature, all invertebrate MTs should be referred to as metallothionein (Fowler et al. 1987). The first invertebrate MTs were demonstrated in the American oyster by Casterline and Yip (1975) and in the limpet by Howard and Nickless (1977) and Noel-Lambot (1976). Later Ridlington and Fowler (1979) purified and partially characterized the oyster protein, and showed that the amino acid composition contained only 7% cysteine (subsequently this has been changed to 15% (Fowler et al. 1986) and 25% (Frazier 1986)). Many other invertebrate MTs have been isolated (for a listing see Kagi and Nordberg 1979; Kagi et al 1984; Hamer 1986) and purified to varying degrees (Fig. 1). Most of the proteins have the same general functional characteristics as the horse kidney MT, but their amino acid composition may be lower in cysteine and minor amounts of aromatic amino acids may be present. One of the most exhaustively purified and characterized invertebrate MTs is that from the crab, *Scylla serrata* (Olafson et al. 1979; Lerch et al. 1982; Otvos et al. 1982). The crab MT has two isoforms and the amino acid sequence is similar to the mammalian form, especially in the spacing of the cysteines. Recently Brouwer et al. (in press) have determined the sequence of the MT isolated from the American lobster *Homarus americanus*, and showed that it is very similar to the crab sequences.

In addition to the metal-binding proteins classified as MTs, a number of proteins and peptides have been isolated that classically are not MT, but appear to function in a similar manner. Stone and Overnell (1985) reviewed the occurrence of these compounds and showed them to include a variety of forms, in plant as well as in animal sources. For example, bacteria (*Pseudomonas putida*), fission yeast (*Schizosaccharomyces pombe*), and blue-green algae (*Synchococcus* sp.), all have metal-binding proteins and peptides that are not homologous to MT, but bind cadmium (Olafson et al. 1980; Higham et al. 1986a; Hayashi et al. 1986). Another class of compounds that bind metals are the phytochelatins. These low-molecular weight heavy metal-binding peptides do not resemble MT structurally but do functionally (Grill et al. 1985, 1987).

Ever since the initial isolation and characterization of MT, one of the primary questions concerning the protein's existence has been its biological/biochemical function in the living cell or organism. Kagi and Vallee (1960, p. 3464) made the following comment which to this day is not entirely resolved:

> "Even though metallothionein represents 1 to 2% of the total weight of soluble protein in kidney cortex, efforts to identify its biological function have not been successful thus far. Its physical characteristics are consistent with a role in a wide range of potential homeostatic mechanisms, either in catalysis, storage, immune phenomena, or detoxication."

It is the last mechanism, detoxication, that has received the most attention over the intervening years. This emphasis is undoubtedly due to the facility with which MT binds metals such as cadmium and the fact that synthesis of the protein can be induced by metals. The preponderance of research on MT has been directed both toward the elucidation of the structure of the molecule and clarification of its role in detoxification processes. In addition, natural occurrence of MT has been studied in organisms both from uncontaminated and metal-contaminated environments. These investigations have shown MT to be present in all organisms that have been tested. In recent experiments using the blue crab, *Callinectes sapidus*, native MT levels were shown to be significantly affected by the molt cycle. Additional evidence suggests that during this cycle MT acts as a donor molecule for copper (+I) and zinc in the biosynthesis of metalloproteins (Engel and Brouwer 1987). We are making progress, therefore, toward understanding MT's natural function in the homeostatic control of metals at the cellular and organismal levels.

3 Chemical and Biochemical Characteristics of Metallothioneins

MTs are thought to be involved in interrelated processes associated with metal metabolism, such as metal detoxification, regulation of zinc and copper, and donation of metals to apo-metalloproteins (Kagi and Nordberg 1979; Foulkes 1982; Brady 1982; Karin 1985; Cousins 1985; Hamer 1986; Engel 1987; Engel and Brouwer 1987). On a molecular level, however, the mechanisms by which metals are sequestered by or transferred from MT to other cellular ligands (apo-metalloproteins) are unknown. It is imperative, therefore, to demonstrate: (1) the metal affinities for the binding sites in MT, (2) the metal transfer rates from MT to other ligands, (3) the thermodynamics and stoichiometry of the MT-ligand interaction, and (4) the intermolecular interface of the MT-ligand complex. Since the structure of the MT molecule defines these properties, the available information on the structure of mammalian and invertebrate MTs is examined below. Available data on the thermodynamics and kinetics of metal binding/dissociation of MT are discussed in relation to metal transfer from MT to apo-metalloproteins in vitro and in vivo. In addition, attempts are made to correlate observed functional properties with structural properties of the MT molecule.

3.1 Spectroscopic Studies of Metallothionein

^{113}Cd-NMR spectroscopy has demonstrated that mammalian Cd-MT is an asymmetric molecule, composed of an N-terminal 3-Cadmium/9-Cysteine and a C-terminal 4-Cadmium/11-Cysteine domain (Otvos and Armitage 1980; Boulanger et al. 1983; Vasak et al. 1985; Frey et al. 1985; Otvos et al. 1985). The Cd-MT isolated from the portunid crab *Scylla serrata*, however, is a symmetrical molecule composed of two 3-Cadmium/9-Cysteine domains (Otvos et al. 1982). We have recently shown that the amino acid sequence of MT of the lobster *Homarus americanus* (Brouwer et al. in press) shows a strong similarity to that of *Scylla serrata* (Lerch et al. 1982), suggesting that the occurrence of two 3-metal clusters is an intrinsic property of the MTs of marine crustacea. A metal to protein stoichiometry of 6 and the presence of 18–19 cysteines per molecule also has been found in the MT of the crab *Cancer pagurus* (Overnell 1986). However, ^{113}Cd-NMR spectroscopy suggests that the metal-binding sites in this protein are as closely related to rabbit MT as to *Scylla serrata* MT (Higham et al. 1986b).

MT from the blue crab *Callinectes sapidus*, also a portunid, is very similar to that of *Scylla* in terms of its amino acid composition (Brouwer et al. 1984; Engel and Brouwer 1987) and its ^{1}H-NMR spectrum (Armitage and Brouwer, unpublished results). Further studies are required to establish whether the mode of metal binding to *Scylla* MT is typical of all marine crustaceans.

Much simpler MTs (isolated from the fungi *Neurospora crassa* and *Agaricus bisporus*) have only 25 amino acids and bind 6 Cu(I) ions to 7 cysteinyl residues (Lerch 1980; Beltramini et al. 1984; Munger and Lerch 1985). *Neurospora* MT is capable of binding transition metals in a fashion analogous to that of the mammalian and crustacean MTs. The first two metal ions are bound exclusively to thiol groups as a coupled binuclear complex. The third metal ion, however, is bound in a different environment, reflecting the smaller number of thiol ligands in *Neurospora* MT as compared to the 3-metal cluster of the mammalian and crustacean MTs (Beltramini et al. 1984).

3.2 X-Ray Diffraction Studies

Results of X-ray diffraction studies on (Cd_5Zn_2)-MT from rat liver confirmed the two-domain structure derived by spectroscopic means (Furey et al. 1986). The N-terminal domain enfolds a 3-metal cluster of one Cd and two Zn atoms coordinated by nine cysteine thiolate ligands. The C-terminal domain (residues 30 to 61) enfolds a 4-cadmium cluster coordinated by 11 cysteinyl thiolates. The domains are roughly spherical and there is limited contact between the domains (Furey et al. 1986).

The metal to cysteine assignments obtained by X-ray crystallography differ from those derived by NMR spectroscopy (Frey et al. 1985), suggesting that the conformation of the metal-reconstituted Cd_7-MT (used for NMR spectroscopy) differs from that of the native Cd_5Zn_2-MT (used for X-ray crystallography). This unresolved discrepancy may arise from the fact that pure mammalian Cd_7-MT, which is necessary for NMR spectroscopy, does not exist in vivo (Nettesheim et al. 1985), and hence has to be made artificially by metal reconstitution of the Cd,Zn-protein.

Marine invertebrate MTs differ from the mammalian MTs in the structure of their metal-binding clusters, amino acid composition, and binding of metals. Crustacean MTs, for example, contain five or six proline residues (Engel and Brouwer 1987) and mammalian MTs only two per polypeptide chain. Since proline readily accommodates turns and bends in a polypeptide chain, this difference may cause the three-dimensional structures of the two classes of MTs to be distinctly different. In fact, lobster MT does not cross-react with anti-mammalian-MT antibodies (Garvey, Winge and Brouwer, unpublished results). The more simple symmetrical marine invertebrate MTs can be prepared in vivo as Cd-, Zn-, or Cu-MT (Otvos et al. 1982; Brouwer et al. 1984, 1986; Engel and Brouwer 1984a, b). It seems possible, therefore, to perform both X-ray diffraction and NMR spectroscopy on native invertebrate MTs. Such studies may help to resolve the existing discrepancy between the NMR- and X-ray-derived structure of the mammalian MTs.

Copper binds to mammalian MT to form a Cu_{12}-MT complex (Nielson et al. 1985), but the mode of binding is still disputed. Spectral studies on $Cu(I)_6Cys_7$-MT from *Neurospora crassa* suggest that the seven cysteine residues act as a single polymeric Cu(I)μ-thiolate structure (Lerch 1980; Beltramini and Lerch 1983). It seems likely that the $(Cd,Zn)_7$-MT and Cu_{12}-MT will have different tertiary structures, which may impart metabolic specificity to Zn-MT and Cu-MT by allowing the cell to discriminate between the two forms (Nielson et al. 1985). Thus, for a better understanding of the structure/function relationships in these metal-binding proteins, it will be necessary to determine the three-dimensional structure of Cu(I)-MT.

With the background structural data on MTs discussed above it is now possible to address in the following sections the topics that are relevant to the thermodynamics and kinetics of metal binding by MT: (1) the metal-binding affinities of the two metal-binding clusters in mammalian MTs, (2) comparison of the metal-binding affinities of the two 3-metal domains of the marine invertebrate MTs to those of the mammalian MTs, (3) the order (if any) of cluster formation upon reconstitution of apo-MT with different metals, (4) preferential binding of metals to one domain or another, (5) the kinetics of metal dissociation from the different domains, and (6) the physiological importance of two (different) metal-binding domains.

4 Metal-Binding Affinities of Mammalian MTs

Despite our understanding of MT structure, relatively little is known about the affinities of Cd, Zn, or Cu for the multiple binding sites in this protein. On a semiquantitative level, pH titrations have shown that the metal-MT bond strength decreases in the order of $Cu(I) > Cd(II) > Zn(II)$ (Rupp and Weser 1978). The stability of the Cd(II)- and Zn(II)-thiolate complexes in MT, which absorb strongly around 250 and 220 nm, respectively, has been studied by spectrophotometric titrations. Based on the assumptions that (1) the metals are bound to equivalent and independent binding sites, (2) the pK of all 20 cysteines involved in metal binding are the same (pK = 8.9), and (3) the molar absorptivities of the different metal-thiolate charge-transfer transitions are the same, it was estimated that $K_{Cd} = 10^{22}\,M^{-1}$ and $K_{Zn} = 10^{18}\,M^{-1}$ (Vasak and Kagi 1983). However, the identification of thiolate clusters in metallothionein renders the simplifying assumptions listed above somewhat doubtful. In fact, several lines of evidence suggest that the two clusters of mammalian MT exhibit differential affinities for cadmium, with the tighter binding occurring at the 4-metal cluster (Otvos and Armitage 1980; Armitage et al. 1982; Nielson and Winge 1983).

4.1 Metal-Binding Affinities of Nonmammalian MTs

Little is known about the metal-binding affinities of nonmammalian MTs. During a recent workshop on “High Affinity Metal-Binding Proteins in Non-Mammalian

Species" (National Institutes of Environmental Health Sciences, Research Triangle Park, NC, 1986) the general consensus was that invertebrate MTs bind cadmium less tightly than mammalian MT (Petering and Fowler 1986). It is evident that the lack of information on the values of metal-MT stability constants for invertebrates requires more careful studies. Such fundamental information is needed to critically evaluate the potential physiological function(s) of MT.

4.2 Order of Metal Binding in Metallothioneins

To understand the biological function of MT it is necessary to determine the mechanism of both metal binding and release by MT. Reconstitution of mammalian MT by cadmium, zinc, and copper has been extensively studied by Winge and his group (Nielson and Winge 1983, 1984; Nielson et al. 1985). In the case of Cd and Zn it was found that metal binding initially occurs in the 4-metal domain. Upon saturation of the four sites in the C-terminal cluster, binding occurs in the N-terminal 3-metal domain. Both clusters are filled in a cooperative fashion. In the case of copper, the N-terminal 3-metal domain fills prior to the C-terminal 4-metal domain. The formation of the N-terminal metal cluster occurs in a cooperative fashion yielding a saturated cluster with approximately six Cu(I) ions bound (Nielson and Winge 1984). Vasak and Kagi (1981) have renatured rabbit liver MT with Co(II). Their electronic and EPR spectroscopic data show that the first four Co(II) are distributed between the two clusters, in complete contrast with the data for Cd, Zn, and Cu reported above. Differential chemical modification of cysteines in MT reconstituted with varying amounts of cadmium show that cluster formation does not occur until the addition of more than 3 equiv. of Cd (Bernhard et al. 1986). At Cd-to-protein ratios between 3–5 mol of Cd/mol of MT, formation of the 4-metal cluster occurs.

Metal reconstitution experiments with nonmammalian MTs are limited to *Neurospora crassa* MT (Beltramini et al. 1984). Since this protein consists of a single domain, the reported results do not shed light on the controversy concerning the pathway of metal reconstitution of the double-domain mammalian MTs.

4.3 Domain-Selective Metal Binding in Metallothioneins

Since MTs bind a variety of metal ions, it is important to know whether different metals may bind to the same MT molecule and whether binding to the two domains is random or specific. Cu,Zn-MT has been isolated from calf livers. The Zn could be replaced by ^{113}Cd and subsequent NMR spectroscopy showed Cd to be bound to the C-terminal 4-metal cluster and Cu(I) to the N-terminal 3-metal cluster (Briggs and Armitage 1982). This observation has been corroborated by metal reconstitution experiments which showed that Cu(I) binds preferentially to the 3-metal domain in rat liver MT (Nielson and Winge 1984).

Measurements of metal concentrations on hepatic MT isolated from cadmium-exposed animals showed the presence of both zinc and cadmium. The metal ions in Cd, Zn-MT are distributed so that the C-terminal cluster contains four Cd ions and the N-terminal cluster two Zn and one Cd (Otvos and Armitage 1982; Furey et al. 1986). Before addressing the functionally relevant question of why mammalian cells produce mixed-metal, Cd,Zn-MT, in response to Cd exposure, it is first necessary to understand how these proteins are produced in vivo (Nettesheim et al. 1985). It was discovered that the native cluster composition (four Cd in the C-terminal, two Zn, one Cd in the N-terminal cluster) could be reproduced exactly by simply mixing together the appropriate amounts of Cd_7-MT and Zn_7-MT and allowing intermolecular metal exchange to occur. The driving force for this reaction appears to be the relative thermodynamic instability of 3 metal clusters containing Cd (Nettesheim et al. 1985). In this light, it seems somewhat unexpected that the symmetrical marine invertebrate MTs, composed of two 3-metal-binding clusters can be prepared in vivo as a pure Cd_6-MT, without containing Zn (Otvos et al. 1982; Brouwer et al. 1984). Studies on invertebrate Cd_6-MT and Zn_6-MT in solution must be carried out to determine whether metal exchange is an intrinsic property of metallothioneins and hence may be of physiological significance.

4.4 Kinetics of Metal Dissociation from Horse, Crab, and Fish Metallothioneins

The kinetics of the removal of Zn and Cd by EDTA from Cd,Zn-MT from horse kidney has been studied by Li et al. (1980). Zinc removal from Cd,Zn-MT, under pseudo first-order conditions for EDTA, was a triphasic process, and was much more rapid than the removal of Cd. The pseudo first-order rate constants for the removal of both metals are given in Table 2.

Winge and Miklossy (1982) showed that the reactivity of native Cd,Zn-MT from rat liver toward EDTA differed from that of the purified C-terminal 4-Cd metal domain. The slower reactivity of the native MT with EDTA strongly suggests that the juxtaposition of the two domains exerts a stabilizing effect on each other. Such observations need to be substantiated for different MTs and different metals, because they may indicate that the occurrence of two-metal binding domains in MTs is functionally and physiologically important.

Table 2. Metal removal from metallothionein by EDTA (Li et al. 1980; Higham et al. 1986b)

Source	Protein	Metal	Pseudo first-order rate constants (s^{-1})			Metal removed (%)
			k_1	k_2	k_3	
Horse kidney	Cd, Zn-MT	Zn	Fast	8.9×10^{-4}	1.7×10^{-4}	100
	Cd:Zn = 5:2	Cd	Fast	3.7×10^{-6}	–	100
Plaice	Cd, Zn-MT	Zn	Fast	1.8×10^{-3}	1.3×10^{-4}	100
	Cd:Zn = 4:1	Cd	Fast	Very slow		5
Crab	Cd, Zn-MT	Zn	Fast	8.1×10^{-3}	4.0×10^{-4}	100
	Cd:Zn = 16:1	Cd	Fast	1.8×10^{-2}	2.2×10^{-4}	12

MTs isolated from the hepatopancreas of the crab, *Cancer pagurus*, and the plaice *Pleuronectes platessa*, after cadmium injection, are predominantly cadmium proteins containing only small amounts of zinc. The rates of removal of cadmium- and zinc by EDTA were monitored directly from the intensity of the proton NMR resonances due to the cadmium and zinc-EDTA complexes (Higham et al. 1986b), and are given in Table 2.

Comparison of the rate constants of metal removal from the three different MTs is difficult since the results were obtained with proteins carrying different amounts of cadmium and zinc, but some conclusions are possible. First, zinc removal from all three proteins occurs in three steps, and the rate of the first step has the same order of magnitude for all three MTs. These similarities suggest that the proteins contain one or two sites that readily exchange zinc with ligands in solution. It seems likely therefore that this site is involved in the donation of zinc to apoenzymes, as will be discussed in greater detail in the following section. Second, the rate constant of the second step of zinc transfer to EDTA is similar for the vertebrate MTs ($k_2 = 9 \times 10^{-4}\,s^{-1}$ for horse and $1.8 \times 10^{-3}\,s^{-1}$ for plaice) and faster for the crab protein ($k_2 = 8 \times 10^{-3}\,s^{-1}$), and the same is true of the third step: $k_3 = 1.7 \times 10^{-4}\,s^{-1}$ for horse, $1.3 \times 10^{-4}\,s^{-1}$ for plaice, and $4 \times 10^{-4}\,s^{-1}$ for crab. Third, all three MTs contain a very reactive cadmium-binding site, but only crab MT contains additional cadmium-binding sites, from which cadmium can be easily removed. It seems evident that the crab MT with its two 3-metal domains is less stable than the vertebrate MTs with their 3- and 4-metal domains.

The limited data presented above makes it clear that it is necessary to carry out kinetic studies on the removal of zinc and cadmium from well-defined, single-metal, and mixed-metal mammalian and invertebrate metallothioneins. No data are available on the thermodynamic and kinetic stability of Cu-MTs. This lack of information stands in the way of a critical evaluation of the potential biological function of Cu-MT.

4.5 Metal Transfer from Metallothionein to Apo-Metalloproteins in Vitro

As mentioned above, Zn-MTs contain reactive zinc that may be transferred from MT to the active site of enzymes that require Zn for their biological activity. Several studies have shown that this reaction does indeed take place (Udom and Brady 1980; Li et al. 1980; Winge and Miklossy 1982). Zn-MT from rat liver, for example, has been used to reactivate the apo-forms of the zinc enzymes alcohol dehydrogenase, yeast aldolase, thermolysin, alkaline phosphatase, and carbonic anhydrase. It was concluded that metals were directly transferred from MT to the apoenzymes, since zinc chelators had minimal effects on the reactivation when added after the incubation of MT+apoenzyme, but inhibited reactivation when added to MT prior to incubation with apoenzyme (Udom and Brady 1980). Similarly, Li et al. (1980) have demonstrated that apocarbonic anhydrase abstracts zinc from MT in second-order processes that are two to three orders of magnitude more rapid than those involving EDTA. The temperature dependence of the second-order rate constants suggested a direct interaction, possibly binding, of the two proteins in the rate-determining transition state of this reaction (Li et al. 1980).

It is well established that two major polymorphic forms of mammalian MT exist in various tissues and cell lines and that the isoforms differ to a limited extent in their primary and tertiary structure (Winge and Miklossy 1982). It is of considerable interest to determine whether these differences affect the metal-binding sites, which may be of significance in the physiological functioning of metallothionein. Winge and Miklossy (1982) studied the kinetics of metal removal from the two polymorphic forms of Zn-MT isolated from rat liver. They found that isoform I invariably restored greater esterase activity to apocarbonic anhydrase than did isoform II. This kinetic variation was also apparent when the reactivity of MT with EDTA was studied. EDTA removed Zn from Zn-MT isoform I at a greater rate compared to isoform II. The significance of these differences is unclear, but it is conceivable that the dissimilarities in conformation and Zn-binding affinity of MT may correlate with a functional distinction (Winge and Miklossy 1982).

The mechanism by which copper is incorporated into metalloproteins has not been elucidated (Cousins 1985), but free Cu(II) ions do not seem to be involved in the process (Holtzman and Gaumnitz 1970; Keyhani and Keyhani 1975). The hypothesis that MT plays a role in copper insertion in metalloproteins is therefore tenable. Reports in the literature on this putative function of copper-MT seem conflicting. Cu(I)-MT, isolated from rat liver, cannot reactivate either bovine or rat superoxide dismutase (Geller and Winge 1982), and Cu(I)-MT from *Neurospora* cannot reactivate *Neurospora* tyrosinase (Beltramini and Lerch 1982), but both apoenzymes can be reconstituted with Cu(II) in solution (Kertesz et al. 1972; Rigo et al. 1978). Hemocyanins, the blood oxygen carrier of many molluscs and arthropods, contain two Cu(I) ions in their active site and can only be reconstituted with Cu(I) (Konings et al. 1969). These proteins are therefore potential substrates for Cu(I)-MT. Apohemocyanin from the crab *Carcinus maenas* cannot be reactivated with Cu(I)-MT from *Neurospora* (Beltramini and Lerch 1982), but apohemocyanin from the American lobster *Homarus americanus* can be partially reactivated by Cu(I)-MT isolated from lobster hepatopancreas (Brouwer et al. 1986; Engel and Brouwer 1987). Recently we have demonstrated that Cu(I)-MT from the lobster can be separated into three components. The first two are related in primary structure to equine renal MT, but the third is not. Only the third, chemically distinct form of MT can transfer its Cu(I) to the active site of apohemocyanin (Brouwer et al. in press). The latter process only takes place in the absence of oxygen, presumably because the newly formed active sites are initially distorted and oxidized by oxygen. Cu(I) transfer from MT to apohemocyanin is followed by a rearrangement of the reconstituted binuclear-copper site into the native stereochemical configuration. This slow conformational change is the rate-limiting step in the formation of active oxygen binding sites (Engel and Brouwer 1987; Brouwer and Brouwer-Hoexum 1989).

4.6 In Vivo Studies

As discussed above, Zn-MT participates in ligand substitution reactions in vitro. A few studies have appeared lately that examined such reactions in vivo. When

Ehrlich cells were placed in metal-free culture media they rapidly lost zinc from the MT pool to other parts of the cell. The loss of zinc is much more rapid than can be accounted for by the biodegradation of MT. Apparently, the protein directly donates zinc to other sites in the cell by ligand substitution reactions (Villalobos et al. 1986). MT seems to mediate the interaction of both zinc and lead with d-aminolevulinic acid dehydratase in rat liver in vitro and in vivo (Goering and Fowler 1987).

Evidence that both Zn-MT and Cu-MT play an important role in metal metabolism of the blue crab will be presented in a following section of this review (Engel 1987; Engel and Brouwer 1987).

5 Aspects of Metallothionein Function

The major objective, and one of the most challenging aspects of this review, is to discuss the function(s) of MT in the different phyla and kingdoms. Kagi and Vallee (1960) and Vallee (1987) have indicated that the physiological function of MT is still eluding us, even though great advances have been made in describing and quantifying the biochemical and physical characteristics of the molecule. While the structural and metal-binding characteristics of MT have been demonstrated using established chemical and biochemical techniques, it has been far more difficult to show how MT participates in routine metal metabolism. This situation seems somewhat paradoxical, since it is obvious that MT must participate in the regulation and buffering of copper and zinc in all organisms where MT has been demonstrated. At the present time, about the only function that has been demonstrated concerns the detoxification of elevated intracellular metal concentrations that are the result of either acute exposures in the laboratory or chronic environmental exposures.

The existence of at least two interactive pools of MT in the cytosol of organisms has been postulated (Fig. 2) (Engel and Roesijadi 1987). These two

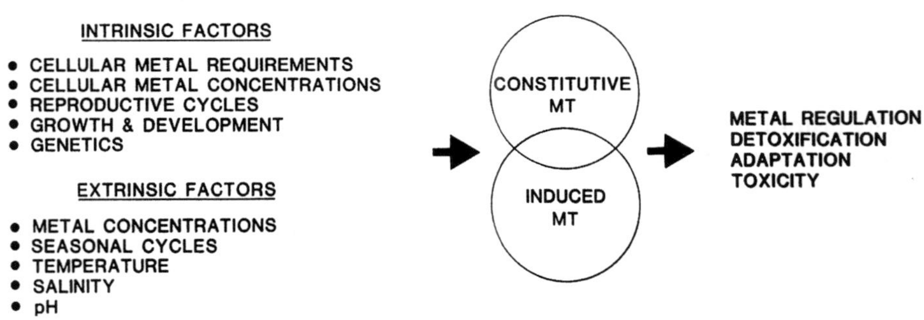

Fig. 2. A diagram describing the presence of two hypothesized intracellular pools of metallothionein (*MT*), constitutive and induced, and the factors, both intrinsic and extrinsic, that control the concentrations of these proteins. The intrinsic factors tend to be those that exist in metabolizing tissues or organisms, while the extrinsic factors are those that are contributed by the environment (After Engel and Roesijadi 1987)

pools are the induced (e.g., Cd-MT) and constitutive MT (e.g., Cu,Zn-MT). As the diagram indicates, the two pools are interrelated and their contributions to the total cytosolic concentration can be influenced by different factors (i.e., intrinsic and extrinsic). The intrinsic factors are those that are related to the regulation of metals for normal routine metabolism, growth reproduction, and development. The extrinsic factors are those that are primarily related to the environment in which the organism lives and are highly variable and episodic. The functions of the two pools also may be different, one concerned with transient changes in the cytosolic metal concentration due to environmental fluctuations, and the other with normal metal regulatory processes. Such an orientation leaves us with two dissimilar forcing functions for MT activity: metal contamination and normal physiology. However, during normal cellular activity the processes of regulation and detoxification may be very closely associated and impossible to separate, since detoxification is simply an extreme case of metal regulation.

5.1 Function in Mammals

From the first description and demonstration of cadmium-MT in the horse kidney (Margoshes and Vallee 1957; Kagi and Vallee 1960, 1961) the majority of the research effort has been devoted to the cadmium-induced form of the protein and its function in cadmium detoxification. In an industrialized society this research focus is reasonable since cadmium is an important industrial metal and may be released to the environment during the refining of other metals, such as zinc (Hiatt and Huff 1975; Uthe and Zitko 1980; Mohlenberg and Jensen 1980). The primary exposure pathways for man are through the inhalation of aerosols that contain cadmium or through the ingestion of foods that contain the metal in significant concentrations (Hiatt and Huff 1975; Friberg 1984). The actual efficiency with which mammalian MT sequesters and stores cadmium is well documented (Cherian and Goyer 1978; Kagi et al. 1984; Hamer 1986). In mammals the sequestration of cadmium may occur either in the gut or liver, and then the cadmium-MT complex is transported to and is stored in the kidney, and turns over at a slow rate (Ridlington et al. 1981). The build up of cadmium in the kidney over time can cause serious physiological problems that can result in pathological conditions (Friberg 1984). These particular problems were discussed in detail at a meeting entitled "Metallothionein and Cadmium Nephrotoxicity Conference" at the National Institutes of Environmental Health Sciences, Research Triangle Park, North Carolina (1984) (see *Environmental Health Perspectives*, Vol. 54).

While the primary focus of mammalian MT research has been on the processes of cadmium detoxification, some efforts have been made to demonstrate the functions of constitutive MT in routine metal metabolism (Brady 1982; Webb and Cain 1982; Cousins 1985). The two areas that have been explored most extensively concern the relationship between MT induction and stress, and the involvement of MT in neonatal metal metabolism. Since stress reactions constitute a complex area of research, we will only mention briefly how MT may be involved in such reactions. Oh et al. (1978) and Cousins (1982, 1985) discussed the role of MT in the response to stress and in the metabolism of copper and zinc in the rat.

Since glucocorticoid concentrations are correlated with stress, the observation that these steroid hormones alter zinc metabolism and increase MT synthesis (Karin et al. 1980a, b, 1981) in the liver without the administration of exogenous zinc, supports the hypothesis that MT is active in normal stress-related metal metabolism (Karin 1985). In another investigation, Oh et al. (1978) demonstrated that stresses such as cold, heat, exercise, and CCl_4 intoxication cause an increase in the zinc on MT, but do not significantly increase the liver or kidney zinc concentrations.

In regard to the second area of research, metal metabolism in the gestating and neonatal mammal, a number of investigators (Hartmann and Weser 1977; Johnson and Evans 1980; Brady and Webb 1981; Ebadi et al. 1984) have shown that during gestation and at birth animals have a high requirement for copper and zinc. Templeton et al. (1985) showed that liver MT concentrations increase dramatically at day 18 of gestation in the rat and continue to increase through birth. In another investigation a copper/zinc thionein was demonstrated in the distal portion of the gut in the neonatal rat, which is presumably involved in the absorption of copper and zinc from the maternal milk (Johnson and Evans 1980). It appears, therefore, that MT has a role in the metabolism of copper and zinc during normal mammalian development. Such evidence is not irrefutable, but is suggestive of a primary regulatory function of MT. However, the functional linkage showing direct involvement of MT in the donation of zinc or copper to apo-metalloproteins has not been demonstrated conclusively.

MT has also been shown to be an effective free-radical scavenger which can be important both in normal cellular metabolism and in modifying radiation sensitivity. Thornalley and Vasak (1985) demonstrated that rabbit liver MT scavenged free-hydroxyl and superoxide radicals from the xanthine/xanthine oxidase reaction more effectively than bovine serum albumin. During the radiolysis of aqueous aerobic solutions, both hydroxyl and superoxide radicals are formed that can cause damage at the cellular and biochemical levels in organisms or cells. Thornalley and Vasak (1985) also demonstrated that MT was an effective scavenger of radiation-produced free radicals and, therefore, may have radiation protective capabilities. Bakka et al. (1982) and Matsubara et al. (1986, 1987) have demonstrated both in cell cultures and in mice that the induction of MT produces a significant level of radiation protection. Both metal injection and skin excision caused the induction of MT synthesis in mice, which led to increased radiation resistance (Matsubara et al. 1987). The apparent radiation protection conferred by MT probably involves the same mechanisms that are involved in other sulfhydryl radioprotective agents. Whether the radiation protective effect is due to free-radical scavenging or the mobilization of zinc for wound healing is unclear, but the results support the earlier observations of Oh et al. (1978).

5.2 Function in Lower Animals

While the primary concern of mammalian MT research has been related to human health, research on the nonmammalian species has been oriented toward understanding how MT may alter the effects of environmental metal contamination. A wide variety of organisms has been examined in the laboratory, including sea urchins, polychaetes, mussels, crabs, insects, fish, frogs, turtles, birds, and lizards (Weser et al. 1973; Brown and Chatel 1978; Ridlington and Fowler 1979; Rainbow and Scott 1979; George 1980; Lerch et al. 1982; Frazier and George 1983; Suzuki and Tanaka 1983; Klaverkamp et al. 1984; Yamamura and Suzuki 1984a, b; Nemer et al. 1985; Flos et al. 1986; Frazier et al. 1985; Hamilton and Mehrle 1986; Jenkins and Sanders 1986; Maroni et al. 1986; Mokdad et al. 1987). The primary emphasis of the studies on the induction and structure of MT has been on the metabolism of cadmium. This research direction has been predicated on the fact that cadmium is toxic to both aquatic and terrestrial vertebrate and invertebrate organisms, and is usually a component of industrial effluents (Hiatt and Huff 1975; Uthe and Zitko 1980; Mohlenberg and Jensen 1980). Among the nonmammalian species that have been examined, cadmium has been shown to be an effective inducer of MT synthesis (Kagi and Nordberg 1979; Roesijadi 1981; Engel and Brouwer 1984a, b; Hamer 1986; Petering and Fowler 1986). One of the most extensively studied organisms is the mussel, *Mytilus edulis*. George and co-workers have demonstrated both the kinetics of MT induction by cadmium in the mussel and its intracellular and cytosolic partitioning (George et al. 1979; George 1980; George and Viarengo 1985; Frazier 1986). Preexposure to low concentrations of metals has been shown to induce MT and provide protection against acute metal toxicity (Cherian and Nordberg 1983). Roesijadi et al. (1982) and Roesijadi and Fellingham (1987) have shown that preexposure of mussels to low concentrations of copper, cadmium, zinc, and mercury decreases the acute toxic effects of mercury. Buckley et al. (1982) and McCarter et al. (1982) demonstrated that salmon can acclimate to chronic low concentrations of copper that result in a subsequent increased tolerance of copper. Recently, Klerks and Weis (1987) reviewed the genetic adaptation of aquatic organisms to environments that contain elevated concentrations of metals other than just cadmium. Their review clearly illustrates that adaptation to elevated concentrations of metals occurs in all phyla tested and that this adaptation involves many different processes including the induction of MT. Some of the best evidence for adaptation comes from research with annelids. Klerks and Levington (1988), using an oligochaete, demonstrated that worms from an area heavily contaminated with cadmium are more resistant to metal challenges. Young and Roesijadi (1983) demonstrated that copper-induced damage is repaired in the tentacles of a tube-dwelling polychaete after the induction of MT synthesis.

Up to this point we have been mainly concerned with cadmium, but two other environmentally important metals, copper and zinc, have been shown to be effective inducers of MT in aquatic organisms. The induction of MT by these metals, however, has not been studied as exhaustively as its induction by cadmium. Since both copper and zinc are also components of domestic and industrial effluents, their toxicological relevance may be more important to aquatic organisms than cadmium. There have been many reports that elevated environmental metal concentrations (i.e., cadmium, copper, and zinc) are correlated with the presence or concentrations of metals bound to MT (Talbot and Magee 1978; Overnell and Trewhella 1979; Noel-Lambot et al. 1980; Ray and White 1981; Weidow et al. 1982; Engel and Brouwer 1984a; Langston and Zhou 1986; Hamilton and Mehrle

1986; Olsson and Haux 1986; Brown et al. 1987). As a result of such correlations, MT has been offered as a likely candidate to monitor environmental metal concentrations (Hennig 1986; Engel and Roesijadi 1987). However, until the normal physiological function of MT is demonstrated and the variables that control its synthesis are explained, its usefulness as a monitoring tool is questionable (Engel and Roesijadi 1987).

To examine how natural processes might affect MT turnover and synthesis we have selected growth in crustaceans as a model system for study. The process of growth and molting is a physiologically complex progression of events. During molting the organism sheds and reforms its exoskeleton, breaks down and rebuilds tissues, undergoes significant changes in hormonal balance, and in some cases breaks down about two-thirds of the respiratory protein, hemocyanin, in the hemolymph. The degradation of hemocyanin is of particular interest, because it is a copper-containing protein. When this protein is degraded, copper is released and must be sequestered to protect the organism, and when the protein is resynthesized there is a metabolic requirement for this metal. Since this process involves the regulation of copper for normal metabolism, it should provide a system in which it is possible to investigate the dynamics of copper metabolism and the involvement of copper-MT in the process.

A series of experiments were conducted using the blue crab, *Callinectes sapidus*, to determine what effect the molting process had on copper and zinc tissue concentrations and their binding to MT (Engel 1987). These experiments show significant decreases of copper in the hemolymph and digestive gland of blue crabs at ecdysis, followed by an increase during the soft and paper shell stages. Zinc followed a similar pattern of change. During the premolt stages, MT primarily binds zinc (9 : 1), but at ecdysis there is an abrupt change from a zinc-dominated protein to a copper-dominated one. These changes in the metal composition of MT are positively correlated with the changes occurring in the concentrations of hemocyanin in the hemolymph. Therefore, a direct linkage seems to exist between copper-MT and the resynthesis of hemocyanin. Such a relationship would suggest that copper-MT is acting as the donor of copper (+I) for the synthesis of hemocyanin. This is an attractive model since the copper bound to both MT and hemocyanin is copper (+I).

Since the MT isolated from the American lobster, *Homarus americanus*, is similar to the MT isolated from the blue crab (Engel and Brouwer 1987), it was used as the test system for the donation of copper (+I) for hemocyanin reconstitution. Brouwer et al. (1986) demonstrated that lobster copper-MT was a potential donor of copper for the activation of lobster apohemocyanin. These experiments have been expanded (Engel and Brouwer 1987), and it has been demonstrated that copper-MT can donate copper to apohemocyanin to reconstitute its oxygen-binding properties. The rate of copper insertion into the apoprotein is more rapid than the rate of the reestablishment of oxygen binding. This difference seems to be the result of conformational changes in the molecule that are required before it becomes functional. Such data further support the hypothesis that crustacean MT has an active role in the metabolism and mobilization of metals during the molting process.

At molt there is a rapid catabolism of hemocyanin and the subsequent release of significant amounts of copper. The decrease in hemocyanin is accompanied by a decrease in the total copper concentration in the digestive gland (Engel 1987). Al-Mohanna and Nott (in press), using electron probe techniques (EDAX) to examine the metal concentrations of subcellular organelles, have demonstrated the presence of copper- and sulfur-containing, membrane-bound granules in the digestive glands of newly molted shrimp. They hypothesize that these inclusions are excreted from the digestive gland and pass out through the gut. Since these inclusions contain both copper and sulfur, it is conceivable that they contain MT-bound copper. In the blue crab, a similar process would account for the observed decrease in both total copper and MT in the digestive glands at molt. Under such circumstances MT might be acting as a detoxifying protein to sequester the elevated copper concentration in the digestive gland due to hemocyanin degradation.

It is apparent, therefore, that the function of MT in marine crustaceans appears to be dictated by metal availability and physiological need, and the ultimate control of MT turnover and synthesis may be through the hormonal complex that coordinates the molting process.

6 Concluding Remarks

The presence of MT in tissues of metazoan organisms at all stages of development and under different environmental conditions, strongly suggests that these small proteins are intimately involved in constitutive metabolism of metals. Two important topics concerning the complete elucidation of the natural function of MT are: (1) the functional significance of cadmium-MT in evolution and (2) a model system in which intracellular metal concentration changes are dramatic and rapid but do not require external sources of metal to force the changes in mobilization.

Throughout the literature on MT the overwhelming body of chemical, biochemical, and toxicological evidence concerns the sequestration and metabolism of cadmium. Since the structure of MT has been conserved throughout evolutionary time (Kagi et al. 1984), it seems likely that the ability of cadmium to bind to and induce MT must have adaptive significance. On the other hand, it might be argued that the ability of cadmium to induce MT synthesis is merely a chemical "coincidence" by virtue of the fact that both Cd^{2+} and the biologically relevant Zn^{2+} have a d^{10} electronic configuration. However, this does not explain why Cd^{2+} is so much more efficient in the induction of MT synthesis in terrestrial vertebrates than in marine invertebrates. It seems likely, therefore, that at some point in the past the ability of organisms to sequester cadmium must have had some selective advantage or the trait would have been lost from the gene pool. While this type of discussion is a bit like tilting at windmills, the biological significance of cadmium is still an enigma.

Metals in mammals are generally under tight regulatory control and therefore, the MT pool and the associated metals zinc and copper turn over slowly. It would seem, therefore, that mammals would not be the best model for the study of MT function, but during gestation or in neonatal animals these pools are more active and amenable to investigation (Brady 1982; Templeton et al. 1985). A difficulty of this system, however, is that zinc is a necessary part of many important metalloenzymes which makes it difficult to select or trace a single protein to study with respect to changes in Zn-MT. A great advantage of the crustacean system is that it is possible to follow Cu-MT and its changes and the changes in a single protein, hemocyanin (Engel and Brouwer 1987).

Another problem in delineating the constitutive function of MT is in determining whether all MTs synthesized under different conditions are the same molecule within a given species. For example, the fruit fly *Drosophila melanogaster* has been shown to have two structurally different MTs that are induced by different metals (Lastowski-Perry et al. 1985; Mokdad et al. 1987). In the intermolt blue crab the MT is dominated by copper, but during the premolt stages the MT is a zinc-dominated protein. At present it is unknown whether these MTs are the same protein and whether both metals induce the same gene. Irrespective of the answers to these questions, we may conclude that MT is a constitutive protein that is directly involved in the synthesis of copper and zinc metalloproteins at many levels in phylogeny, as well as a protein that is induced to high concentrations by metal exposure.

References

Al-Mohanna SY, Nott JA (in press) The accumulation of metals in the hepatopancreas of the shrimp *Penaeus semisulcatus* de Hann (Crustacea; Decapoda) during the molt cycle. In: Proceedings of the First Arabian Conference on Environment and Pollution, pp 195–207. Allen Press, Oxford

Armitage IM, Otvos JD, Briggs RW, Boulanger Y (1982) Structure elucidation of the metal-binding sites in metallothionein by ^{113}Cd NMR. Fed Proc 41:2974–2980

Bakka A, Johnson AS, Endresen L, Rugstad HE (1982) Radioresistance in cells with high content of metallothionein. Experientia 38:381–383

Barnes RD (1968) Invertebrate zoology. Saunders, Philadelphia, 743 pp

Beltramini M, Lerch K (1982) Copper transfer between *Neurospora* copper metallothionein and type 3 copper apoproteins. FEBS Lett 142:219–222

Beltramini M, Lerch K (1983) Spectroscopic studies on *Neurospora* copper metallothionein. Biochemistry 22:2043–2048

Beltramini M, Lerch K, Vasak M (1984) Metal substitution of *Neurospora* copper-metallothionein with bivalent metal ions. Biochemistry 23:3422–3427

Bernhard WR, Vasak M, Kagi JHR (1986) Cadmium binding by and metal cluster formation in metallothionein: a differential modification study. Biochemistry 25:1975–1980

Boulanger Y, Goodman CM, Forte CP, Fesik SW, Armitage IM (1983) Model for mammalian metallothionein structure. Proc Natl Acad Sci USA 80:1501–1505

Brady FO (1982) The physiological function of metallothionein. Trends Biochem Sci 7:143–145

Brady FO, Webb M (1981) Metabolism of zinc and copper in the neonate (zinc, copper)-thionein in the developing rat kidney and testis. J Biol Chem 256:3931–3935

Briggs RW, Armitage IM (1982) Evidence for site-selective metal binding in calf liver metallothionein. J Biol Chem 257:1259–1262

Brouwer M, Brouwer-Hoexum TM (1989) Reactivation of apohemocyanin by Cu(I)-metallothionein. In: D. Winge and D. Hamer (eds) Metal ion homeostasis: molecular biology and chemistry. UCLA Symp Molecular and cellular biology, 98. Liss, New York, pp 247–256

Brouwer M, Winge DR, Gray WR (in press) Structural and functional diversity of copper-metallothionein from the American lobster Homarus americanus. J Inorg Biochem

Brouwer M, Brouwer-Hoexum TM, Engel DW (1984) Cadmium accumulation by the blue crab, *Callinectes sapidus*: involvement of hemocyanin and characterization of cadmium-binding proteins. Mar Environ Res 14:71–88

Brouwer M, Whaling P, Engel DW (1986) Copper-metallothioneins in the American lobster, *Homarus americanus*: potential role as Cu(I) donors to apohemocyanin. Environ Health Perspect 65:93–100

Brown DA, Chatel KW (1978) Interactions between cadmium and zinc in cytoplasm of duck liver and kidney. Chem Biol Interact 22:271–279

Brown DA, Bay SM, Greenstein DJ, Szalay P, Hershelman GP, Ward CF, Westcott AM, Cross JN (1987) Municipal wastewater contamination in the Southern California Bight. Part II. Cytosolic distribution of contaminants and biochemical effects in fish livers. Mar Environ Res 21:135–161

Buckley JT, Roch M, McCarter JA, Rendell CA, Matheson AT (1982) Chronic exposure of coho salmon to sublethal concentrations of copper-I. Effects on growth, on accumulation and distribution of copper, and on copper tolerance. Comp Biochem Physiol 72C:15–19

Casterline JI, Yip G (1975) The distribution and binding of cadmium in oyster, soybean, and rat liver and kidney. Arch Environ Contamin Toxicol 3:319–329

Cherian MG, Goyer RA (1978) Metallothioneins and their role in the metabolism and toxicity of metals. Life Sci 23:1–10

Cherian MG, Nordberg M (1983) Cellular adaptation in metal toxicology and metallothionein. Toxicology 28:1–15

Cousins RJ (1982) Relationship of metallothionein synthesis and degradation to intracellular zinc metabolism. In: Foulkes EC (ed) Biological roles of metallothionein. Elsevier/North Holland Biomedical Press, Amsterdam New York, pp 251–260

Cousins RJ (1985) Absorption, transport, and hepatic metabolism of copper and zinc: special reference to metallothionein and ceruloplasmin. Physiol Rev 65:238–309

Ebadi M, White RJ, Swanson S (1984) The presence and function of zinc-binding in developing and mature brains. In: The neurobiology of zinc, pt. A: physiochemistry, anatomy, and techniques. Liss, New York, pp 39–57

Engel DW (1987) Metal regulation and molting in the blue crab, *Callinectes sapidus*: copper, zinc, and metallothionein. Biol Bull 172:69–82

Engel DW, Brouwer M (1984a) Trace metal-binding proteins in marine molluscs and crustaceans. Mar Environ Res 13:177–194

Engel DW, Brouwer M (1984b) Cadmium-binding proteins in the blue crab, *Callinectes sapidus*: laboratory-field comparison. Mar Environ Res 14:139–151

Engel DW, Brouwer M (1987) Metal regulation and molting in the blue crab, *Callinectes sapidus*: metallothionein function in metal metabolism. Biol Bull 173:237–249

Engel DW, Roesijadi G (1987) Metallothioneins: a monitoring tool. In: Vernberg FJ, Thurberg FP, Calabrese A, Vernberg WB (eds) Pollution and physiology of estuarine organisms. Univ South Carolina Press, Columbia, pp 421–438

Flos R, Bas J, Hidalgo J (11986) Metallothionein in the liver of the small lizard *Podarcis muralis*. Comp Biochem Physiol 83C:93–98

Foulkes EC (ed) (1982) The biological roles of metallothionein. Elsevier/North-Holland Biomedical Press, Amsterdam New York

Fowler BA, Engel DW, Brouwer M (1986) Purification and characterization studies of cadmium-binding proteins from the American oyster, *Crassostrea virginica*. Environ Health Perspect 65:63–69

Fowler BA, Hildebrand CE, Kojima Y, Webb M (1987) Nomenclature of metallothionein. In: Kagi JHR, Kojima Y, (eds) Metallothionein II. Birkhäuser, Basel, pp 19–22

Frazier JM (1986) Cadmium-binding proteins in the mussel *Mytilus edulis*. Environ Health Perspect 65:39–43

Frazier JM, George SG (1983) Cadmium kinetics in oysters, a comparative study of *Crassostrea gigas* and *Ostrea edulis*. Mar Biol 76:55–61

Frazier JM, George SG, Overnell J, Coombs TL, Kagi JHR (1985) Characterization of two molecular weight classes of cadmium-binding proteins for the mussel, *Mytilus edulis*. Comp Biochem Physiol 80C:257–262

Frey MH, Wagner G, Vasak M, Sørensen OW, Neuhaus D, Worgotter E, Kagi JHR, Ernst RR, Wutrich K (1985) Polypeptide-metal cluster connectivities in metallothionein 2 by novel H-^{113}Cd heteronuclear two-dimensional NMR experiments. J Am Chem Soc 107:6847–6851
Friberg L (1984) Cadmium and the kidney. Environ Health Perspect 54:1–11
Furey WF, Robbins AH, Clancy LL, Winge DR, Wang BC, Stout CD (1986) Crystal structure of Cd, Zn metallothionein. Science 231:704–710
Geller BL, Winge DR (1982) Metal binding sites of rat liver Cu-thionein. Arch Biochem Biophys 213:109–117
George SG (1980) Correlation of metal accumulation in mussels with the mechanisms of uptake, metabolism and detoxification: a review. Thalass Jug 16:347–365
George SG, Viarengo A (1985) A model of heavy metal homeostasis and detoxification in mussels. In: Vernberg FJ, Thurberg FP, Calabrese A, Vernberg WB (eds) Marine pollution and physiology: recent advances. Univ South Carolina Press, Columbia, pp 125–144
George SG, Carpene E, Coombs TL, Overnell J, Youngson A (1979) Characterisation of cadmium-binding proteins for mussels, *Mytilus edulis* (L.) exposed to cadmium. Biochim Biophys Acta 580:225–233
Goering PL, Fowler BA (1987) Liver Zn-thionein regulates the interaction of Zn and Pb with *a*-amino-levulinic acid dehydratase. Fed Proc 46:2239
Grill E, Winnacker E-L, Zenk MH (1985) Phytochelatins: the principal heavy-metal complexing peptides of higher plants. Science 230:674–676
Grill E, Winnacker E-L, Zenk MH (1987) Phytochelatins, a class of heavy-metal-binding peptides from plants, are functionally analogous to metallothioneins. Proc Natl Acad Sci USA 84:439–443
Hamer DH (1986) Metallothionein. Annu Rev Biochem 55:913–951
Hamilton SJ, Mehrle PM (1986) Metallothionein in fish: review of its importance in assessing stress from metal contaminants. Trans Am Fish Soc 115:596–609
Hartmann H-J, Weser U (1977) Copperthionein from fetal bovine liver. Biochim Biophys Acta 491:211–222
Hayashi Y, Nakagawa CW, Murasugi A (1986) Unique properties of Cd-binding peptides induced in fission yeast, *Schizosaccharomyces pombe*. Environ Health Perspect 65:13–19
Hennig HF-KO (1986) Metal-binding proteins as metal pollution indicators. Environ Health Perspect 65:175–187
Hiatt V, Huff JE (1975) The environmental impact of cadmium: an overview. Int J Environ Stud 7:277–285
Higham DP, Sadler PJ, Scawen MD (1986a) Cadmium-binding proteins in *Pseudomonas putida*: pseudothioneins. Environ Health Perspect 65:5–11
Higham DP, Nicholson JK, Overnell J, Sadler PJ (1986b) NMR studies of crab and plaice metallothioneins. Environ Health Perspect 65:157–165
Holtzman NA, Gaumnitz BM (1970) Studies on the rate of release and turnover of ceruloplasmin and apoceruloplasmin in rat plasma. J Biol Chem 245:2354–2358
Howard AG, Nickless G (1977) Heavy metal complexation in polluted molluscs. 1. Limpets (*Patella vulgata* and *Patella intermedia*). Chem Biol Interact 16:107–114
Jenkins KD, Sanders BM (1986) Relationships between free cadmium ion activity in seawater, cadmium accumulation and subcellular distribution, and growth in polychaetes. Environ Health Perspect 65:205–210
Johnson WT, Evans GW (1980) Isolation of a (copper, zinc)-thionein from the small intestine of neonatal rats. Biochem Biophys Res Commun 96:10–17
Kagi JHR, Kojima Y (eds) (1987) Chemistry and biochemistry of metallothionein. In: Metallothionein II. Birkhäuser, Basel, pp 25–61
Kagi JHR, Nordberg M (eds) (1979) Metallothionein. Birkhäuser, Basel, 378 pp
Kagi JHR, Vallee BL (1960) Metallothionein: a cadmium- and zinc-containing protein from equine renal cortex. J Biol Chem 235:3460–3465
Kagi JHR, Vallee BL (1961) Metallothionein: a cadmium- and zinc-containing protein from equine renal cortex. II. Physicochemical properties. J Biol Chem 236:2435–2442
Kagi JHR, Vasak M, Lerch K, Gilg DEO, Hunyiker P, Bernhard WR, Good M (1984) Structure of mammalian metallothionein. Environ Health Perspect 54:93–103
Karin M (1985) Metallothioneins: proteins in search of function. Cell 41:9–10

Karin M, Herschman HR, Weinstein D (1980a) Primary induction of metallothionein by dexamethasone in cultured rat hepatocytes. Biochem Biophys Res Commun 92:1052–1059
Karin M, Anderson RD, Slater E, Smith K, Herschman HR (1980b) Metallothionein mRNA induction in HeLa cell in response to zinc or dexamethasone is a primary induction response. Nature (London) 286:295–297
Karin M, Anderson RD, Herschman HR (1981) Induction of metallothionein mRNA in HeLa cells by dexamethasone and by heavy metals. Eur J Biochem 118:527–531
Kertesz O, Rotilio G, Brunori M, Zito R, Antonini E (1972) Kinetics of reconstitution of polyphenoloxidase from apoenzyme and copper. Biochem Biophys Res Commun 49:1208–1214
Keyhani E, Keyhani J (1975) Cytochrome c oxidase biosynthesis and assembly in *Candida utilis* yeast cells: function of copper in the assembly of active cytochrome c oxidase. Arch Biochem Biophys 167:596–602
Klaverkamp JF, MacDonald WA, Duncan DA, Wagermann R (1984) Metallothionein and acclimation to heavy metals in fish: a review. In: Cairns VW, Hodson PV, Nriagu JO (eds) Contaminant effects on fisheries. John Wiley & Sons, New York, pp 99–114
Klerks PL, Levington J (1988) Effects of heavy metals in a polluted aquatic environment. In: Levin SA, Kelly CR, Harwell MA (eds) Ecotoxicology: problems and approaches. Springer, Berlin Heidelberg New York Tokyo, pp 57–84
Klerks PL, Weis JS (1987) Genetic adaptation to heavy metals in aquatic organisms: a review. Environ Pollut 45:173–205
Kojima Y, Berger C, Vallee B, Kagi JHR (1976) Amino-acid sequence of equine renal metallothionein-1B. Proc Natl Acad Sci USA 73:3413–3417
Konings WN, van Driel R, van Bruggen EFJ, Gruber M (1969) Structure and properties of hemocyanins. V. Binding of oxygen and copper in *Helix pomatia* hemocyanin. Biochim Biophys Acta 194:55–66
Langston WJ, Zhou M (1986) Evaluation of the significance of metal-binding proteins in the gastropod *Littorina littorea*. Mar Biol 92:505–515
Lastowski-Perry D, Otto E, Maroni G (1985) Nucleotide sequence and expression of a *Drosophila* metallothionein. J Biol Chem 260:1527–1530
Lerch K (1980) Copper metallothionein, a copper-binding protein from *Neurospora crassa*. Nature (London) 284:368–370
Lerch K, Ammer D, Olafson RW (1982) Crab metallothionein: primary structures of metallothionein 1 and 2. J Biol Chem 257:2420–2426
Li TY, Kraker AJ, Shaw CF, Petering DH (1980) Ligand substitution reactions of metallothioneins with EDTA and apo-carbonic anhydrase. Proc Natl Acad Sci USA 77:6334–6338
Margoshes M, Vallee BL (1957) A cadmium protein from the equine kidney cortex. J Am Chem Soc 79:4813–4814
Maroni G, Otto E, Lastowski-Perry D (1986) Molecular and cytogenetic characterization of a metallothionein gene of *Drosophila*. Genetics 112:493–504
Matsubara J, Shida T, Ishioka K, Egawa S, Inada T, Machida K (1986) Protective effect of zinc against lethality in irradiated mice. Environ Res 41:558–567
Matsubara J, Tajima Y, Karasawa M (1987) Promotion or radioresistance by metallothionein induction prior to irradiation. Environ Res 43:66–74
McCarter JA, Matheson AT, Roch M, Olafson RW, Buckley JT (1982) Chronic exposure of coho salmon to sublethal concentrations of copper-II, distribution of copper between high- and low-molecular weight proteins in liver cytosol and the possible role of metallothionein in detoxification. Comp Biochem Physiol 72C:21–26
Mohlenberg F, Jensen A (1980) The ecotoxicology of cadmium in fresh and sea water and water pollution with cadmium in Denmark. Jensen, Kobenhavn, 42 pp
Mokdad R, Debec A, Wegnez M (1987) Metallothionein genes in *Drosophila melanogaster* constitute a dual system. Proc Natl Acad Sci USA 84:2658–2662
Munger K, Lerch K (1985) Copper metallothionein from the fungus *Agaricus bisporus*: chemical and spectroscopic properties. Biochemistry 24:6751–6755
Nemer M, Wilkinson DG, Travaglini EC, Sternberg EJ (1985) Sea urchin metallothionein sequence: key to an evolutionary diversity. Proc Natl Acad Sci USA 82:4992–4994

Nettesheim DG, Engeseth HR, Otvos JD (1985) Products of metal exchange reactions of metallothionein. Biochemistry 24:6744–6751
Nielson KB, Winge DR (1983) Order of metal binding in metallothionein. J Biol Chem 258:13063–13069
Nielson KB, Winge DR (1984) Preferential binding of copper to the β-domain of metallothionein. J Biol Chem 259:4941–4946
Nielson KB, Atkin CL, Winge DR (1985) Distinct metal-binding configurations in metallothionein. J Biol Chem 260:5342–5350
Noel-Lambot F (1976) Distribution of cadmium, zinc and copper in the mussel *Mytilis edulis*. Existence of cadmium-binding proteins similar to metallothionein. Experientia 32:324–326
Noel-Lambot F, Bouquegneau JM, Frankenne F, Desteche A (1980) Cadmium, zinc, and copper accumulation in limpets (*Patella vulgata*) from the Bristol Channel with special reference to metallothioneins. Mar Ecol Progr Ser 2:81–89
Oh SH, Deagen JT, Whanger PD, Weswig PH (1978) Biological function of metallothionein. V. Its induction in rats by various stresses. Am J Physiol 3:E282–E285
Olafson RW, Sims RG, Boto KG (1979) Isolation and characterization of the heavy metal-binding protein metallothionein from marine invertebrates. Comp Biochem Physiol 62B:407–416
Olafson RW, Loya S, Sims RG (1980) Physiological parameters of prokaryotic metallothionein induction. Biochem Biophys Res Commun 95:1495–1503
Olsson PE, Haux C (1986) Increased hepatic metallothionein content correlates to cadmium accumulation in environmentally exposed perch (*Perca fulviatilis*). Aqu Toxicol 9:231–242
Otvos JD, Armitage IM (1980) Structure of the metal clusters in rabbit liver metallothionein. Proc Natl Acad Sci USA 77:7094–7098
Otvos JD, Olafson RW, Armitage IM (1982) Structure of an invertebrate metallothionein from *Scylla serrata*. J Biol Chem 257:2427–2431
Otvos JD, Engeseth HR, Wehrli S (1985) Preparation and ^{113}Cd NMR studies of homogeneous reconstituted metallothionein: reaffirmation of the two-cluster arrangement of metals. Biochemistry 24:6735–6740
Overnell J (1986) The occurrence of cadmium in crabs (*Cancer pagurus*) and the isolation and properties of cadmium metallothionein. Environ Health Perspect 65:101–106
Overnell J, Trewhella E (1979) Evidence for natural occurrence of (cadmium, copper)-metallothionein in the crab *Cancer pagurus*. Comp Biochem Physiol 64C:69–76
Petering DH, Fowler BA (1986) Discussion summary. Roles of metallothionein and related proteins in metal metabolism and toxicity: problems and perspectives. Environ Health Perspect 65:217–224
Rainbow PS, Scott AG (1979) Two heavy metal-binding proteins in the midgut gland of the crab, *Cancer maenas*. Mar Biol 55:143–150
Ray S, White M (1981) Metallothionein-like protein in lobsters (*Homarus americanus*). Chemosphere 10:1205–1213
Ridlington JW, Fowler BA (1979) Isolation and partial characterization of a cadmium-binding protein from the American oyster (*Crassostrea virginica*). Chem Biol Interact 25:127–138
Ridlington JW, Chapman DC, Goeger DE, Whanger PD (1981) Metallothionein and Cu-chelatin: characterization of metal-binding proteins from tissues of four marine animals. Comp Biochem Physiol 70B:93–104
Rigo A, Viglino P, Brunori M, Cocco D, Calabrese L, Rotilio G (1978) The binding of copper ions to copper-free bovine superoxide dismutase. Biochem J 169:277–280
Roesijadi G (1981) The significance of low molecular weight, metallothionein-like proteins in marine invertebrates: current status. Mar Environ Res 4:167–179
Roesijadi G, Fellingham GW (1987) Influence of Cu, Cd, and Zn preexposure on Hg toxicity in the mussel *Mytilus edulis*. Can J Fish Aqu Sci 44:680–684
Roesijadi G, Drum AS, Thomas JT, Fellingham GW (1982) Enhanced mercury tolerance in marine mussels and relationship to low molecular weight, mercury-binding proteins. Mar Pollut Bull 13:250–253
Rupp H, Weser U (1978) Circular dichroism of metallothioneins. A structural approach. Biochim Biophys Acta 533:209–226
Stone H, Overnell J (1985) Non-metallothionein cadmium binding proteins. Comp Biochem Physiol 80C:9–14

Suzuki KT, Tanaka Y (1983) Induction of metallothionein and effect on essential metals in cadmium-loaded frog *Xenopus laevis*. Comp Biochem Physiol 74C:311–317
Talbot V, Magee RJ (1978) Naturally-occurring heavy metal binding proteins in invertebrates. Arch Environ Contamin Toxicol 7:73–81
Templeton DM, Banerjee D, Cherian MG (1985) Metallothionein synthesis and localization in relation to metal storage in rat liver during gestation. Can J Biochem Cell Biol 63:16–22
Thornalley PJ, Vasak M (1985) Possible role for metallothionein in protection against radiation-induced oxidative stress. Kinetics and mechanism of its reaction with superoxide and hydroxyl radicals. Biochim Biophys Acta 827:36–44
Udom AO, Brady FO (1980) Reactivation in vitro of zinc-requiring apo-enzymes by rat liver zinc-thionein. Biochem J 187:329–335
Uthe JF, Zitko V (eds) (1980) Cadmium pollution of Belledune Harbor, New Brunswick, Canada. Can Tech Rep Fish Aqu Sci 963, 107 pp
Vallee BL (1979) Metallothionein: historical review and perspectives. In: Kagi JHR, Nordberg M (eds) Metallothionein. Birkhäuser, Basel, pp 19–40
Vallee BL (1987) Implications and inferences of metallothionein structure. In: Kagi JHR, Kojima Y (eds) Metallothionein II. Birkhäuser, Basel, pp 5–16
Vasak M, Kagi JHR (1981) Metal thiolate clusters in cobalt (II)-metallothionein. Proc Natl Acad Sci USA 78:6709–6713
Vasak M, Kagi JHR (1983) Spectroscopic properties of metallothionein. Metal Ions Biol Syst 15:213–273
Vasak M, Kawkes GE, Nicholson JK, Sadler PJ (1985) ^{113}Cd NMR studies of reconstituted seven-cadmium metallothionein: evidence for structural flexibility. Biochemistry 24:740–747
Villalobos J, Krezoski S, Shaw CF, Petering DH (1986) Probable ligand substitution reactivity of zinc-metallothionein in Ehrlich cells. Fed Proc 45:1663
Webb M, Cain K (1982) Functions of metallothionein. Biochem Pharmacol 31:137–142
Weidow MA, Kneip TJ, Garte SJ (1982) Cadmium-binding proteins from the blue crab (*Callinectes sapidus*) environmentally exposed to cadmium. Environ Res 28:164–170
Weser U, Rupp H, Donay F, Linnemann F, Voelter W, Voetsch W, Jung G (1973) Characterization of Cd, Zn-thionein (metallothionein) isolated from rat and chicken liver. Eur J Biochem 39:127–140
Winge DR, Miklossy KA (1982) Domain nature of metallothionein. J Biol Chem 257:3741–3746
Yamamura M, Suzuki KT (1984a) Induction and characterization of metallothionein in the liver and kidney of Japanese quail. Comp Biochem Physiol 77A:101–106
Yamamura M, Suzuki KT (1984b) Isolation and characterization of metallothionein from the tortoise *Clemmys mutica*. Comp Biochem Physiol 79C:63–69
Young JS, Roesijadi G (1983) Reparatory adaptation to copper-induced injury and occurrence of a copper-binding protein in the polychaete, *Eudistylia vancouveri*. Mar Pollut Bull 14:30–32

Chapter 4

Nuclear Magnetic Resonance Spectroscopic Techniques for the Study of Cellular Function

W. Ross Ellington and *Robert W. Wiseman*[1]

Contents

1 Noninvasive Approaches to the Study of Cellular Metabolism

The tremendous strides in metabolic research that have taken place over the last 50 years have largely been accomplished through studies of subcellular systems. For instance, metabolite levels and enzyme activities have been determined in extracts of cells. Enzymes have been highly purified and kinetic properties analyzed

[1] Department of Biological Science, The Florida State University, Tallahassee, FL 32306, USA

Advances in Comparative and Environmental Physiology, Vol. 5

in in vitro systems where substrate concentrations far exceed enzyme concentrations. All of these approaches involve physical disruption of cells resulting in relatively homogeneous extracts devoid of the ordered microstructure found in living cells.

As Clegg (1984) has eloquently pointed out, the existence of this ordered microstructure has important impact on metabolic function in living cells. For instance, Clegg (1984) has suggested that most of cellular metabolism takes place in the phase of cellular water in intimate contact with the cytoarchitecture, implying that the bulk of cell water is ametabolic. In addition, most cells contain a variety of dynamic and static compartments of enzymes and/or metabolites, the existence of which depend upon cellular microarchitecture (Friedrich 1985). There is evidence for compartmentation of metabolic pathways as in the case of carbohydrate metabolism in vascular smooth muscle (Lynch and Paul 1987). There is also evidence that enzyme concentrations in vivo are often higher than corresponding substrate concentrations (Srivastava and Bernhard 1986). Under cellular conditions, enzymes of specific metabolic sequences may interact, passing substrates from active site to active site (Srivastava and Bernhard 1986). Enzymes may reversibly bind with various intracellular structures often displaying dramatic alterations in activity (Clarke et al. 1984). Specific metabolites may be compartmentalized. In vertebrate muscle, extracellular pyruvate equilibrates directly with the mitochondrial pyruvate pool rather than with cytosolic pyruvate (Peukhurinen et al. 1983; Schadewaldt et al. 1983; Bunger 1985). High energy phosphates are clearly compartmented in a variety of systems including skeletal muscle (Hebisch et al. 1984).

It should be obvious that the existence and control of cellular metabolic functions depends on the microarchitecture of the cells. Although useful, metabolic techniques which disrupt cellular structure may obscure subtle details of metabolic organization and control. Thus, to probe aspects of cellular biochemistry there has been considerable interest in developing noninvasive techniques. First used by Moon and Richards (1973) to study red blood cell metabolism, nuclear magnetic resonance (NMR) spectroscopy has become an extremely powerful, noninvasive tool for studying metabolic processes in vivo. Considerable advances have been made over the past decade. In this essay, we will endeavor to give a practical introduction to NMR techniques, illustrate the kinds of information that can be derived from these methods, and will highlight some recent studies of interest to comparative physiologists and biochemists.

2 Principles of Nuclear Magnetic Resonance

In this section we intend to give a theoretical as well as practical treatment of NMR and NMR techniques. More thorough discussions of this can be found in Gadian (1982), Meyer et al. (1982), and Budinger and Margolis (1986).

2.1 Nuclear Spin and Magnetism

Nuclear magnetic resonance spectroscopy obtains from the fact that certain atomic nuclei, many of biological interest, possess the property of spin. That is, they exhibit angular momentum which causes them to spin very much as a gyroscope. Because the nuclei have charge, a magnetic field is produced along the axis of spin. Thus, the nuclei behave as tiny bar magnets. ^{1}H, ^{31}P, ^{14}N, and ^{23}Na are naturally abundant nuclei which are commonly used in NMR analyses of biological systems. Furthermore, rare isotopes of certain nuclei, such as ^{13}C, are commonly used as metabolic tracers in biological systems. The angular momentum of spin is given by

$$p = \hbar[I(I+1)]^2 \ , \quad \text{where } p = \text{angular momentum}, \ \hbar = \text{Plank's constant}/2\pi, \ I = \text{spin quantum number.}$$

Some nuclei of odd mass numbers, such as ^{1}H, ^{31}P, and ^{13}C, have spin quantum numbers (I) equal to $\frac{1}{2}$.

When placed in a magnetic field, termed H_0 (also known as B_0), nuclei with spin become aligned along the axis of the magnetic field (Z-axis in Fig. 1A). Because of their spin properties, these nuclei precess at a small angle around the magnetic field (Fig. 1A). The frequency of precession (v) is given by

$$v = \gamma H_0/2\pi \ , \quad \text{where } \gamma = \text{gyromagnetic ratio, a constant specific for each nucleus in question;} \ H_0 = \text{strength of the external magnetic field.}$$

In a powerful superconducting magnet with a field strength of 5 Tesla (T), ^{1}H, ^{31}P, and ^{13}C would precess at around 213, 86.2, and 53.5 MHz, respectively (Gadian 1982). The electrons in a given nucleus also influence the frequency of precession in an external field. This has to be taken into account by modifying the above expression,

$$v = \frac{\gamma H_0}{2\pi}(1-s) \ , \quad \text{where } s = \text{chemical shift} \ .$$

Chemical shift is typically in the range of 10^{-5}. Differences in electronic configurations of ^{31}P nuclei in inorganic phosphate, creatine phosphate, and the three phosphorus nuclei of ATP result in slightly different frequencies of precession for each nucleus type. As we will see shortly, this has important practical considerations in NMR spectroscopy.

For nuclei such as ^{1}H, ^{31}P, and ^{13}C, which have spin quantum numbers of $\frac{1}{2}$, nuclei precess about the external field (H_0) at two discrete energy levels: a lower energy state precessing around the +Z axis and a high energy state precessing around the −Z axis (Fig. 1A). The difference in energy levels between these two states is directly proportional to γ, $\hbar$, and H_0. At room temperature, under equilibrium conditions, there is a slight excess (1 in 10^6) of nuclei in the low energy state. Because of the small difference in populations at the two energy levels, these systems are capable of absorbing only a small amount of energy to produce transitions from low to high energy states. NMR spectroscopic tech-

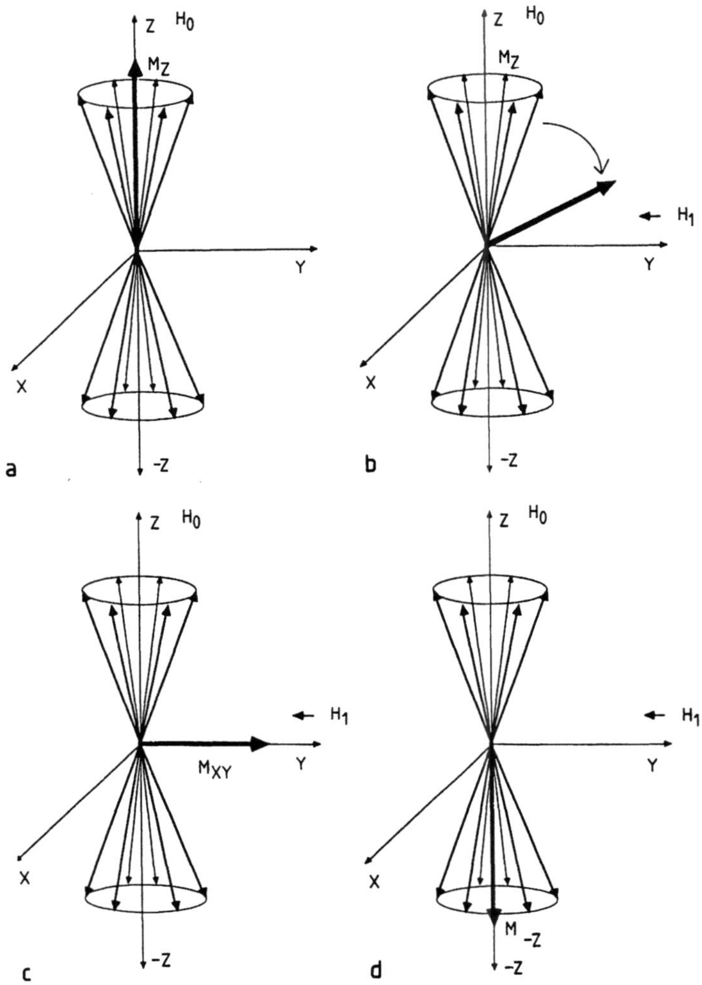

Fig. 1 a–d. The effect of application of a rf (radio frequency) pulse (H_1) on net magnetization in an applied magnetic field (H_0). See text for explanation of figure

niques are inherently insensitive and under most conditions are only able to detect nuclei in the 100 μM range in in vivo studies (Gadian 1982).

Because there is a slight excess of nuclei in the lower energy state, a macroscopic or net magnetization vector (M_z) develops around the +Z axis (Fig. 1 A). The magnitude of M_z is directly related to the total number of nuclei within the field. In a typical, modern NMR spectrometer, the sample is placed in the vicinity of a coil of wire. This coil acts as a transceiver, delivering an oscillating magnetic field, the lines of which run parallel to the X-axis. This coil later receives an oscillating magnetic field from the sample. If an oscillating magnetic field (H_1) is delivered to the sample at precisely the frequency of precession (v), some of the nuclei in the low energy state will absorb energy (resonate) and make transi-

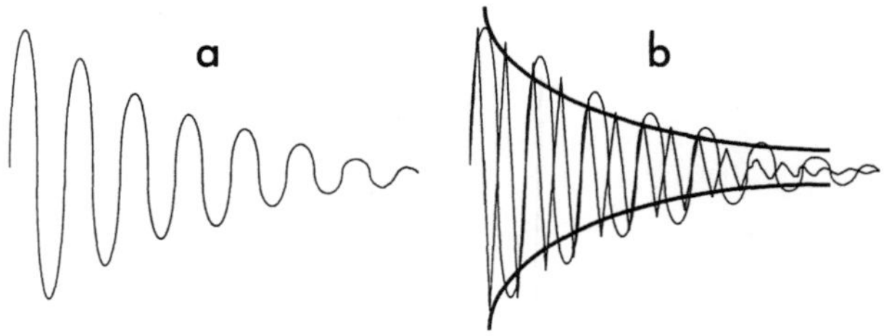

Fig. 2a, b. Simulated free induction decay (FID) signals

tions to the higher energy state (Fig. 1B). The net magnetization vector will tip away from the Z-axis and develop a magnetization vector in the XY plane (M_{xy}) (Fig. 1B). M_{xy} will rotate around the Z-axis and induce a current in the transceiver coil. If H_1 is applied long enough for populations at the two energy levels to become equal, M_{xy} is rotated completely into the XY plane and the induced current in the transceiver coil is maximal (Fig. 1C). Under these circumstances application of H_1 produces a 90° shift in the magnetization vector and M_z will be zero under these conditions. By the same token, if H_1 is applied for a longer period, a slight excess of nuclei in the high energy state develops and the magnetization vector is shifted toward the $-Z$ axis. M_{xy} will decrease and reach zero when H_1 produces a 180° shift in the net magnetization vector (Fig. 1D).

In a typical NMR experiment, the H_1 field is delivered as a pulse of a fixed duration at a specific frequency. Usually, the pulse is delivered over a specific range of frequencies (broad band pulse) covering a range of $\pm 100-15000$ Hz. If a given pulse is sufficiently long to maximize M_{xy}, we say that it is a 90° pulse. After a 90° pulse, the precession of nuclei in the XY plane will induce an oscillating current, known as the free induction decay (FID), in the transceiver coil. The FID is digitized for computer storage. On the oscilloscope, the FID appears as a decaying sine wave; the frequency being the frequency of precession (v) (Fig. 2). The presence of several nuclei differing in chemical shift results in a complex FID consisting of multiple sine waves (Fig. 2). The FID, in effect, is represented by a signal envelope (Meyer et al. 1982) (Fig. 2). After a sufficient number of FIDs have been collected, the data are converted by Fourier transform (FT) into an NMR spectrum (Fig. 3). Peak intensity (height) and area are directly related to the number of nuclei in the sample. Nuclei differing in chemical shift are resolved in the frequency domain (abscissa) of the spectrum.

Let us return to the decaying nature of the FID signal (Fig. 2). This is a manifestation of the phenomenon of relaxation. Immediately after a 90° pulse, M_{xy} will be maximal as the nuclei precess in the XY plane around the Z-axis (Fig. 4A). The precession frequencies of individual nuclei of each nucleus type (ca. β-P of ATP) may differ slightly. As a consequence, these nuclei become desynchronized resulting in the fanning out of the individual magnetization vectors (Fig. 4B) and M_{xy} will decrease. When the individual magnetization vectors are completely randomized in the XY plane, M_{xy} will be zero (Fig. 4C). This process is known a transverse or spin-spin relaxation. The decay in M_{xy} is exponential with a time constant T_2, the intrinsic spin-spin relaxation time. T_2 values depend on the electronic characteristics of the nucleus.

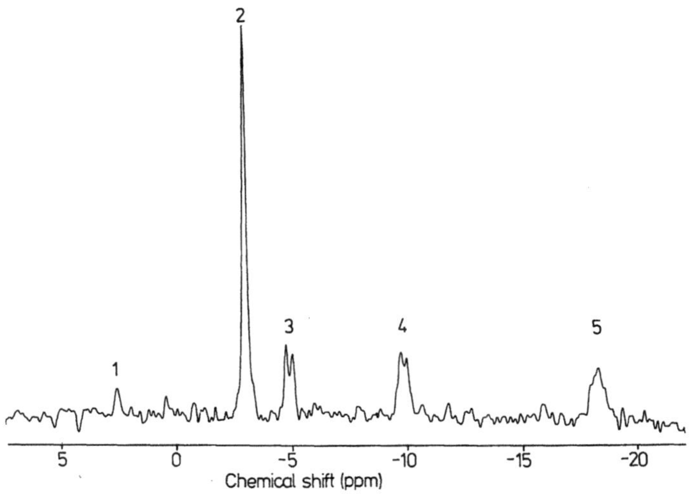

Fig. 3. ^{31}P-NMR spectrum of a perfused ventricle of the marine gastropod, *Busycon contrarium*. Chemical shift values are expressed relative to an external standard of 85% O-phosphoric acid. *Numbers* correspond to the following metabolites: *1* inorganic phosphate; *2* arginine phosphate; and *3–5* phosphates of ATP. Spectrum is from Ellington (1983)

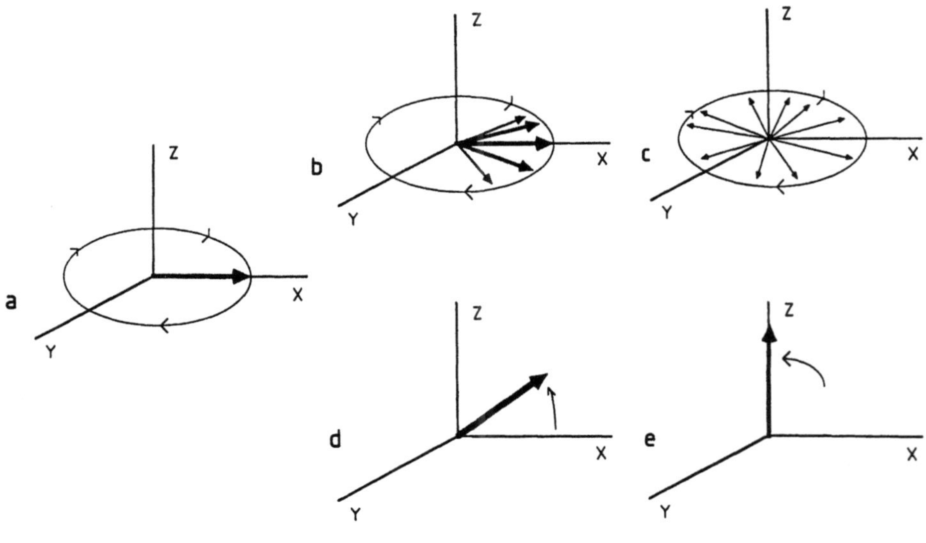

Fig. 4a–e. Spin-spin (*b–c*) and longitudinal (*d–e*) behavior. See text for explanation of figure

An additional kind of relaxation takes place after the H_1 pulse. With time, some nuclei in the high energy state will lose energy and return to the low energy state (Fig. 4D). This results in a lowering of M_{xy} and a development of a slight magnetization vector in the +Z axis (M_z). When the system is at equilibrium once again, there will be a slight surplus of nuclei in the low energy state and M_z will be maximal while M_{xy} will be zero (Fig. 4E). This return to the low energy state is known as longitudinal or spin lattice relaxation. During longitudinal relaxation, M_{xy} also decays in an exponential time course with a time constant T_1, the spin lattice relaxation time. Typically, T_1 values of key ^{31}P nuclei in vivo range from 0.2 to 5.0 s, while T_2 values are considerably shorter. Relaxation effects are extremely important from the standpoint of spectral resolution and sensitivity of biological NMR.

2.2 Practical Considerations on the NMR Experiment

2.2.1 Resolution and Line Width

Consider the individual peaks in the ^{31}P-NMR spectrum in Fig. 3. They are well resolved in the frequency domain. The differences in v are expressed as chemical shift (δ) given by the following expression:

$$\delta = \frac{v_{sample} - v_{reference}}{v_{reference}} \times 10^6$$

where δ = chemical shift expressed as ppm,
v_{sample} = resonance frequency of sample in Hz,
$v_{reference}$ = resonance frequency of a reference in Hz.

In the case of Fig. 3, the reference was 85% o-phosphoric acid. Chemical shift values are independent of field strength, even though the actual resonance frequencies change.

Note in Fig. 3 that the arginine phosphate peak appears much sharper than the β-ATP peak. Sharpness is quantified as line width, frequency span at half-peak height. Thus, the line width for arginine phosphate is less than that of β-ATP. Line width is largely determined by the T_2 of a given nucleus by the following:

$$\text{line width} = 1/\pi\ T_2.$$

Thus, the longer the spin-spin relaxation time, the narrower the line width. This phenomenon is manifested as a prolonged FID. Different nuclei may intrinsically have longer T_2 values as is the case of arginine phosphate vs β-ATP.

Spin-spin relaxation behavior also depends on the kind of molecule a particular nucleus is found in. Decreased molecular mobility results in a shorter T_2^*, hence broader lines. T_2^* is the observed spin-spin relaxation time of a nucleus under specific chemical/physical conditions. Thus, resonances even in randomly diffusing macromolecules can be extremely broad in contrast to those of low-molecular weight compounds in solution. Nuclei in molecules which bind to relatively immobile structures in cells may have extremely short T_2^* values. Immediately after an H_1 pulse, transverse relaxation will proceed so rapidly that M_{xy} will decay to zero with minimal FID development. In effect, resonances are rendered so broad that they are invisible. As we will see in a later section, nearly all of cellular ADP is NMR invisible in vivo due to its binding to actin.

In addition to intrinsic T_2 and the molecular mobility of a given nucleus, the homogeneity of the applied magnetic field has dramatic effects on the line width. In a sample containing a homogeneous distribution of a given nucleus, as might be the case of a simple aqueous solution, slight inhomogeneities in the applied magnetic field will result in nuclei showing slightly different precession frequencies depending on their position in the field. This will tend to enhance dephasing of the M_{xy} vector resulting in line broadening. Thus, inhomogeneity tends to reduce T_2^* values. In routine NMR, samples are spun in the magnetic field. This process exposes the sample to all portions of the magnetic field which tends to dampen the effects of the slight inhomogeneities of the magnetic field. In NMR of biological samples of such things as cells, tissues, and whole organs, the problems of inhomogeneity are further exacerbated by the sample geometry. In contrast to a simple solution, nuclei are very likely going to be distributed in an inhomogeneous manner in living systems. Furthermore, the presence of perfusion or superfusion lines prohibits spinning of samples. To compensate for sample inhomogeneity, the shape of the applied field is adjusted to fit the unique characteristics of a given sample. This process, called shimming, involves the application of minute electronic adjustments in the magnetic field prior to the taking of real data. The methods for magnet shimming remain beyond the technical scope of this essay. Shimming of magnets is also critical in high resolution NMR of homogeneous samples.

2.2.2 Sensitivity and Signal to Noise Considerations

As previously indicated, NMR spectroscopy is an inherently insensitive technique due to the limited capacity of the system to absorb energy. The magnitude of the FID is directly proportional to the number of nuclei within the transceiver system. Thus, the experimenter often attempts to maximize the FID by placing as much sample as possible in the detector field. At equilibrium, the difference in energy levels between the low and high energy states is directly proportional to the applied field strength (H_o). The energy difference for protons in a 400-MHz (9.4 Tesla) magnet will be greater than the energy difference in a 150-MHz magnet. Therefore, sensitivity can be enhanced by the use of a higher applied field strength.

In biological samples signal from a single FID may not be substantially greater in amplitude than electronic noise in the system. Electronic noise is often random with time. Whereas FIDs produced by repetitive pulses on a sample should be identical, given a few assumptions which we will discuss in the next section. Therefore, repetitive FIDs should sum, while the random noise will not be directly additive. In practice a single NMR spectrum is based on the Fourier transformation of the summed FIDs from a number of repetitions or as it is known in jargon, data acquisitions. It can be shown that the signal to noise ratio (S/N) is proportional to the square root of the number of data acquisitions. An experimenter programs the spectrometer to deliver an H_1 pulse, record the FID, and then delay. The entire loop is repeated until the S/N ratio is adequate for generation of a well-resolved spectrum. The number of data acquisitions required is dependent on a variety of factors including concentration of nuclei in sample, field strength, and duration of delay between pulses.

2.2.3 Saturation Effects

The summing of repetitive FIDs, although essential for enhanced S/N, introduces a major problem for the experimenter. Immediately after a 90° pulse, M_{xy} will be maximal while M_z will be zero (see Fig. 1 C). Relaxation results in a decay of M_{xy} to zero in an exponential manner with a time constant of T_1. Full relaxation (where M_z = equilibrium magnetization) takes roughly $5 \times T_1$. For ^{31}P in biological systems, complete relaxation after a 90° pulse will take from 1 to 18 s. M_{xy} really is an index of the difference in populations at the high and low energy levels. If a second H_1 pulse is delivered prior to complete relaxation of the system, the FID generated from this second pulse will not be proportional to the number of nuclei in the sample. This is because at the time of the second pulse the system was already partially displaced from equilibrium by the first pulse. This phenomenon is called the saturation effect. Saturation effects become most pronounced when the intervals between pulses are short, especially for nuclei with relatively long T_1 values. This is readily demonstrated in Fig. 5 showing the integrals of arginine phosphate and β-ATP in living muscle from spectra generated at various intervals between pulses. The T_1 values for arginine phosphate and β-ATP in the heart of a marine gastropod are 3.0 and 0.25 s, respectively (Ellington 1983). It should be obvious that the integrals for both compounds, especially arginine phosphate, differ substantially depending on the interval between pulses.

Saturation effects pose severe problems when quantifying the levels of nuclei in samples. The most straightforward way to avoid these problems is to lengthen the interval between pulses such that it is $5 \times T_1$ of the nucleus displaying the longest T_1 value. In the case of the system in Fig. 5, this would mean that the inter-

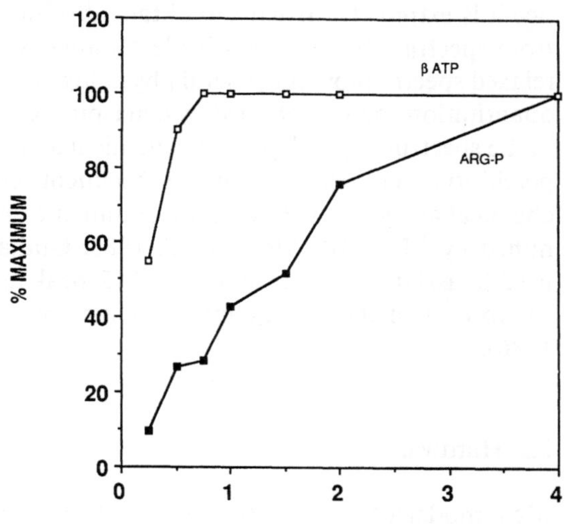

Fig. 5. The effect of various pulse intervals on the integrals for the β-ATP and arginine phosphate in a superfused preparation of the radula protractor muscle of the marine gastropod, *Busycon canaliculatum* (R.W. Wiseman unpublished observations)

val would have to be 15 s for the system to be entirely relaxed prior to the next pulse. For samples with very low concentrations of nuclei, fully relaxed spectra would require a prohibitively large amount of time to accumulate sufficient number of data acquisitions to achieve an adequate S/N ratio. Alternatively, it is possible to empirically determine "saturation factors" to compensate for saturation effects in spectra that are not fully relaxed. First used by Dawson et al. (1977), this procedure involves dividing the integrals for nuclei in a fully relaxed spectrum by integrals from a spectrum generated using a short but convenient pulse interval. The resulting saturation factor is then multiplied times integrals from subsequent spectra generated using the short pulse interval. Saturation factors for arginine phosphate and β-ATP in spectra of molluscan hearts taken with a pulse interval of 3 s are 1.55 and 1.0, respectively (Ellington 1983). The underlying assumption in the use of saturation factors is that the T_1 values of nuclei in a given tissue are constant under most physiological conditions. Recent ^{31}P-NMR studies of the rat heart have shown that T_1 values for creatine phosphate change dramatically during hypoxic insults (Cranney et al. 1987). Thus, saturation factors should be used with some caution.

It is also possible to reduce saturation effects by reducing the duration of the H_1 pulse. This is done by applying a pulse which does not achieve a 90° flip in the magnetization vector.

2.2.4 Quantitation of Levels of Compounds in Living Tissues

Prior to discussing examples of quantitation of specific metabolites in living tissues, it would be useful to make some general comments about the process of quantitation using NMR spectra. Integrals of peaks in fully relaxed NMR spectra are directly proportional to the concentrations of compounds in that tissue. Again, it must be emphasized that peaks in NMR spectra correspond to NMR visible nuclei which reflect free or unbound compounds (see Sect. 2.2.1). In practice it is extremely difficult to obtain absolute values for levels or concentrations from spectra. However, relative levels are easily estimated from integrals of fully relaxed spectra providing useful physiological information. Furthermore, absolute quantitation can be obtained by determining the absolute levels of one compound in the spectrum by independent chemical analysis (e.g., preparation of neutralized perchloric acid extracts and subsequent enzymatic analysis). For instance, chemical assays of ATP levels in a number of muscle types agree with levels determined by ^{31}P-NMR (Meyer et al. 1985; Graham et al. 1986). Thus, ATP can be used to assign absolute values to NMR peak areas (Meyer et al. 1985). Procedures for absolute metabolite quantitation have recently been discussed (Wray and Tofts 1986).

2.3 Hardware

Most modern NMR spectrometers utilize a superconducting magnet consisting of a coil of wire immersed in a Dewar flask of liquid helium which is itself immersed in a Dewar of liquid nitrogen. Once brought up to field, the magnets maintain field strength as long as lost cryogens are replaced and Dewars remain functional. The lines of the magnetic field run parallel to the bore of the magnet. Typically, spectrometers used for biological NMR experiments have relatively wide vertical bores (ca. 10 cm) in order to accommodate relatively large samples. Horizontal bore magnets are available which have sufficiently large bores to accommodate small mammals (e.g., rabbits) and human limbs.

In the simplest case, samples are placed in a probe which is inserted into the bore of the magnet. At minimum, a probe has transceiver coil wrapped around the volume occupied by the sample. The sample is placed in such a way as to be as close as possible to the coil. Provision for heating/cooling and superfusion of samples may be available. Probe design and construction are discussed in detail in Hoult (1978) and Gadian (1982). A jet of compressed gas (either air or nitrogen) is present to cause samples to spin when needed.

The remaining portions of the spectrometer consist of a radio frequency synthesizer, a preamplifier, and a computer. The frequency synthesizer generates pulses of a specific frequency or range of frequencies which are delivered for specified intervals. The preamplifier amplifies the FID returning after the pulse. Finally, the computer controls the experiments and stores and processes data. An oscilloscope is usually available to display raw and transformed data. Hard data copy is made using an XY plotter.

The experimenter has a number of options for temperature control. A thermister may be located near the transceiver coil. Air or nitrogen of a specified temperature is passed over the thermister and sample. A feedback loop activates a resistive heater located in the probe which heats the air to maintain set-point temperature. When maintaining subambient temperatures, nitrogen can be passed through a heat exchanger immersed in a propanol-dry ice bath. Even more precise temperature control can be maintained in samples superfused or perfused by passing the fluid through a heat exchanger built directly into the probe. An external circulating bath provides fluid of a constant temperature to the heat exchanger.

2.4 Samples and Sample Preparations

2.4.1 Cell-Free Samples

By far the easiest biological samples to deal with are cell-free extracts such as neutralized perchloric acid extracts of tissues. These samples are placed directly in standard NMR tubes. When spun at high speeds, samples will develop a vortex which effectively reduces the total amount of sample in the detector field. To prevent this Teflon vortex plugs are inserted into the tube and advanced down to the meniscus of the sample, permitting high speed spinning (10–20 Hz) with no vortex development.

2.4.2 Spectroscopy of Isolated Cells

In vivo NMR studies can be performed on isolated cells by packing them into NMR tubes as has been done with sea urchin sperm (Christen et al. 1983) and eggs (Inoue and Yoshioka 1980). Because of the dense packing, these packed cell suspensions show poor long-term viability and are of limited value. The major problems with the maintenance of cells in NMR systems are twofold: (1) to keep the cells in suspension and provide continuous movement of fluid over the cell surface and (2) to provide oxygen and other nutrients while at the same time removing wastes. As we shall see, a number of solutions to these problems have been developed, some more successful than others.

To maintain suspensions, there must be continuous agitation of the medium without lysing the cells. Air-driven turbines moving a small propeller have been used with limited success in maintenance of suspensions of red blood cells (Ogino et al. 1985) and cultured renal epithelial cells (Jan et al. 1987). Alternatively, a "double bubble" method (Gillies et al. 1982) is available where two small gas tubes pass into the cell suspension. One tube penetrates to near the bottom and delivers minute gas bubbles to the system. The other tube is inserted into the upper part of the cell suspension where gas is delivered at a higher flow rate. The net effect is to produce a gentle movement of cells in the suspension.

The major drawback to the above methods is that it is difficult to add or remove substances from the medium surrounding the cells. To accomplish this the cells must be superfused and immobilized in such a way as to prevent them from being washed out. Certain cells in secondary culture can be induced to attach and grow on 150 μm Cytodex (Pharmacia) beads which can be superfused in NMR tubes. Well-resolved ^{31}P-NMR spectra have been generated using embryonic mouse fibroblasts superfused in this manner (Ugurbil et al. 1981). The use and limitations of such cell spheroids has recently been reviewed (Lin et al. 1987). It is also possible to extrude cells embedded in 0.86-mm-thick agarose threads into NMR tubes. These cells show good viability and can be easily superfused as has been shown for human breast cancer cells (Neeman et al. 1987). Both approaches permit convenient control of perfusate characteristics.

We would be remiss in not mentioning the fact that NMR spectroscopy can be done on single cells, namely, the giant muscle fibers from the barnacle *Balanus nubilus* (Hoyle et al. 1973). Well-resolved ^{31}P-NMR spectra have been obtained on single fibers which were not superfused (Dubyak and Scarpa 1983; Hansen et al. 1986). Hamm and Yue (1987) recently published work utilizing a special single cell holder which permitted superfusion and maintenance of cells in excellent physiological condition during data acquisition.

2.4.3 Perfusion of Tissues and Organs

Isolated tissues are easily superfused in tubes mounted in NMR spectrometers. Inflow lines can be connected to peristaltic pumps with outflow either recycled or removed from the system. Superfusate can be gassed to a particular gas atmosphere prior to entry into the pump. Tissues must be relatively thin providing minimum diffusion distances for gases and metabolites. Skin (Lin et al. 1982) and bladder epithelia (Bond et al. 1981) are good examples of tissues which can be superfused for extended time periods.

Whole organs present a number of problems in terms of in vivo NMR spectroscopy. In the first place, there is a limit to the size of the organs that can be accommodated in the magnet. Typically, the maximal diameter of most sample chambers is around 25 mm. Secondly, an even more important consideration is a mechanism for getting perfusate to deeper cells in the organs. This latter problem is resolved by direct perfusion of the particular organ. In the case of the ventricle of the marine mollusc, *Busycon contrarium*, the ventricle is cannulated at the atrioventricular junction (Ellington 1983, 1985; Graham and Ellington 1986). Internal perfusion of the ventricle with sea water results in direct contact of a large fraction of the muscle cells with the medium because the fibers are arranged as a trabeculum. The inorganic ion content of the blood is identical to that of sea water.

Isolated working mammalian hearts have been studied extensively by NMR-spectroscopic techniques (for instance, see Gard et al. 1985). Typically, perfusion is done in the Langendorf mode, cannulation of the aorta, and retrograde flow of perfusate into the heart. Outflow is sucked out of the sample tube. It is possible to observe cardiac performance by suturing into the ventricle a fluid-filled balloon connected to a pressure transducer. Cardiac work is increased by simply raising the perfusion inflow pressure. Vascular perfusion has been used to maintain a number of other systems including rabbit bladder and uterine smooth muscle (Kushmerick et al. 1986), cat skeletal muscle (Meyer et al. 1985), and rat liver (Desmoulin et al. 1987).

2.4.4 Whole Organisms

Animals that are sufficiently small can be placed directly in NMR tubes. Large numbers of embryos of the brine shrimp *Artemia salina* have been superfused in NMR tubes and ^{31}P-NMR data collected under various physiological conditions (Busa et al. 1982). If carefully positioned in the NMR tube, spectra can be generated from much larger, intact organisms as has been done with helmith parasites (Mansour et al. 1982; Rohrer et al. 1986), sea anemones (Steen 1986), insects (Storey et al. 1984; Ziegler and Roth 1985), and crustaceans (Butler et al. 1985; Kamp and Juretschke 1987; Thebault et al. 1987). ^{31}P-NMR spectra of the tail musculature of intact, conscious newts have recently been published (Hitzig et al. 1987a). Generally, this approach is less than optimal due to physical constraints imposed by the size of the particular sample tube. Furthermore, it is difficult to position the animals in such a way to have only a single tissue/organ type within the limits of the transceiver coil. Thus, NMR spectra may be obscured by signal contributions from a number of tissues.

The inherent problems of spatial localization of signals in whole animal NMR spectroscopy have been somewhat alleviated by the use of surface coil systems. A surface coil is simply a flat coil of wire which is placed directly on or in the immediate vicinity of the organ system of interest. The coil functions as a transceiver device. Typically, a surface coil is mounted in a probe of sufficient size to accommodate the entire animal. Spatial localization is further enhanced by electronic adjustments of the probe as well as alterations in data acquisition parameters (Gadian 1982). Surface coils have been extensively utilized in NMR studies of organ metabolism in mammals. For instance, surface coils have been used to generate NMR spectra of rat brains in situ (Behar et al. 1983; Devel et al. 1985). Additional in situ studies have been done on liver (Shulman et al. 1985), hearts (Neurohr et al. 1983), and muscle (Kuwabara et al. 1986). A surface coillike system has been used with at least one invertebrate. ^{31}P-NMR studies have been conducted on the common shore crab *Carcinus maenas* (Briggs et al. 1985). In this study, individual crabs were placed in a probe consisting of a plastic chamber containing recirculating sea water. One walking leg was lifted up out of the water and inserted through a seven-turn copper NMR coil. High resolution NMR spectra of the leg musculature in situ were obtained for a period of up to 48 h (Briggs et al. 1985). It is likely that surface coil methods may be widely applicable to nonmammalian and other invertebrate systems. Gadian (1982) extensively discusses surface coil technology.

3 In Vivo Measurement of Proton and Inorganic Ion Levels

3.1 Intracellular pH

3.1.1 General Considerations

Over the past 5 years there has been increased interest in the role of the intracellular pH (pH_i) in the regulation of cellular function. Changes in pH_i often accompany or are associated with dramatic changes in cellular metabolism (see the review by Busa and Nuccitelli 1984). Mechanisms of regulation of pH_i have been reviewed by Roos and Boron (1981). Considerable efforts have been directed at developing accurate methods for estimation of intracellular H^+ concentrations. Each method tends to have inherent limitations. Use of the weak acid ^{14}C-DMO (5,5-dimethyloxazolidine-2,4-dione) (Waddell and Butler 1959) often involves extensive equilibration times as well as destruction of the tissue in question. Intracellular pH electrodes either of the recessed glass tip type (Thomas 1978) or the liquid ion exchanger type (Kurkdjian and Barbier-Brygoo 1983) are generally only applicable to relatively large cells. Entrapped pH-sensitive fluorometric probes, such as 6-carboxyfluorescein (Piwnica-Worms and Lieberman 1983), have been used in the microfluorometric measurement of pH_i in isolated cell preparations but cannot be used for large tissue masses. NMR provides a noninvasive alternative for the measurement of intracellular H^+ concentrations. In this section, we will discuss the use of NMR-spectroscopic techniques, using a number of nuclei and pH probes, to measure pH_i in a variety of different systems.

3.1.2 ^{31}P-NMR Measurement of pH_i Using the Inorganic Phosphate (Pi) Resonance

Inorganic phosphate (Pi) is typically a well-defined resonance in ^{31}P-NMR spectra of living cells and tissues (see Fig. 3). Under physiological conditions Pi has a pK_a of around 6.8. Thus, under physiological pH conditions, Pi exists as two distinct species, $H_2PO_4^{-1}$ and HPO_4^{-2}. As the pH is increased the relative proportion of the latter species increases, while the opposite is true when the pH is decreased. The unique electronic environment of the phosphorus nuclei in each species produces different precession frequencies when a Pi solution is placed in an applied magnetic field. In solution, $H_2PO_4^{-1}$ and HPO_4^{-2} are in rapid chemical exchange. Thus, NMR spectra of Pi solutions do not display two resonances, but rather show a single peak whose resonance frequency is the algebraic sum of the frequencies of the two Pi species based on their relative proportions. As pH changes, there is a systematic shift in the resonance frequencies of Pi, the bulk of these changes taking place ± 1.0 pH unit around the pK_a (Fig. 6).

Given the above phenomenon, it is possible to estimate the pH_i from the chemical shift of Pi in ^{31}P-NMR spectra using the following expression:

$$pH_i = pK'_a + \log \frac{{}^{\delta}H_2PO_4^{-1} - {}^{\delta}obs}{{}^{\delta}obs - {}^{\delta}HPO_4^{-2}} ,$$

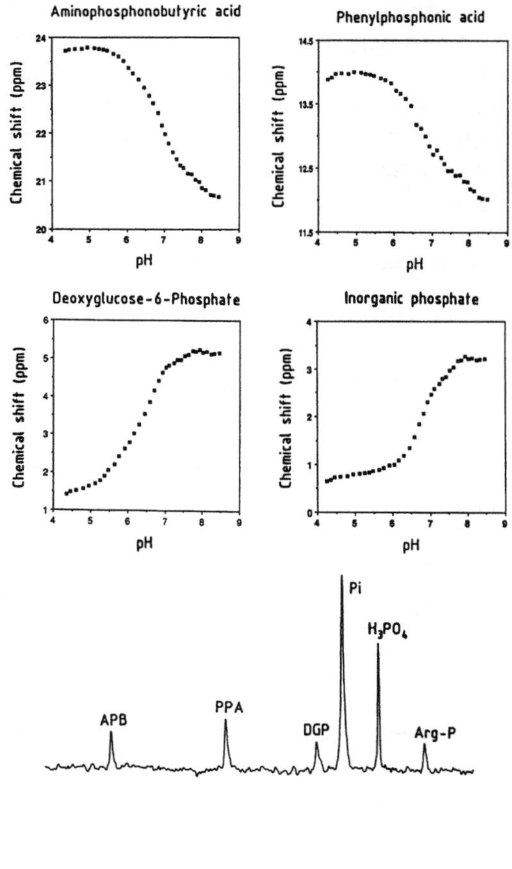

Fig. 6. The effect of pH on the chemical shift of aminophosphonobutyric acid (*APB*), phenylphosphonic acid (*PPA*), 2-deoxyglucose-6-phosphate (*DGP*), and inorganic phosphate (*Pi*). Chemical shift values are expressed relative to 85% o-phosphoric acid. *ARG-P* corresponds to arginine phosphate

where pK'_a = apparent pK_a of Pi under physiological conditions; $^{\delta}obs$ = chemical shift of Pi in spectrum; $^{\delta}H_2PO_4^{-1}$, $^{\delta}HPO_4^{-2}$ = chemical shift of each species alone under physiological conditions.

The pK'_a and chemical shift values for $H_2PO_4^{-1}$ and HPO_4^{-2} have to be determined empirically under conditions (temperature, ionic strength, and presence of organic ions) similating the physiological state. The usual approach is to prepare a model solution consisting of inorganic and organic ions present in relative concentrations similar to what might be found in vivo. For instance, a model solution consisting of 5 mM Mg-ATP, 5 mM creatine phosphate, 5 mM EGTA, 5 mM potassium phosphate, and potassium propionate has been used to simulate the cytoplasmic environment of mammalian smooth muscle (Kushmerick et al. 1986). The model solution is then titrated and sequential ^{31}P-NMR spectra obtained. The relevant parameters can then be obtained by plotting chemical shift vs pH (Fig. 6). Titration curves for Pi in centrifuged, isotonic KCl extracts of tissue samples are virtually identical to those from model solutions, indicating that binding of cytoplasmic proteins has a minimal effect on pK'_a (Jacobus et al. 1982).

The Pi observed in NMR spectra is very likely of cytoplasmic origin. In most systems, the mitochondrial contribution is trivial. Furthermore, extracellular Pi concentrations are typically low in most systems. Perfusion or superfusion with media containing Pi is usually avoided as this will complicate interpretation of the Pi signal. However, if the pH_i and extracellular pH (pH_e) are very different, two distinct resonances will be observed corresponding to the two compartments. In some cases, very broad Pi peaks are observed which may correspond to heterogeneity of cell type in a given tissue. Time resolution of measurement of pH_i depends primarily on the characteristics of the samples and the number of data acquisitions required to achieve a S/N ratio sufficient to resolve resonances. Figure 7 shows intracellular pH changes in the adductor muscle of a scallop during a brief bout of muscle contraction (Chih and Ellington 1985). Each pH_i point corresponds to data acquired over a 40-s time interval. Thus, time resolution can be quite good in certain systems.

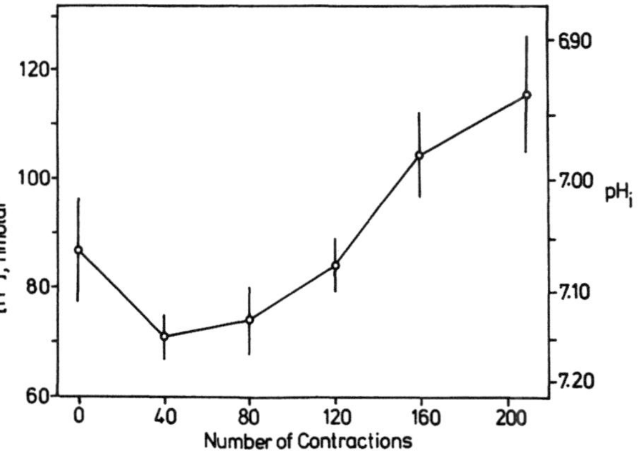

Fig. 7. Changes in intracellular pH (*pH_i*), as measured using ^{31}P-NMR spectroscopy, in the adductor muscle of the scallop *Argopecten irradians concentricus* during the time course of contractile activity (one contraction per second). Data are from Chih and Ellington (1985)

A final consideration must be mentioned concerning the use of Pi as a probe of pH_i in living tissues. This consideration is also true of other pH probes. Occasionly, the field strength of a superconducting magnet may change resulting in alterations in the precession frequency of nuclei in the applied field. If this change should occur during data acquisition, erroneous estimates of pH_i might result. To compensate for this possibility, workers often estimate the chemical shift of Pi relative to an internal standard resonance which shows no change in chemical shift with pH or for that matter with any physiological change of state. In muscles, creatine phosphate or arginine phosphate are ideal internal standards. The pK_a's for these compounds are less than 5.0 which means that under most physiological pH conditions they do not display a change in chemical shift. The problem is complicated in tissues lacking appreciable phosphagen levels. In some tissues, endogenous glycerophosphorylcholine (GPC = 0.49 ppm relative to H_3PO_4) is used an internal reference. Often, an external standard (H_3PO_4, methylene diphosphonic acid) in a small, sealed capillary is placed in the immediate vicinity of the sample in question.

3.1.3 Use of Artificial ^{31}P Probes

Inorganic phosphate levels in certain cells are so low as to render Pi a poor probe of pH_i. Furthermore, in systems such as red blood cells or organs perfused with red cell suspensions, where there is a large 2,3-diphosphoglycerate (2,3-DPG) signal, Pi often overlaps with the 2-P of 2,3-DPG (Labotka and Kleps 1983). Thus, there has been considerable interest in developing other phosphate probes of intracellular pH. Perfusion of muscles with 2-deoxyglucose results in the generation of a new resonance corresponding to 2-deoxyglucose-6-phosphate. The pK_a' of this compound is 6.13 (Meyer et al. 1985; Fig. 6, this Chap.). Estimates of pH_i in mammalian skeletal muscle using the chemical shift of 2-deoxyglucose-6-phosphate are identical to pH_i values determined via the Pi probe (Meyer et al. 1985). The major disadvantage of 2-deoxyglucose-6-phosphate is that it is not a particularly effective probe in the alkaline pH range (Fig. 6).

Methylphosphonate has a pK_a' of 7.6, is permeable across the plasma membrane, and is well resolved from other resonances in ^{31}P-NMR spectra (Labotka and Kleps 1983). This compound has been used as a pH_i probe but has some inherent limitations due to its high permeability (DeFronzo and Gillies 1987). A series of novel phosphonate pH_i indicators have been discovered by Swergold et al. (1987). These compounds (2-amino-4-phosphonobutyric acid, 2-amino-5-phosphovaleric acid, and 2-amino-6-phosphonohexanoic acid) are readily taken up by mammalian cells presumably by amino acid transporters (Swergold et al. 1987). These compounds have pK_a's ranging from 6.9 to 7.6 (see Fig. 6) and have used to estimate pH_i in fibroblasts, cells which have very low Pi levels (Swergold et al. 1987). The general applicability of these probes to other systems remains to be evaluated.

3.1.4 Fluorinated Probes of pH_i

A rather different strategy for NMR measurement of pH_i involves the use of fluorinated α-methylamino acids as indicators of pH_i (Taylor and Deutsch 1983). Fluorine is not present in any large amounts in biological systems. However, ^{19}F is a NMR visible nucleus. Taylor and Deutsch (1983) showed that fluoromethylalanines (or their methyl esters) were readily taken up by lymphocytes but were not metabolized except for conversion of the methyl esters to acids by esterases. The fluoromethyl amino acids were not cytotoxic. ^{19}F-NMR spectra showed that all probes displayed considerable change in chemical shift above and below their respective pK'_as and yielded estimates of pH_i fully compatible with estimates obtained by other methods (Taylor and Deutsch 1983). Again, the general applicability of this method has not been fully explored.

3.1.5 Estimation of pH_i Using ^1H-NMR

Frog muscles contain relatively high concentrations of the dipeptide carnosine (Seo et al. 1983). Using ^1H-NMR (see Sect. 4.3 for details) it is possible to resolve the 2-H and 4-H resonances of the histidine moiety of carnosine in intact muscle preparations. The pK'_a of this histidine portion is around 7.0 and both 2-H and 4-H resonances show dramatic changes in chemical shift as the pH_i is changed (Seo et al. 1983). Using this approach, Seo (1984) was able to measure changes in pH_i in frog sartorius muscle during anoxia and contractile activity.

3.1.6 Measurement of pH_e and Transmembrane pH Gradients by NMR

Under many experimental circumstances it is useful to be able to measure the pH_e by NMR. Phenylphosphonic acid is a useful indicator in this regard. This compound is not permeable, has a pK'_a of 7.01, and has a range of chemical shift values which are far displaced from the typical phosphate-containing compounds in ^{31}P-NMR spectra (Meyer et al. 1985). The presence of 10 mM phenylphosphonic acid in the perfusion media of cat biceps and soleus muscles had no effect on muscle viability or function. Taylor and Deutsch (1983) have used the fluoromethylalanines and their methyl esters as indicators of the transmembrane pH gradient. If lymphocytes are superfused with a medium containing fluoromethylalanine methyl ester, this compound is transported into the cell where it is immediately hydrolyzed to the acid by endogenous esterases. In effect, the extracellular compartment contains fluoromethylalanine methyl ester *only*, while the intracellular compartment contains fluoromethylalanine *only* (Taylor and Deutsch 1983). The ^{19}F resonances for both compounds are well-resolved in NMR spectra and show characteristic changes in chemical shift with pH. Thus, it is possible to simultaneously measure pH_i and pH_e in this system via ^{19}F-NMR (Taylor and Deutsch 1983).

3.2 Intracellular Free Mg^{2+} Levels

The intracellular free Mg^{2+} concentration critically influences a variety of cellular processes including energy metabolism and the modulation of Ca^{2+}-dependent processes (Garfinkel et al. 1986). Methods for measurement of intracellular Mg^{2+} have recently been reviewed (Garfinkel et al. 1986). ^{31}P-NMR measurement of free Mg^{2+} is based on an early observation by Cohn and Hughes (1962) that the β-phosphate of ATP displayed dramatic changes in chemical shift when ATP was titrated with Mg^{2+}. This property was first exploited by Gupta et al. (1978) in the measurement of free Mg^{2+} in erythrocytes by ^{31}P-NMR. The approach is similar to NMR measurement of pH_i. ATP is titrated with Mg^{2+} in a model solution while sequential NMR spectra are taken. Free Mg^{2+} is then estimated by the following equation (Kushmerick et al. 1986):

$$pX = pK' + \log \frac{^{\delta}\text{com} - {}^{\delta}\text{obs}}{^{\delta}\text{obs} - {}^{\delta}\text{dis}} ,$$

where pX = −log free Mg^{2+}; pK′ = −log apparent association constant of ATP and Mg^{2+}; $^{\delta}$com = chemical shift of β-ATP fully complexed Mg-ATP; $^{\delta}$dis = chemical shift of β-ATP of ATP alone; $^{\delta}$obs = chemical shift of β-ATP in a given in vivo spectrum.

Intracellular free Mg^{2+} concentrations ranging from 0.25 − 0.8 mM have been observed by ^{31}P-NMR in a variety of in vivo systems including skeletal muscle (Gupta and Moore 1980) and hearts (Gupta et al. 1984).

As Gupta et al. (1983) have stated, the validity of ^{31}P-NMR determinations of free Mg^{2+} is highly dependent on the accuracy of the Mg-ATP association constant. These workers have measured the association constant of Mg-ATP under physiological conditions using optical absorbance spectroscopy as well as ^{31}P-NMR. They found complete concordance in pK′ values (Gupta et al. 1983). Kushmerick et al. (1986) described a more elaborate procedure for estimation of the association constant.

Recently Morris et al. (1987) have shown that the nontoxic (+) isomer of fluorocitrate can be used as a probe of intracellular free Mg^{2+} by ^{19}F-NMR. Fluorocitrate has an association constant for magnesium of 0.84 mM. Morris et al. (1987) showed that the free Mg^{2+} in the perfused ferret heart was around 0.8 mM.

3.3 Measurement of Other Cation Levels

3.3.1 Intracellular and Extracellular Sodium Levels

^{23}Na is the natural isotope of sodium and is ubiquitously distributed at high concentrations in biological systems. ^{23}Na-NMR is quite sensitive. However, in biological systems sodium is found in both intracellular (Na_i) and extracellular (Na_e) compartments and the ^{23}Na-NMR resonances completely overlap. To circumvent these problems, paramagnetic chemical shift reagents (dysprosium [III] tripolyphosphate)(dys PPP) were developed (Gupta and Gupta 1982). The chemical shift reagent will not readily penetrate cells and literally interacts with Na_e to produce a displacement in its chemical shift relative to that of Na_i (Gupta and Gupta 1982). The Na_i and Na_e resonances become fully resolved in NMR spectra, and it is then possible to estimate sodium levels in both compartments. Using this approach, Na_i concentrations have been measured noninvasively in yeast cells (Balschi et al. 1982), bacteria (Castle et al. 1986), rabbit proximal tubule (Gullans et al. 1985), and perfused rat hearts (Pike et al. 1985). A newer shift reagent, dysprosium triethylenetetraminehexaacetate, has been gradually infused into anesthesized rats and Na_i and Na_e levels estimated in brain and muscle using ^{23}Na-NMR techniques (Naritomi et al. 1987). Use of these chemical shift reagents in whole animals is limited by their toxicity.

3.3.2 Intracellular Potassium Measurements

^{39}K is the abundant potassium isotope in biological systems and is also an NMR-visible species as is ^{23}Na. However, sensitivity for ^{39}K is much less. Ogino et al. (1985) were able to measure K_i and K_e in erythrocyte suspensions in the presence of dysPPP using ^{39}K-NMR. More recently this approach has been applied to K_i measurements in perfused frog hearts (Burstein and Fossel 1987) and rat brain, muscle liver and kidney in situ (Adam et al. 1987).

3.3.3 Intracellular Calcium Measurements

Recently, Levy et al. (1987) have synthesized a ^{19}F calcium chelator known as quinMF [2-(2-amino-4-methyl-5-fluorphenoxy)-methyl-8 aminoquinidine-N,N,N′,N′-tetraacetic acid]. This chelator has a high affinity for Ca^{2+} with an association constant of 63 nM. This compound can become entrapped in red blood cells when introduced as the methyl ester (see Sect. 3.1.6) (Levy et al. 1987). Ca-quinMF and quinMF display different resonances in ^{19}F-NMR spectra, thus it is possible to estimate Ca_i based on the association constant and the relative integrals of the chelator/Ca-chelator complex (Levy et al. 1987). The use of quinMF is somewhat complicated by the fact that it does chelate Mg^{2+} to some extent and appropriate corrections must be made for this. Levy et al. (1987) have measured internal Ca^{2+} concentrations as low as 25–30 nM in erythrocytes.

4 In Vivo Measurement of Intracellular Metabolite Levels

4.1 ^{31}P-NMR Spectroscopy

In addition to yielding information concerning changes in pH_i, ^{31}P-NMR can provide noninvasive measurements of the relative levels of key phosphate-containing metabolites in living tissue. Given the inherent limitations of sensitivity, NMR

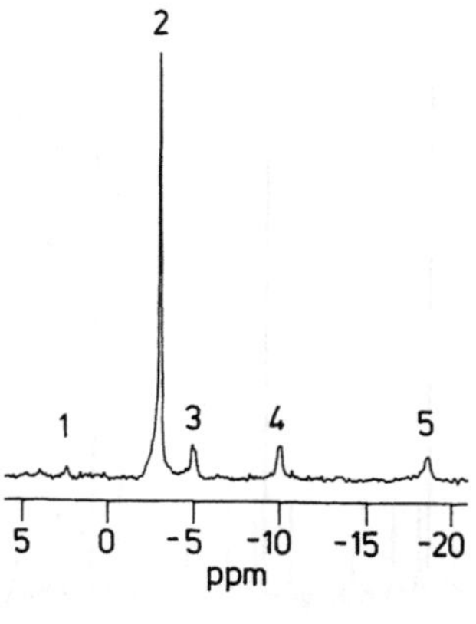

Fig. 8. ^{31}P-NMR spectrum of the adductor muscle of the scallop *Argopecten irradians concentricus*. *Numbers* correspond to the following resonances: *1* Pi; *2* arginine phosphate; *3* γ-ATP; *4* α-ATP; and *5* β-ATP. Data are from Graham et al. (1986)

Table 1. A comparison of the levels of key phosphate-containing compounds in the adductor muscle of the scallop *Argopecten irradians concentricus* as determined using ^{31}P-NMR (Graham et al. 1986) and conventional enzymatic assay of neutralized perchloric acid extracts (Chih and Ellington 1986)

Metabolite	NMR (μmol g^{-1})	Enzyme assay (μmol g^{-1})
Arginine phosphate	22.45	19.50
ATP	5.71	5.85
ADP	<0.05[a]	0.52
Inorganic phosphate	6.03	10.50

[a] The ratio of the integrals of the β-ATP and γ-ATP peaks was not significantly different from unity, indicating ADP levels are not detectable by NMR.

techniques are capable of measuring only metabolites present in concentrations exceeding 0.2 mM. However, in vivo quantitation of higher sensitivity is possible as cyclic phosphates at levels of around 70 µM in perfused tumors have been observed by NMR (R.A. Graham, personal communication). Figure 8 is a typical ^{31}P-NMR spectrum of the adductor muscle of a scallop. This spectrum displays the major peaks present in virtually every muscle type: Pi, phosphagen (arginine phosphate in this case), γ-ATP, α-ATP, and β-ATP. The γ-ATP resonance contains signal from ADP, while the α-ATP resonance contains contributions from both ADP and NAD. The β-ATP resonance contains signal from ATP alone. If the integral from the β-ATP peaks is subtracted from the integral of the γ-ATP peak, one obtains an estimate of the relative level of ADP in the system.

As indicated before, one of the great virtues of the NMR technique is that it detects *free* metabolites while *bound* compounds will not be detected. Whereas chemical techniques, which involve destruction of the tissue, assess *total* metabolite levels. Discrepancies between NMR and chemically determined estimates of phosphate metabolite levels may indicate significant binding of a particular metabolite. Table 1 shows a sample data set for scallop adductor muscle where such a discrepancy is readily evident. Arginine phosphate and ATP levels are in virtual agreement. In contrast, ADP levels are below the level of detection by

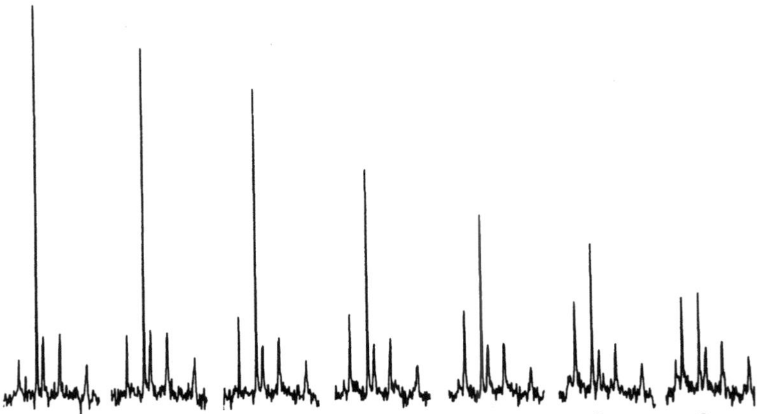

Fig. 9. Time course of changes in the ^{31}P-NMR spectrum of a perfused ventricle of the marine gastropod *Busycon contrarium* subjected to anoxia/ischemia for 6 h. Data are from Ellington (1983)

NMR while total ADP levels are at least an order of magnitude higher (Table 1). This suggests that free ADP levels are extremely low in vivo, which is very likely due to the binding of ADP to F-actin (Seraydarian et al. 1962). Table 1 also shows that Pi levels by NMR are lower than corresponding levels determined by conventional techniques. These data indicate some binding of Pi to subcellular structures (Korn et al. 1987) or perhaps reflect the artifactual hydrolysis of some arginine phosphate during the perchloric acid extraction procedure. Similar discrepancies have been observed in vertebrate skeletal muscle (Meyer et al. 1985).

The major conclusion from the above discussion of ^{31}P-NMR data is that free ADP levels are extremely low in these muscles. This also appears to be true in other muscles. This has important implications from the standpoint of regulation of metabolic processes as well as the effective energy yield of ATP hydrolysis determined by the relative free concentrations of ADP, ATP, and Pi. The discrepancy for Pi is also important. Meyer et al. (1986) demonstrated that changes in Pi levels determine the maximum extent of activation of glycogenolysis in vertebrate skeletal muscle.

Perhaps the greatest virtue of NMR techniques is the ability to follow the time course of the effect on an experimental treatment on the *same* experimental preparation. This is readily demonstrated in Fig. 9 where a gastropod ventricle was subjected to anoxia/ischemia and the time course observed by taking sequential ^{31}P-NMR spectra. Note that there are reciprocal changes in the arginine phosphate and inorganic phosphate resonances (Fig. 9). To accomplish this task by a conventional approach, it would have been necessary to use different preparations for each point in time! As might be expected, ^{31}P-NMR techniques have been used extensively to look at the effect of various physiological treatments on energy metabolism in tissues: acidosis (Garlick et al. 1979), ischemia (Lavanchy et al. 1984a), and anoxia (Barbour et al. 1984) in mammalian hearts, contractile activity and increased work loads in vertebrate skeletal muscle

(Dawson et al. 1978; Chance et al. 1985; Meyer et al. 1985), and recovery following contractile activity in crustacean muscle (Thebault et al. 1987). These studies represent but a few examples from a large body of work that has been done using ^{31}P-NMR.

^{31}P-NMR data can also be used to detect cellular heterogeneity in a given tissue. This is readily evident when looking at spectra from vertebrate skeletal muscles. Line widths for Pi in mammalian biceps muscle (Busby et al. 1978; Kushmerick et al. 1983) are usually broader than those for creatine phosphate which is not surprising since this muscle consists of a heterogeneous population of fiber types. Whereas soleus muscle, which has a more homogeneous fiber population, has a Pi line width roughly comparable to that of creatine phosphate (Kushmerick et al. 1983).

4.2 ^{13}C-NMR Spectroscopy

The natural abundance of ^{13}C is low (1.1%) and NMR sensitivity for this nucleus is considerably below that of ^{31}P. Because of this low natural abundance, ^{13}C-NMR spectroscopy is generally not practical in the study of tissue metabolism. However, the recent availability of a diverse array of ^{13}C-labeled substrates has made ^{13}C-NMR an attractive method for tracer studies of metabolism in tissues. ^{13}C-NMR techniques afford several advantages over conventional methods using ^{14}C-labeled substrates (Cohen et al. 1981 b). In the first place, ^{13}C-NMR provides information about the distribution of tracer in the carbon skeleton of each of the carbons of the metabolite in question. This capability is readily evident in Fig. 10 which depicts NMR spectra showing the distribution of ^{13}C in gastropod ventricles after the anoxic catabolism of aspartate-3-^{13}C and aspartate -4-^{13}C. To gain this information using ^{14}C radioisotope techniques, it would be necessary to isolate each metabolite, fragment the molecule into carbons, and count each carbon separately. The other major advantage of ^{13}C-NMR, of course, is that it is possible in some experimental systems to follow noninvasively the time course of incorporation of label into the tissue. The major disadvantage of the ^{13}C-NMR approach stems from the insensitivity of the overall method.

There are two general approaches to ^{13}C-NMR. The first is to do in vivo spectroscopy of living cells, tissues, and organs. Alternatively, NMR spectroscopy can be done on neutralized, perchloric acid extracts of tissues that have previously been incubated with a ^{13}C-labeled substrate. In vivo ^{13}C-NMR is limited primarily by the ability to detect poorly enriched carbons as well as the capacity to resolve carbons with similar resonance frequencies. Early studies demonstrated that it was possible to follow gluconeogenesis in rat hepatocytes using a variety of ^{13}C substrates (Cohen et al. 1979, 1981 a). More recently, carbohydrate metabolism has been studied in perfused livers (Cohen 1987 a) and hearts (Lavanchy et al. 1984 a, b). Additionally, Cohen (1987 b) was able to measure in vivo flux through pyruvate kinase in perfused rat livers using ^{13}C-NMR. ^{13}C-NMR studies can also be done in situ yielding information about glycogen metabolism (Neurohr et al. 1983; Reo et al. 1984; Shulman et al. 1985).

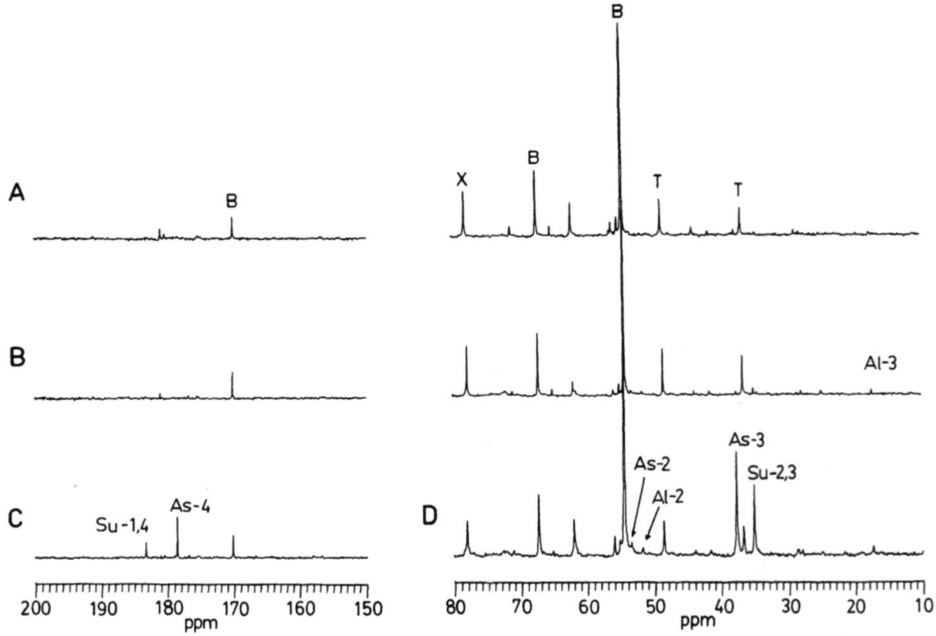

Fig. 10A–D. ^{31}P-NMR spectra of neutralized perchloric acid extracts of the ventricles of the marine gastropod ***Busycon contrarium***. ***A*** Extract of freshly dissected tissue; ***B*** extract of tissue first incubated under anoxic conditions for 3 h; *C* extract of tissue incubated under anoxic conditions for 3 h in the presence of aspartate-4-^{13}C (150–200 ppm range of spectrum depicted); and *D* extract of tissue incubated under anoxic conditions for 3 h in the presence of aspartate-3-^{13}C (10–80 ppm range of spectrum depicted). Abbreviations are as follows: betaine (*B*), chloroform external standard (*X*), taurine (*T*), C-3 of alanine (*Al-3*), C-2,3 of succinate (*Su-2,3*), C-3 of aspartate (*As-3*), C-2 of alanine (*Al-2*), C-2 of aspartate (*As-2*), C-4 of aspartate (*As-4*), and C-1,4 of succinate (*Su-1,4*). Data are from Graham and Ellington (1985)

If ^{13}C-NMR spectroscopy is coupled with direct enzymatic measurement of the concentration of specific metabolites in neutralized, perchloric acid extracts of tissues, an entirely new dimension is added to the analyses of metabolism. Since enzymatic analyses yield total metabolite pool size and ^{13}C-NMR provides data on ^{13}C content at specific carbon sites, it is possible to estimate the percent enrichment at each carbon site of a particular key metabolite. In an elegant mathematical analysis, Chance et al. (1983) have shown that it is possible to estimate metabolic flux in very complex pathways by following the fractional enrichment of key metabolites of the particular pathway. To this end, citric acid cycle flux in perfused hearts has been evaluated using a number of ^{13}C-labeled substrates (Chance et al. 1983; Malloy et al. 1987). Similarly, combined ^{13}C-NMR and GC-MS approaches have been used to evaluate hepatic glucose synthesis (Kalderon et al. 1987). Furthermore, it is possible to use these techniques to assess substrate preferences in perfused hearts (Sherry et al. 1985).

4.3 ^{1}H-NMR Spectroscopy

Sensitivity for protons is the highest of all NMR-visible nuclear species. Since protons (^{1}H) are present in all organic compounds, considerable interest has focused on the in vivo use of ^{1}H-NMR. However, the use of this approach is severely limited by the fact that the protons of water are present at 55 M while the protons of compounds of interest exist in mM concentrations at maximal. Thus, without appropriate experimental modifications, a typical biological ^{1}H-NMR spectrum consists of a massive water peak. The remaining proton resonances of interest ride on the shoulders of the water peak making resolution and quantification virtually impossible.

A number of approaches have been utilized to suppress or eliminate the water resonance in biological samples. A rather straightforward approach is to suppress the water signal by a presaturation technique (Yoshizaki et al. 1981). In brief, the selective 90° rf (radio frequency) pulse is applied directly on the water resonance. This is followed by the broad band pulse and data acquisition. Since the water protons are already displaced from equilibrium by the presaturation pulse, the development of magnetization of the water signal by the broad band pulse is substantially reduced. Therefore, the net effect is to reduce the amplitude of the water signal. More sophisticated water suppression approaches are currently used (Jue et al. 1985; Williams et al. 1985), but the details are beyond the scope of this essay.

^{1}H-NMR spectroscopy has been applied extensively to the study of muscle metabolism. Typical spectra are dominated by the single resonance from the three protons of the methyl group (CH_3) of lactate (Fig. 11). As a consequence, it is possible to measure intracellular lactate levels via ^{1}H-NMR. Furthermore, if pH_i is measured using the chemical shift of the histidine protons of carnosine (Sect. 3.1.5), it is possible to correlate lactate production and intracellular pH change in amphibian muscles (Seo et al. 1983; Seo 1984; Arus and Barany 1986). It is also possible to distinguish free creatine from creatine phosphate in certain muscle preparations (Arus et al. 1984) and therefore noninvasively measure the simultaneous utilization of phosphagen and production of lactate, and correlate these events with the change in pH_i. ^{1}H-NMR spectra of muscles in whole animals

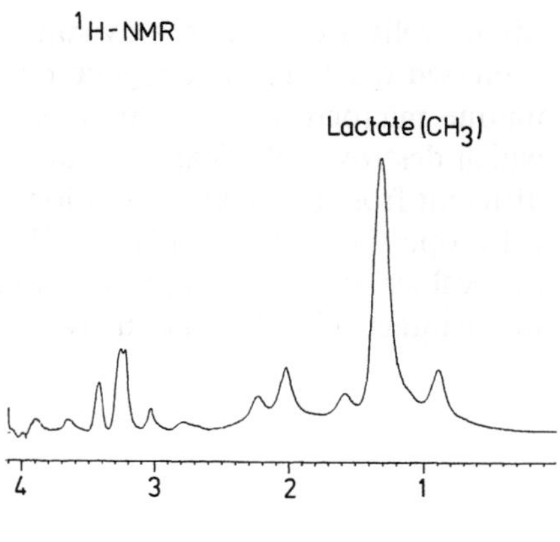

Fig. 11. ^{1}H-NMR spectrum of a portion of turtle ventricular muscle (W. R. Ellington unpublished results)

suffer from resolution problems. However, changes in muscle lactate levels are easily observed by this approach (Williams et al. 1985). It is also possible to detect lactate in perfused heart (Ugurbil et al. 1984) and liver (Jue et al. 1985) preparations as well as in the brain of the rat in situ (Behar et al. 1984). Recently, Jue et al. (1987) have shown that it is possible, using sophisticated data editing routines, to observe the resonances of lactate, pyruvate, β-hydroxybutyrate, and acetoacetate by ^{1}H-NMR. This leaves open the possibility for the noninvasive estimate of the cytoplasmic NADH/NAD$^+$ ratio (lactate: pyruvate redox couple) and the mitochondrial NADH/NAD$^+$ ratio (β-hydroxybutyrate: acetoacetate redox couple) in perfused livers (Jue et al. 1987). ^{1}H-NMR has not been applied to invertebrate systems to any extent.

4.4 ^{14}N-NMR-Spectroscopy

^{14}N-NMR has not been used extensively as a probe of nitrogen-containing compounds in biological systems. This is primarily due to the low sensitivity for this nucleus (0.1% of ^{1}H) and the fact that ^{14}N nuclei in biologically important compounds typically have very short T_2 values leading to very broad lines (Richards and Thomas 1974). However, Balaban and Knepper (1983) have shown that it is possible to use ^{14}N-NMR to measure urea and trimethylamine concentrations in isolated rat kidneys. Wray and Wilkie (1987) have used this approach to follow the time course of accumulation of betaine in the frog *Xenopus laevis* during acclimation to salty water. ^{14}N-NMR spectroscopy may be an appropriate method to follow the time course of changes in free amino acids in the tissues of marine invertebrates during osmotic stress. The chemical shifts of the amino groups of alanine, glycine, taurine, glutamate, aspartate, and other free amino acids are different and can be resolved by NMR (Balaban and Knepper 1983).

5 Measurement of Kinetic Processes by NMR Spectroscopy

5.1 General Considerations

Living cells are extremely dynamic systems consisting of hundreds of enzyme-catalyzed reactions. The regulation of these reactions is often dependent on the unique microenvironment in which the enzyme is found. Invasive approaches, which destroy biological structure and place enzymes in environments radically different from the in vivo situation, may not give a true picture of the physiological properties of the system. NMR spectroscopy provides a unique approach to the evaluation of the properties of key enzymes in that kinetics can be monitored in the intact cell. Thus, an in vivo analysis of kinetic properties can be developed.

5.2 Magnetization or Spin Transfer Studies of Enzyme Kinetics

The most commonly used approach to measurement of in vivo kinetics is by magnetization or spin transfer. These methods have been reviewed extensively (Alger and Shulman 1984; Brindle and Campbell 1987). To understand how this approach works we will use the arginine phosphokinase reaction (below) as an example:

$$\text{Arginine phosphate} + \text{ADP} + \text{H}^+ \leftrightarrow \text{Arginine} + \text{ATP}.$$

For the purposes of kinetic analysis, the above reaction can be simplified to a two-site chemical exchange model (Forsen and Hoffman 1963) of exchange of phosphate between the γ-P of ATP and arginine phosphate:

$$\text{ARG-P} \underset{k_r}{\overset{k_f}{\rightleftharpoons}} \text{ATP} .$$

The constants k_f and k_r represent the pseudo-first-order unidirectional rate constants for the forward (ATP synthesis) and reverse (ARG-P synthesis) directions, respectively. Magnetization transfer can be understood using the analogy of radioisotope tracer methodology (Hitzig et al. 1987b). In this specific case, the phosphorus becomes magnetically labeled and kinetics are monitored by following the transfer of magnetism to the other site.

Two general approaches for measurement of kinetics are typically used. Saturation transfer (SATRAN) involves the selective irradiation of a particular resonance such that it becomes saturated and its magnetization (M_{xy}) is zero. Chemical exchange with another nucleus will result in a decrease in the magnetization of that nucleus which is dependent, among other things, on the rate of chemical exchange. Using the example above, if the γ-P of ATP is saturated, there will be a reduction in the magnetization of ARG-P which will be manifested as a decreased integral in the ARG-P resonance (see Fig. 12b). The ratio of the integral of ARG-P, in the absence and the presence of saturation of ATP, can be shown to be related to the k_f and the T_1 for ATP (Campbell et al. 1977). The magnetization ratio as well as the T_1 values are determined experimentally and the k_f calculated by substitution. The reciprocal experiment for estimation of k_r is done by the selective irradiation of ARG-P which then results in the diminution of the γ-ATP integral (Fig. 12a). A practical discussion of the above SATRAN method can be found in Graham et al. (1986).

The alternative to SATRAN is a procedure known as inversion transfer. Inversion transfer involves the selective inversion of a resonance by a 180° pulse and the transfer of this inversion by chemical exchange to the other site. Using the above arginine phosphokinase example, γ-ATP is selectively inverted by a 180° pulse. This is then followed by 90° broad band pulses with variable delays after the inversion pulse. The inverted γ-ATP resonance will show a recovery of original magnetization with an exponential time course. In contrast, the ARG-P resonance will initially decrease, reach a minimum, and then increase back to the equilibrium magnetization. It can be shown that the initial rate of decrease of ARG-P is a measure of k_f (Meyer et al. 1982). By the same token, k_r can then be

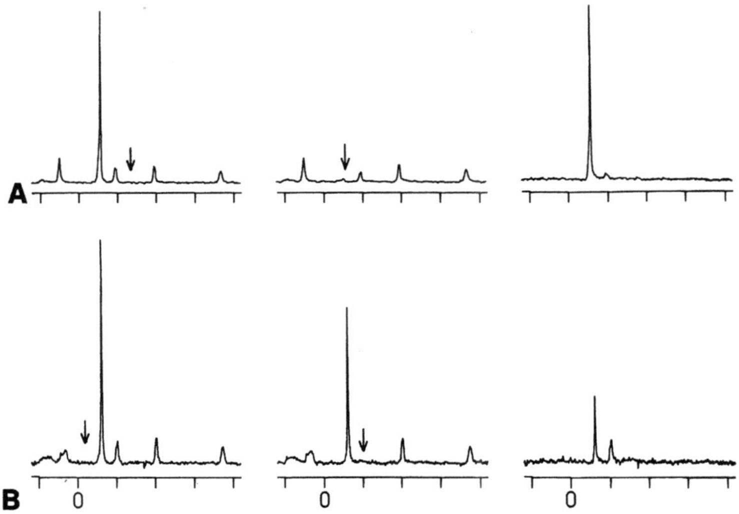

Fig. 12A, B. Saturation transfer ^{31}P-NMR experiments on the adductor muscle of the scallop *Argpecten irradians concentricus*. *Arrows* correspond to the site of the selective, saturating irradiation: arginine phosphate (*A*, middle spectrum) and γ-ATP (*B*, middle spectrum). The spectra on *left-hand side* represent control spectra. The spectra on the *right-hand side* correspond to the difference spectra. Data are from Graham et al. (1986)

estimated by the selective inversion of ARG-P. Degani et al. (1985) provided an excellent example of inversion transfer techniques in the study of creatine phosphokinase kinetics in isolated rat hearts.

The first-order unidirectional rate constants can be used to estimate the forward and reverse fluxes of an enzyme in vivo. In the case of arginine phosphokinase, ^{31}P-NMR spectra provide information about the levels of ARG-P and ATP in a system. Fluxes are then calculated as follows:

$$\text{Forward Flux } (F_f, \text{ ATP synthesis}) = k_f \times \text{ARG-P} \ ;$$
$$\text{Reverse Flux } (F_r, \text{ ARG-P synthesis}) = k_r \times \text{ATP} \ .$$

The forward and reverse fluxes for arginine phosphokinase in scallop adductor muscle are 4.72 and 6.98 $\mu mol\ s^{-1}\ g^{-1}$ wet wt, respectively (Graham et al. 1986).

5.3 Spin Transfer Studies of Creatine Phosphokinase (CPK)

Creatine phosphokinase (creatine phosphate + ADP + H^+ $\leftrightarrow$ creatine + ATP) in muscle is thought to catalyze a reaction near equilibrium with its substrates in vivo. The equilibrium nature of the reaction is important because of the temporal and spatial ATP-buffering roles of the reaction (Meyer et al. 1984). If this enzyme operates near equilibrium, the forward and reverse fluxes of creatine phosphokinase should be equivalent in vivo. Spin transfer studies have shown that this is the case in vertebrate skeletal muscle (Gadian et al. 1981; Haseler et al. 1986) and in perfused hearts (Degani et al. 1985). The equilibrium nature of the CPK reaction permits estimation of the levels of free ADP in these muscles. If the levels of creatine phosphate, creatine, and ATP are known, then it is possible to estimate the free ADP levels (ADP_{free}) by the following equation:

$$ADP_{free} = \frac{\text{Creatine} \times \text{ATP}}{\text{Creatine phosphate} \times K'},$$

The K′ represents the apparent equilibrium constant for the CPK reaction.

A discrepancy between the forward and reverse fluxes of CPK has been noted in some heart preparations. For instance, Matthews et al. (1982) found that the F_f was consistently greater than the F_r. In working hearts, there was an increase in both F_f and F_r as the work load was increased, but the discrepancy between the two fluxes also increased (Bittl and Ingwall 1985). The discrepancy between fluxes has been attributed to the participation of ATP in reactions other than simply CPK (Matthews et al. 1982). Bittl and Ingwall (1985) have suggested that this is a manifestation of the compartmentalization of ATP in heart such that there are two kinetically distinct ATP pools, mitochondrial and cytosolic, and have used this as evidence supporting the existence of a phosphocreatine shuttle system in heart muscle.

5.4 Spin Transfer Studies of Other Enzyme Systems

Spin transfer techniques have been used to assess the in vivo activities of a variety of other enzymes including the following: arginine phosphokinase (Butler et al. 1985; Briggs et al. 1985; Graham et al. 1986), adenylate kinase (Brown and Ogawa 1977), ATP synthetase (Matthews et al. 1981; Kingsley-Hickman et al. 1986; Thoma and Ugurbil 1987), glyceraldehyde-3-phosphate dehydrogenase/phosphoglycerate kinase (Brindle and Radda 1987), and lactate dehydrogenase (Brindle et al. 1986).

5.5 Other Applications of Spin Transfer NMR

Spin transfer NMR techniques can be of great utility in the study of transport or diffusion of materials across the membranes of intact cells. The glucose transporter in erythrocytes was recently investigated using ^{13}C-NMR spectroscopy by Kuchel et al. (1987a). This method obtains from the fact that the α and β anomers of glucose are well resolved by NMR. First-order rate constants estimated by this method were comparable to rate constants determined by less direct methods (Kuchel et al. 1987a). A similar approach was applied to the noninvasive measurement of fast membrane transport of $H^{13}CO_3^-$ in erythrocytes (Kuchel et al. 1987b).

6 Special Applications of NMR Spectroscopy in the Investigation of Biological Problems

6.1 Intracellular Myoglobin Diffusion

Vertebrate myoglobin is thought to function in the facilitation of diffusion of oxygen from the peripheral regions of the cell to the mitochondria. For this role to be feasible, the myoglobin must diffuse in the cell at sufficient rates to account for rate of oxygen diffusion in the muscle cell. Using ^{1}H-NMR spectroscopy, Livingston et al. (1983) measured myoglobin line widths in bovine heart muscle as well as myoglobin in free solution. Using this line width analysis, they suggested that myoglobin in situ has a high rate of mobility, roughly one-half of that in free solution. It was concluded that this apparent mobility is consistent with the proposed role of facilitation of oxygen diffusion (Livingston et al. 1983).

6.2 Noninvasive Estimates of Extra- and Intracellular Water Volume

Phenylphosphonic acid is a noncytotoxic NMR-visible compound which is impermeable to cell membranes (Fisher and Dillon 1986). This compound has been used to measure the extracellular volume in isolated, perfused rabbit bladders using ^{31}P-NMR (Fisher and Dillon 1986).

Hoffman and Gupta (1986) have used a combined ^{35}Cl- and ^{2}H-NMR approach to measure intracellular water volume in erythrocyte suspensions. This method takes advantage of the fact that intracellular chloride NMR signals are extremely broad such that they are NMR-invisible. ^{35}Cl- and ^{2}H-NMR spectra are taken of cell suspensions and the cell-free medium. The intracellular water content (f_w) is estimated by the following equation:

$$f_w = \frac{f_D - f_{Cl}}{1 - f_{Cl}},$$

where f_D = ratio of the intensities of water ^{2}H-NMR signals of the cell suspension and cell-free medium;
f_{Cl} = ratio of the intensities in the chloride ^{35}C-NMR signals of the cell suspension and cell-free medium.

Intracellular water content, as estimated by NMR, compares well with estimates derived from gravimetric methods (Hoffman and Gupta 1986).

6.3 Gating Techniques to Observe Very Rapid Physiological Events

As previously indicated, a major disadvantage of NMR methods is their inherent insensitivity. Signal-averaging over a relatively long time period may be necessary to obtain adequate signal to noise. Thus, transient events may be completely overlooked. However, if these transient events can be repeated, NMR data acquisition can be controlled in such a way as to occur precisely over that time period of the short-term event. This process, called "gating", allows the investigator to sum data from a series of repeated, transient events to achieve adequate signal to noise. For instance, Fossel and Ingwall (1982) monitored aortic pressure in perfused hearts and gated ^{31}P-NMR data acquisition to various phases of the cardiac cycle (maximal aortic pressure, minimum aortic pressure). They observed transient changes in all major phosphorus-containing metabolites which were not evident if the data were acquired randomly during the cardiac cycle (Fossel and Ingwall 1982). However, recently Bottomley et al. (1987) did similar studies on human subjects. They suggested that cyclic changes in high energy phosphates were due to ventricular wall thickening and blood volume changes rather than to a specific metabolic effect. Barbour et al. (1984) subjected hearts to repeated cycles of 4.5 s of oxygenated perfusate followed by 1.5 s of anoxic perfusate for periods up to 2 h. Data were then gated to various stages of the aerobic-anoxic recovery cycle revealing interesting changes in pH_i and levels of ATP and Pi taking place over a very short time scale. It is obvious that gating techniques have wide applicability since they can be coupled to the time course of a variety of physiological phenomena.

7 Conclusions

NMR spectroscopy represents a unique research tool for the noninvasive study of the properties of living cells. This essay is intended to give the reader an insight into the full range of possible areas of research afforded by these methods. To this end, it is hoped that NMR spectroscopic techniques will gain widened use in the area of comparative and environmental physiology.

Acknowledgements. We thank R. Rosanske and T. Gedris of the NMR Facility of Florida State University for their assistance in all of our NMR endeavors. We thank R. A. Graham (Fox Chase Cancer Center) for critically looking over the manuscript. Our work cited in this essay was supported by grants PCM-8202370, DCB-8401258, and DCB-8710108 from the U.S. National Science Foundation to W. R. Ellington.

References

Adam WR, Koretsky AP, Weiner MW (1987) Measurement of tissue potassium in vivo using ^{39}K nuclear magnetic resonance. Biophys J 51:265–271

Alger JR, Shulman RG (1984) NMR studies of enzymic rates in vitro and in vivo by magnetization transfer. Q Rev Biophys 17:83–124

Arus C, Barany M (1986) Application of high field ^{1}H-NMR spectroscopy for the study of perfused amphibian and excised mammalian muscles. Biochim Biophys Acta 886:411–424

Arus C, Barany M, Westler WM, Markley JL (1964) ^{1}H NMR of intact muscle at 11 T. FEBS Lett 165:231–236

Balaban RS, Knepper MA (1983) Nitrogen-14 nuclear magnetic resonance spectroscopy of mammalian tissues. Am J Physiol 245:C439–C444

Balschi JA, Cirillo VP, Springer CS (1982) Direct high resolution nuclear magnetic resonance studies of cation transport in vivo. Biophys J 38:323–326

Barbour RL, Sotak CH, Levy GC, Chan HP (1984) Use of gated perfusion to study early effects of anoxia on cardiac energy metabolism: a new ^{31}P NMR method. Biochemistry 23:6054–6062

Behar KL, den Hollander JA, Stromski ME, Ogino T, Shulman RG, Petroff OAC, Pritchard JW (1983) High resolution ^{1}H nuclear magnetic resonance study of cerebral hypoxia in vivo. Proc Natl Acad Sci USA 80:4945–4948

Behar KL, Rothman DL, Shulman RG, Petroff OAC, Pritchard JW (1984) Detection of cerebral lactate in vivo during hypoxemia by ^{1}H NMR at relatively low field strengths (1.9 T). Proc Natl Acad Sci USA 81:2517–2519

Bittl JA, Ingwall JS (1985) Reaction rates of creatine kinase and ATP synthesis in the isolated rat heart. J Biol Chem 260: 3512–3517

Bond M, Shporer M, Peterson K, Civan MM (1981) ^{31}P nuclear magnetic resonance analysis of toad urinary bladder. Mol Physiol 1:243–263

Bottomley PA, Herfkeus RJ, Smith SL, Bashmore TM (1987) Altered phosphate metabolism in myocardial infarction: P-31 MR spectroscopy. Radiology 165:703–707

Briggs RW, Radda GK, Thulborn KR (1985) ^{31}P-NMR saturation transfer study of the in vivo kinetics of arginine kinase in *Carcinus* crab leg muscle. Biochim Biophys Acta 845:343–348

Brindle KM, Campbell ID (1987) NMR studies of kinetics in cells and tissue. Q Rev Biophys 19:159–182

Brindle KM, Radda GK (1987) ^{31}P-NMR saturation transfer measurements of exchange between Pi and ATP in the reactions catalyzed by glyceraldehyde-3-phosphate dehydrogenase and phosphoglycerate kinase in vivo. Biochim Biophys Acta 928:45–55

Brindle KM, Campbell ID, Simpson RJ (1986) A ^{1}H-NMR study of the activity expressed by lactate dehydrogenase in the intact human erythrocyte. Eur J Biochem 158:299–305

Brown TR, Ogawa S (1977) ^{31}P nuclear magnetic resonance kinetic measurements on adenylate kinase. Proc Natl Acad Sci USA 74:3627–3631

Budinger TF, Margolis AR (1986) Medical magnetic resonance imaging and spectroscopy. Soc Magn Reson Med, Berkeley, Cal

Bunger R (1985) Compartmented pyruvate in perfused working heart. Am J Physiol 249:H439–H449.

Burstein D, Fossel ET (1987) Nuclear magnetic resonance studies of intracellular ions in perfused frog heart. Am J Physiol 252:H1138–H1146

Busa WB, Nuccitelli R (1984) Metabolic regulation via intracellular pH. Am J Physiol 246:R409–R438

Busa WB, Crowe JH, Matson GB (1982) Intracellular pH and the metabolic status of dormant and developing *Artemia* embryos. Arch Biochem Biophys 216:711–718

Busby SJW, Gadian DG, Radda GK, Richards RE, Seeley PJ (1978) Phosphorus nuclear magnetic resonance studies of compartmentation in muscle. Biochem J 170:103–114.

Butler KW, Deslauriers R, Geoffrion Y, Storey JM, Storey KB, Smith CP, Somorjai RL (1985) ^{31}P nuclear magnetic resonance studies of crayfish (*Orconectes virilis*). Eur J Biochem 149:79–83

Campbell ID, Dobson CM, Ratcliffe RG, Williams RJD (1977) Fourier transform NMR pulse methods for the measurement of slow exchange rates. J Magn Reson 29:397–405

Castle AM, MacNab RM, Shulman RG (1986) Measurement of intracellular sodium concentration and sodium transport in *Escherichia coli* by ^{23}Na nuclear magnetic resonance. J Biol Chem 261:3288–3294

Chance B, Leigh JS, Clark BJ, Maris J, Kent J, Noika S, Smith D (1985) Control of oxidative metabolism and oxygen delivery in human skeletal muscle: a steady state analysis of the work/energy cost transfer function. Proc Natl Acad Sci USA 82:8384–8388

Chance EM, Seeholzer SH, Kobayashi K, Williamson JR (1983) Mathematical analysis of isotope labeling in the citric acid cycle with applications to ^{13}C NMR studies in perfused rat hearts. J Biol Chem 258:13785–13794

Chih CP, Ellington WR (1985) Metabolic correlates of intracellular pH change during rapid contractile activity in a molluscan muscle. J Exp Zool 236:27–34

Christen R, Schackmann RW, Dahlquist FW, Shapiro BM (1983) ^{31}P-NMR analysis of sea urchin sperm activation. Exp Cell Res 149:289–294

Clarke FM, Stephan P, Huxham G, Hamilton D, Morton DJ (1984) Metabolic dependence of glycolytic enzyme binding in rat and sheep heart. Eur J Biochem 138:643–649

Clegg JS (1984) Properties and metabolism of the aqueous cytoplasm and its boundaries. Am J Physiol 246 :R133–R151

Cohen SM (1987a) ^{13}C and ^{31}P NMR study of gluconeogenesis: utilization of ^{13}C-labeled substrates by perfused liver from streptozotocin-diabetic and untreated rats. Biochemistry 26:563–572

Cohen SM (1987b) Effects of insulin on perfused liver from streptozotocin-diabetic and untreated rats: ^{13}C NMR assay of pyruvate kinase flux. Biochemistry 26:573–580

Cohen SM, Ogawa S, Shulman RG (1979) ^{13}C NMR studies of gluconeogenesis in rat liver cells: utilization of labeled glycerol by cells from euthyroid and hyperthyroid rats. Proc Natl Acad Sci USA 76:1603–1607

Cohen SM, Glynn P, Shulman RG (1981a) ^{13}C NMR study of gluconeogenesis from labeled alanine in hepatocytes from euthyroid and hyperthyroid rats. Proc Natl Acad Sci USA 78:60–64

Cohen SM, Rognstad R, Shulman RG, Katz J (1981b) A comparison of ^{13}C nuclear magnetic resonance and ^{14}C tracer studies of hepatic metabolism. J Biol Chem 256:3428–3432

Cohn M, Hughes TR (1962) Nuclear magnetic resonance spectra of adenosine di- and triphosphate. II. Effect of complexing with divalent metal ions. J Biol Chem 237:176–181

Cranney GB, Evanochko WT, Vaughn JM, Prohost GM (1987) T_1 alterations of ^{31}P metabolites with hypoxic insult. Soc Magn Reson Med, New York Meet 568

Dawson MJ, Gadian DG, Wilkie DR (1977) Contraction and recovery of living muscles studied by ^{31}P nuclear magnetic resonance. J Physiol 267:703–735

Dawson MJ, Gadian DG, Wilkie DR (1978) Muscular fatigue investigated by phosphorus nuclear magnetic resonance. Nature (London) 274:861–866

DeFronzo M, Gillies RJ (1987) Characterization of methylphosphonate as a ^{31}P NMR pH indicator. J Biol Chem 262:11032–11037

Degani H, Laughlin M, Campbell S, Shulman RG (1985) Kinetics of creatine kinase in heart: a ^{31}P NMR saturation- and inversion-transfer study. Biochemistry 24:5510–5516

Desmoulin F, Cozzone PJ, Canioni P (1987) Phosphorus-31 nuclear magnetic resonance study of phosphorylated metabolite compartmentation, intracellular pH and phosphorylation state during normoxia, hypoxia and ethanol perfusion, in the perfused rat liver. Eur J Biochem 162:151–159

Devel PK, Yue GM, Sherman WR, Schickner DJ, Ackerman JJH (1985) Monitoring the time course of cerebral deoxyglucose metabolism by ^{31}P nuclear magnetic resonance spectroscopy. Science 228:1329–1331

Dubyak GR, Scarpa A (1983) Phosphorus-31 nuclear magnetic resonance studies of single muscle cells isolated from barnacle depressor muscle. Biochemistry 22:3531–3536

Ellington WR (1983) Phosphorus nuclear magnetic resonance studies of energy metabolism in molluscan tissues: effect of anoxia and ischemia on the intracellular pH and high energy phosphates in the ventricle of the whelk, *Busycon contrarium*. J Comp Physiol 153:159–166

Ellington WR (1985) Metabolic impact of experimental reductions of intracellular pH in molluscan cardiac muscle. Mol Physiol 7:155–164

Fisher MJ, Dillon PF (1986) Phenylphosphonate (PPA): a ^{31}P-NMR indicator of extracellular pH and volume. Fed Proc 45:765

Forsen S, Hoffman RA (1963) Study of moderately rapid chemical exchange by means of nuclear magnetic double resonance. J Chem Phys 39:2892–2901

Fossel ET, Ingwall JS (1982) Measurement of dynamic aspects of energy metabolism in the heart using ^{31}P NMR magnetic resonance spectroscopy. In: Cohen JS (ed) Noninvasive probes of tissue metabolism. John Wiley & Sons, New York, pp 173–207

Friedrich P (1985) Dynamic compartmentation in soluble multienzyme systems. In: Welch RA (ed) Organized multienzyme systems. Academic Press, New York London, pp 141–176

Gadian DG (1982) Nuclear magnetic resonance and its applications to living systems. Oxford Univ Press, New York

Gadian DG, Radda GK, Brown TR, Chance EM, Dawson MJ, Wilkie DR (1981) The activity of creatine kinase in frog skeletal muscle studied by saturation transfer nuclear magnetic resonance. Biochem J 194:215–228

Gard JK, Kichura GM, Ackerman JJH, Eisenberg JD, Billadello JJ, Sobel BE, Gross RW (1985) Quantitative ^{31}P nuclear magnetic resonance analysis of metabolite concentrations in Langendorf-perfused rabbit hearts. Biophys J 48:803–813

Garfinkel L, Alschuld RA, Garfinkel D (1986) Magnesium in cardiac energy metabolism. J Mol Cell Card 18:1003–1013
Garlick PB, Radda GK, Seeley PJ (1979) Studies of acidosis in the ischemic heart by phosphorus nuclear magnetic resonance. Biochem J 184:547–554
Gillies RJ, Alger JR, den Hollander JA, Shulman RG (1982) Intracellular pH measured by NMR: methods and results. In: Nuccitelli R, Deamer DW (eds) Intracellular pH. Liss, New York, pp 79–104
Graham RA, Ellington WR (1985) Anaerobic aspartate utilization and alanine formation in molluscan cardiac muscle: a ^{13}C-NMR study. J Exp Zool 236:365–370
Graham RA, Ellington WR (1986) Phosphorus nuclear magnetic resonance studies of energy metabolism in molluscan muscle: intracellular pH and the qualitative nature of anaerobic end products. Physiol Zool 58:478–490
Graham RA, Ellington WR, Chih CP (1986) A saturation transfer phosphorus nuclear magnetic resonance study of arginine phosphokinase in the muscle of a marine mollusc. Biochim Biophys Acta 887:157–163
Gullans SR, Avison MJ, Ogino T, Giebisch G, Shulman RG (1985) NMR measurements of intracellular sodium in the rabbit proximal tubule. Am J Physiol 249:F160–F168
Gupta RK, Gupta P (1982) Direct observation of resolved resonances from intra- and extracellular sodium-23 ions in NMR studies of intact cells and tissues using dysprosium (111) tripolyphosphate as paramagnetic shift reagent. J Magn Reson 47:344–350
Gupta RK, Moore RD (1980) NMR studies of intracellular free Mg^{2+} in intact frog skeletal muscle. J Biol Chem 255:3987–3994
Gupta RK, Benovic JL, Rose ZB (1978) The determination of the free magnesium level in the red blood cell by ^{31}P-NMR. J Biol Chem 253:6172–6176
Gupta RK, Gupta P, Vushok WD, Rose ZB (1983) Measurement of the dissociation constant of Mg ATP at physiological nucleotide levels by a combination of ^{31}P NMR and optical absorbance spectroscopy. Biochem Biophys Res Commun 117:210–216
Gupta RK, Gupta P, Moore RD (1984) NMR studies of intracellular metal ions in intact cells and tissues. Annu Rev Biophys Bioeng 13:221–246
Hamm JR, Yue GM (1987) ^{31}P nuclear magnetic resonance measurements of intracellular pH in giant barnacle muscle. Am J Physiol 252:C30–C37
Hansen J, Sharpe T, Bittar EE (1986) Phosphate metabolites in single barnacle muscle fibers investigated by phosphorus-31 nuclear magnetic resonance. Comp Biochem Physiol 83B:875–879
Haseler LJ, Brooks WM, Irving MG, Bulliman BT, Kuchel PW, Doddrell DM (1986) Use of inversion transfer to monitor creatine kinase kinetics in rat skeletal muscle in vivo. Biochem Int 12:613–618
Hebisch S, Soboll S, Schwenen J, Sies H (1984) Compartmentation of high energy phosphates in resting and working rat skeletal muscle. Biochim Biophys Acta 74:117–124
Hitzig BM, Johnson DC, McFarland E, Koutcher JA, Kazemi H, Burt CT (1987a) Unknown phosphate compounds in the tail muscle of intact conscious newts by ^{31}P NMR. Comp Biochem Physiol 86B:537–540
Hitzig BM, Pritchard JW, Kantor HL, Ellington WR, Ingwall JS, Burt CT, Helman SI, Koutchev J (1987b) NMR spectroscopy as an investigative technique in physiology FASEB J 1:22–31
Hoffman D, Gupta RK (1986) NMR measurement of intracellular water volume. J Magn Reson 70:481–483
Hoult DI (1978) The NMR receiver: a description and analysis of design. Proc NMR Res 12:41–77
Hoyle G, McNeill PA, Selerston AI (1973) Ultrastructure of barnacle giant muscle fibers. J Cell Biol 56:74–91
Inoue H, Yoshioka T (1980) Measurement of intracellular pH in sea urchin eggs by ^{31}P-NMR. J Cell Physiol 105:461–468
Jacobus WE, Pores IH, Lucas SK, Kallman CH, Weisfeldt ML, Flaherty JT (1982) The role of intracellular pH in the control of normal and ischemic myocardial contractility: a ^{31}P nuclear magnetic resonance and mass spectrometry study. In: Deamer R, Nuccitelli DW (eds) Intracellular pH. Liss, New York, pp 537–565

Jan AWH, Amsler K, Griewel B, Kinne RKH (1987) Regulation of intracellular pH in LLC-PK_1 cells studied using ^{31}P-NMR spectroscopy. Biochim Biophys Acta 927:203–212
Jue T, Arias-Mendoza F, Gonnella NC, Shulman GI, Shulman RG (1985) A ^{1}H NMR technique for observing metabolite signals in the spectrum of perfused liver. Proc Natl Acad Sci USA 82:5246–5249
Jue T, Chung Y, Shulman RG (1987) Measuring the redox potential in perfused liver with ^{1}H NMR. Soc Magn Res Med, New York, p 276
Kalderon B, Gophev A, Lapiclot A (1987) A quantitative analysis of the metabolic pathways of hepatic glucose synthesis in vivo with ^{13}C-labeled substrates. FEBS Lett 213:209–214
Kamp G, Juretschke HP (1987) An in vivo ^{31}P-NMR study of the possible regulation of glycogen phosphorylase *a* by phosphagen via phosphate in the abdominal muscle of the shrimp *Crangon crangon*. Biochim Biophys Acta 929:121–127
Kingsley-Hickman P, Sako EY, Andreone PA, St Cyr JA, Michurski S, Foker JG, Fromm AHL, Petgein M, Ugurbil K (1986) ^{31}P-NMR measurement of ATP synthesis rate in perfused intact rat hearts. FEBS Lett 198:159–163
Korn ED, Carlier M-F, Pantaloni D (1987) Actin polymerization and ATP hydrolysis. Science 238:638–644
Kuchel PW, Bulliman BT, Chapman BE, Kirk K (1987a) The use of transmembrane differences in saturation transfer for measuring fast membrane transport; application to $H^{13}CO_3^-$ exchange across the human erythrocyte. J Magn Reson 74:1–11
Kuchel PW, Chapman BE, Potts JR (1987b) Glucose transport in human erythrocytes measured using ^{13}C NMR spin transfer. FEBS Lett 219:5–10
Kurkdjian A, Barbier-Brygoo H (1983) A hydrogen ion-selective liquid-membrane electrode for measurement of the vacuolar pH of plant cells in suspension culture. Anal Biochem 132:96–104
Kushmerick MJ, Meyer RA, Brown TR (1983) Phosphorus NMR spectroscopy of cat biceps and soleus muscles. In: Bicher HI, Bruley DF (eds) Oxygen transport to tissue, vol 4. Plenum, New York, pp 303–324
Kushmerick MJ, Dillon PF, Meyer RA, Brown TR, Krisanda JM, Sweeney HL (1986) ^{31}P NMR spectroscopy, chemical analysis, and free Mg^{2+} of rabbit bladder and uterine smooth muscle. J Biol Chem 261:14420–14429
Kuwabara T, Yuasa T, Miyatake T (1986) ^{31}P-NMR studies on an animal model of human defective muscle glycolysis. Muscle Nerve 9:138–143
Labotka RJ, Kleps RA (1983) A phosphate-analogue probe of red cell pH using phosphorus-31 nuclear magnetic resonance. Biochemistry 22:6089–6095
Lavanchy N, Martin J, Rossi A (1984a) Graded global ischaemia and reperfusion of the isolated perfused rat heart. Cardiovasc Res 18:573–582
Lavanchy N, Martin J, Rossi A (1984b) Glycogen metabolism: a ^{13}C-NMR study on the isolated perfused rat heart. FEBS Lett 178:34–38
Levy LA, Murphy E, London RE (1987) Synthesis and characterization of ^{19}F NMR chelators for measurement of cytosolic free Ca. Am J Physiol 252:C441–C449
Lin L-E, Shporer M, Civan MM (1982) ^{13}P nuclear magnetic resonance analysis of frog skin. Am J Physiol 243:C74–C80
Lin P-S, Blumenstein M, Mikkelsen RB, Schmidt-Ullrich R, Bachovchin WW (1987) Perfusion of cell spheroids for study by NMR spectroscopy. J Magn Reson 73:399–404
Livingston DJ, La Mar GN, Brown WD (1983) Myoglobin diffusion in bovine heart muscle. Science 222:71–73
Lynch RM, Paul RJ (1987) Compartmentation of metabolism in vascular smooth muscle. Am J Physiol 252:C328–C334
Malloy CR, Sherry AD, Jeffrey FMH (1987) Carbon flux through citric acid cycle pathways in perfused hearts by ^{13}C NMR spectroscopy. FEBS Lett 212:58–62
Mansour TE, Morris PG, Feeney J, Roberts GCK (1982) A ^{31}P-NMR study of the intact liver fluke *Fasciola hepatica*. Biochim Biophys Acta 721:336–340
Matthews PM, Bland JL, Madiun DG, Radda GK (1981) The steady state rate of ATP synthesis in the perfused rat heart by ^{31}P NMR saturation transfer. Biochim Biophys Res Commun 103:1052–1059

Matthews PM, Bland JL, Gadian DG, Radda GK (1982) A ^{31}P-NMR saturation transfer study of the regulation of creatine kinase in the rat heart. Biochim Biophys Acta 721:312–320
Meyer RA, Kushmerick MJ, Brown TR (1982) Application of ^{31}P NMR spectroscopy to the study of skeletal muscle metabolism. Am J Physiol 242:C1–C11
Meyer RA, Sweeney HL, Kushmerick MJ (1984) A simple analysis of the "phosphocreatine shuttle". Am J Physiol 246:C365–C377
Meyer RA, Brown TR, Kushmerick MJ (1985) Phosphorus nuclear magnetic resonance of fast- and slow-twitch muscles. Am J Physiol 248:C279–C287
Meyer RA, Brown TR, Krilowicz BL, Kushmerick MJ (1986) Phosphagen and intracellular pH changes during contraction of creatine-depleted rat muscle. Am J Physiol 250:C264–C274
Moon RB, Richards JH (1973) Determination of intracellular pH by ^{31}P nuclear magnetic resonance. J Biol Chem 248:7276–7278
Morris PG, Smith GA, Metcalfe JC, Rodrigo GC (1987) A new NMR approach to the measurement of intracellular cations in living systems. Soc Magn Reson Med, New York, p 31
Naritomi H, Kanashiro M, Sasaki M, Kuribayashi Y, Sawada T (1987) In vivo measurements of intra- and extracellular Na^+ and water in the brain and muscle by nuclear magnetic resonance spectroscopy with shift reagent. Biophys J 52:611–616
Neeman M, Rushkin E, Kaye AM, Degani H (1987) ^{31}P-NMR studies of phosphate transfer rates in T470 human breast cancer cells. Biochim Biophys Acta 930:179–192
Neurohr KJ, Barrett EJ, Shulman RG (1983) In vivo carbon-13 nuclear magnetic resonance studies of heart metabolism. Proc Natl Acad Sci USA 80:1603–1607
Ogino T, Shulman GI, Avison MG, Gullans SR, den Hollander JA and Shulman RG (1985) ^{23}Na and ^{39}K NMR studies of ion transport in human erythrocytes. Proc Natl Acad Sci USA 82:1099–1103
Peukhurinen KJ, Hiltunen JK, Hassinen IE (1983) Metabolic compartmentation of pyruvate in the isolated perfused rat heart. Biochem J 210:193–198
Pike MM, Frazer DF, Ingwall JS, Allen PD, Springer CS, Smith TW (1985) ^{23}Na and ^{39}K nuclear magnetic resonance studies of perfused rat hearts. Discrimination of intra- and extracellular ions using a shift reagent. Biophys J 48:159–173
Piwnica-Worms D, Lieberman M (1983) Microfluorometric monitoring of pH_i in cultured heart cells: Na^+-H^+ exchange. Am J Physiol 244:C422–C428
Reo NV, Siegfried BA, Ackerman JJH (1984) Direct observation of glycogenesis and glucagon-stimulated glycogenolysis in the rat liver in vivo by high field carbon-13 surface coil NMR. J Biol Chem 259:13664–13667
Richards RE, Thomas NA (1974) A ^{14}N NMR study of amino acids, peptides and other biologically interesting molecules. J Chem Soc Perkin Trans 2:368–374
Rohrer SP, Saz HP, Howak T (1986) ^{13}P-NMR studies of the metabolisms of the parasitic helminths *Ascaris suum* and *Fasciola hepatica*. Arch Biochem Biophys 248:200–209
Roos A, Boron WF (1981) Intracellular pH. Physiol Rev 61:296–434
Schadewaldt P, Munch U, Staib W (1983) Evidence for the compartmentation of pyruvate metabolism in perfused rat skeletal muscle. Biochem J 216:761–764
Seo Y (1984) Effects of extracellular pH on lactate efflux from frog sartorius muscle. Am J Physiol 247:C175–C181
Seo Y, Yoshizaki K, Morimoto T (1983) A ^{1}H-nuclear magnetic resonance study on lactate and intracellular pH in frog muscle. Jap J Physiol 33:721–731
Seraydarian K, Mommaerts WFHM, Walner A (1962) The amounts and compartmentation of ADP in muscle. Biochim Biophys Acta 65:443–460
Sherry AD, Nunnally RL, Peshock RM (1985) Metabolic studies of pyruvate- and lactate-perfused guinea pig hearts by ^{13}C-NMR. J Biol Chem 260:9272–9279
Shulman GI, Rothman DL, Smith D, Johnson CM, Blair JB, Shulman RG, DeFronzo RA (1985) Mechanism of liver glycogen repletion in vivo by nuclear magnetic resonance spectroscopy. J Clin Invest 76:1229–1236
Srivastava DK, Bernhard SA (1986) Enzyme-enzyme interactions and the regulation of metabolic reaction pathways. Curr Top Cell Reg 28:1–68
Steen GR (1986) Impact of symbiotic algae on sea anemone metabolism: analysis by in vivo ^{31}P nuclear magnetic resonance spectroscopy. J Exp Zool 240:315–326

Storey KB, Miceli M, Butler KW, Smith CP, Deslauriers R (1984) ^{31}P-NMR studies of the freeze-tolerant larvae of the gall fly, *Eurosta solidaginis*. Eur J Biochem 142:591–595
Swergold RB, Freed JJ, Brown TR (1987) Bicarbonate/CO_2 dependent mechanism is important in the regulation of intracellular pH in 3T3 cells. Soc Magn Reson Med, New York Meet, p 511
Taylor JS, Deutsch C (1983) Fluorinated α-methylamino acids as ^{19}F NMR indicators of intracellular pH. Biophys J 43:261–267
Thebault MT, Raffin JP, LeGall JY (1987) In vivo ^{31}P NMR in crustacean muscles: fatigue and recovery in the tail musculature from the prawn *Palasmon elegans*. Biochem Biophys Res Commun 145:453–459
Thoma WJ, Ugurbil K (1987) Saturation-transfer studies of ATP-Pi exchange in isolated perfused rat liver. Biochim Biophys Acta 893:225–231
Thomas RC (1978) Ion-sensitive intracellular microelectrodes. How to make and use them. Academic Press, New York London
Ugurbil K, Geurnsey DL, Brown TR, Glynn P, Tobke N, Edelman IS (1981) ^{31}P-NMR studies of intact anchorage dependent mouse embryo fibroblasts. Proc Natl Acad Sci USA 78:4843–4847
Ugurbil K, Petein M, Maidan R, Michuvski S, Cohn JN, From AH (1984) High resolution proton NMR studies of perfused rat hearts. FEBS Lett 167:73–78
Waddell WJ, Butler TC (1959) Calculation of intracellular pH from the distribution of 5,5-dimethyl-2,4-oxazolidinedione (DMO). Application to skeletal muscle of the dog. J Clin Invest 38:720–729
Williams SR, Gadian DG, Proctor E, Sprague DB, Talbot DF, Brown FF, Young IR (1985) ^{1}H nuclear-magnetic-resonance studies of muscle metabolism in vivo. Biochem Soc Trans 13:839–842
Wray S, Tofts PS (1986) Direct in vivo measurement of absolute metabolite concentrations using ^{31}P nuclear magnetic resonance spectroscopy. Biochim Biophys Acta 886:399–405
Wray S, Wilkie DR (1987) Some novel uses for 14-nitrogen NMR spectroscopy. Soc Magn Reson Med, New York, p 597
Yoshizaki K, Seo Y, Hishikawa H (1981) High-resolution proton magnetic resonance spectra of muscle. Biochim Biophys Acta 678:283–291
Ziegler R, Roth K (1985) ^{13}C-NMR spectroscopy of larvae of *Manduca sexta* in vivo. Naturwissenschaften 72:206–207

Chapter 5

Effects of Temperature on Muscular Function and Locomotory Performance in Teleost Fish

B. D. Sidell[1] and *T. S. Moerland*[2]

Contents

[1] Department of Zoology and Center for Marine Studies, University of Maine, Orono, ME 04469, USA

[2] Department of Radiology, Harvard Medical School, Boston, MA 02115, USA

Advances in Comparative and Environmental Physiology, Vol. 5

1 Introduction

1.1 Propulsion in the Aquatic Medium

All modes of animal locomotion are variations on a theme in which the organism converts chemical potential energy into the kinetic energy of a moving body. The range of permissible variations on this theme is limited by physical constraints of the medium in which locomotion occurs and shaped by locomotory demands of a particular life history.

Several physical properties of water are germane to considerations of locomotion in the aquatic medium. The first of these factors is density. Water has a density that is approximately three orders of magnitude greater than that of air, but very close to that of animal tissue (not surprising given that fish muscle, for example, is 70% – 75% water). As a result, animal tissue requires rather extensive structures to support its mass in air whereas in water it largely is supported by the medium. Submerged fish are approximately neutrally buoyant and generally possess mechanisms which permit adjustment of buoyant mass; for example, gas-filled swim bladders and low-density lipids. A practical consequence of neutral buoyancy is that fish expend little or no energy to overcome the force of gravity; all power available from the body musculature can be dedicated to generation of forward thrust. Common experience tells us that this is clearly not true for life on land, and a substantial portion of the energy costs of human locomotion can be attributed to maintenance of muscular tension as the body moves vertically while subject to the force of gravity (McMahon 1984). Neutral buoyancy with respect to the medium is the principal reason that, for animals of similar body size, swimming is a more economical form of locomotion than either walking/running or flying (Schmidt-Nielsen 1972).

Viscosity is another physical property of water that is important for considerations of locomotion. The dynamic viscosity for water is approximately four orders of magnitude greater than that of air. The greater viscosity and density of water compared to air means that, for animals of similar mass and form traveling at a given speed, the power necessary to overcome drag of the medium is far greater in water than in air. Power to overcome drag can come from added muscle and, given that fish are neutrally buoyant in their locomotory medium, little penalty is incurred by carrying added mass. However, this strategy is necessarily self-limiting because the addition of muscle to overcome drag, in turn, must be tempered by considerations of drag incurred by added bulk. Consequently, we generally find that the musculature of fish is configured in a streamlined body form that minimizes drag.

From the preceding, we can appreciate that viscosity and density of water are important factors in determining body form and locomotory performance of teleost fish. A third consideration stems from the thermal properties of water; this consideration does not focus on parameters that directly affect the hydrodynamics of a swimming fish, but rather on the ability of the organism to act as a chemomechanical converter of energy. The same design features that permit the gills of fish effectively to exchange respiratory gases also permit excellent thermal exchange with the medium. Thermal diffusion in water is approximately tenfold more rapid than gaseous diffusion. Coupled with the high heat capacity of water, rapid thermal diffusion results in extreme difficulty for fish to establish and maintain a thermal disequilibrium with their environment. Exceptions are found in certain lamnid sharks and large tuna which have evolved vascular retia that function as efficient countercurrent heat exchangers. In these species, which we will not consider further, a significant fraction of the metabolic heat energy which results from contractile activity is conserved in the deep body musculature (reviewed by Carey et al. 1971). However, for most teleost fish species and all those considered here, body temperature at any given time is approximately equal to ambient water temperature. This has important implications for locomotory function of fish, as discussed in the next section.

1.2 The Problem of Temperature

In considering the effects of temperature on muscular and locomotory performance of fish, we are attempting to resolve the following paradox. Temperature influences biochemical reaction rates by a factor of two to three for every change of 10 °C and contractile performance of muscle (a biochemical system) should be affected similarly by temperature. In contrast, the power necessary for a given level of locomotory performance in the aquatic medium is essentially independent of temperature. Our a priori expectation, therefore, is that changes in temperature should have a profound effect on locomotion of fish. Representative fish occupy a remarkable range of thermal niches, from equatorial hot springs to the polar oceans, and the life histories of many species in temperate zones include considerable seasonal changes in temperature. Yet, the evidence available to us indicates that locomotory performance of fish from different thermal regimes is relatively similar. Mechanisms must exist, therefore, to confer a degree of independence from intrinsic effects of temperature on biochemical reaction rate.

Our goal in the remaining sections of this assay is to examine the interplay between temperature and the muscular and locomotory performance of teleost fish and to outline our present knowledge of mechanisms that can serve to compensate for the effects of temperature on locomotory function. We are, of course, necessarily limited in scope. There are over 20000 described species of teleosts and this assemblage includes an amazing variety of body forms and locomotory habits; however, data relevant to our purposes are available for relatively few of these species. We will consider only teleosts that swim in the subcarangiform mode, which includes many familiar fish, such as trout, salmon, bass, and carp.

2 The Propulsion Apparatus

2.1 Locomotory Musculature is Composed of Different Fiber Types

The muscle types of fish differ greatly in their content of cytochromes and myoglobin and commonly are referred to by color, rather than by a system of descriptive nomenclature based upon biochemical and contractile characteristics.

Table 1. Enzymatic profile of slow (red), intermediate (pink), and fast (white) skeletal muscle of the carp (*Cyprinus carpio*)[a]

Enzyme	Muscle type		
	Red	Pink	White
Myofibrillar ATPase	0.3	0.6	1.1
Creatine kinase	1970	2800	4240
Glycogen phosphorylase	19	19	13
Hexokinase	3.3	2.2	0.8
Pyruvate kinase	230	460	300
Lactate dehydrogenase	300	780	440
Malate dehydrogenase	1600	1050	690
Succinic dehydrogenase	1330	440	200
Cytochrome oxidase	47	17	6

[a] Data are μmol substrate converted g^{-1} dry wt min^{-1}. Myofibrillar ATPase activity is μmol Pi released mg^{-1} protein min^{-1}. All assays were performed at 20 °C. Data are from Johnston et al. (1977).

Most workers recognize three principal fiber types in teleost fish (Table 1). Red fibers, so-called because of their high content of myoglobin, are comparatively small in diameter, well-vascularized, highly aerobic, and possess relatively low activities of myofibrillar ATPase. In contrast, white fibers of teleosts have little myoglobin, are relatively large in diameter, poorly vascularized, with little aerobic metabolic potential but a high capacity for anaerobic glycolysis, and show high activity of myofibrillar ATPase. A third fiber type, pink, is intermediate in biochemical profile to red and white muscle (Table 1). We mention ATPase catalytic activity of the myofibrils here because this characteristic has been shown to correlate very well with maximum contraction velocities of muscles from a wide variety of animals (Barany 1967). Indeed, myofibrillar ATPase activity of fish white muscle is typically three to five times greater than that of red muscle and the unloaded shortening speed of fibers from white muscle is greater than those from red (Johnston and Tota 1974; Altringham and Johnston 1982; Moerland and Sidell 1986a, b).

Based upon metabolic profiles and contractile characteristics, white and red muscle of teleosts can be considered analogous to the fast glycolytic (type IIb) and slow oxidative (type I) fibers of higher vertebrates, respectively. Similarly, pink muscle may be considered to be analogous to mammalian fast, oxidative muscle (type Ia). Note, however, that red and white muscles of teleosts are very different from the slow and fast muscle of mammals with respect to their patterns of innervation and electrophysiological properties (Sect. 2.1.1). This three fiber-type system of classification is not adequate for all cases and it is very likely that future work will reveal a degree of heterogeneity within each category, as is the case for higher vertebrates (Nemeth et al. 1979). For example, the red fibers of mosaic muscle in rainbow trout (Sect. 2.1.2) are histochemically different from those of the superficial red musculature (Johnston et al. 1975b), and there is evidence of several types of fibers intermediate to red and white in cod (Korneliussen et al. 1978).

2.1.1 Innervation Patterns of Fish Muscle Depart from the Vertebrate Norm

Red fibers of teleosts are multiterminally innervated with nerve endings of the *en grappe* type (Johnston 1980a, 1983a). It is generally thought that red muscle is activated by electrotonic spread of junction potentials, not by propagated action potentials (Stanfield 1972; Johnston 1980a). Red muscle of teleosts is similar to the slow tonic fibers of anurans in this regard.

At least two types of innervation pattern are found in the white musculature of teleosts. Focal innervation, in which fibers are innervated by one large end plate is found among some teleosts, such as herring, and in the taxonomically primitive nonteleost fish (Bone et al. 1978; Johnston 1983a). Activation of focally innervated fibers is by propagated action potential (Hagiwara and Takahashi 1967). However, white musculature of most teleosts is multiply innervated (polyneuronal), a pattern of innervation which is unusual among vertebrates. Each multiply innervated fiber receives many nerve endings that are distributed along the cell surface (10–15 in cod; Altringham and Johnston 1981). Neural stimulation of white muscle can result in activation both by electrotonic spread of junction potentials and by propagated action potentials (Hudson 1969), but it is not clear which mode of activation is employed in situ. The significance of multiterminal innervation is not known, but it has been suggested that stimulation by evenly distributed motor end plates could serve to activate the entire fiber more rapidly, in turn, permitting more rapid tail beat frequencies (Johnsteon 1983a). Alternatively, it is possible that this pattern permits the white musculature to be activated in a graded fashion, resulting in finer control of power development (Johnston 1980a).

2.1.2 Muscle Fiber Types of Fish Are Segregated Anatomically

Fiber types within the locomotory musculature of teleost fish are arranged in anatomically discrete regions (Fig. 1). In contrast, fibers of different types usually are distributed heterogeneously throughout muscles of higher vertebrates. This factor has complicated studies comparing the contractile and biochemical characteristics of different fiber types in the latter groups. The musculature of most teleost fish, therefore, provides a much more manageable system for the investigator interested in the comparative physiology of skeletal muscle. Nature already has performed a major purification step!

Red fibers of fish typically are found as a triangular, superficial strip that lies in the lateral line region extending from the head area to the caudal peduncle. Pink muscle is found in varying quantities immediately beneath the superficial red fibers. The remainder of the trunk musculature is made up of white fibers, which are by far the most abundant of the three muscle types in most teleost species. Individual fibers insert into collagenous sheets, called myoseptae, with the myosept-to-myosept unit being termed a myotome. Myotomes are arranged in a serial fashion along the long axis of the body, but the three-dimensional geometry of a given myotome is complex, so that adjacent myotomes overlap in the deep axial region (Fig. 2). Superficial red fibers near the lateral line region

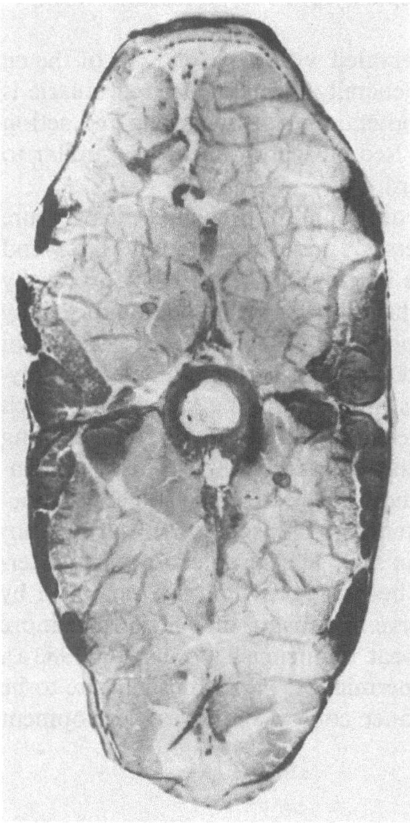

Fig. 1. Cross-section through the trunk of a striped bass, histochemically stained for cytochrome oxidase activity. Section was taken approximately ten myotomes anterior to the caudal peduncle. Note anatomically discrete segregation of dark-staining, slow oxidative (red) muscle located superficially to the surface of the animal and concentrated in the region of the lateral line (Jones and Sidell 1982)

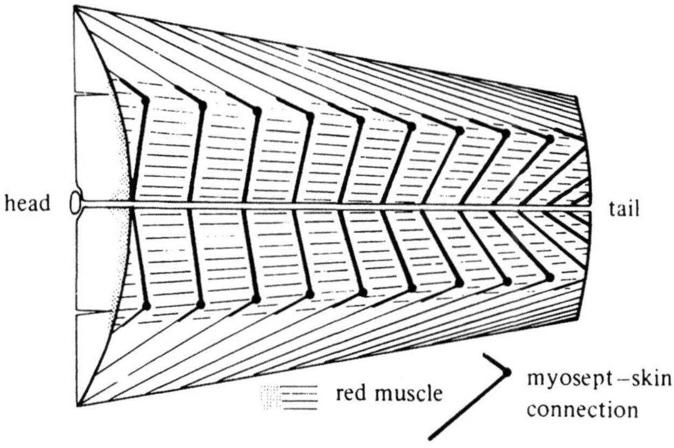

Fig. 2. Diagrammatic representation of myotomal arrangement in the cod (*Gadus morhua*) (Wardle and Videler 1980)

have a simple segmented arrangement and are oriented parallel to the long axis of the body. White fibers in the deep region of the myotome are not oriented as simply and have and angle of insertion that may approach 30° with respect to the body axis. In some species, notably salmonids, red fibers are intermixed with the deep white musculature to give what is termed a mosaic distribution of fibers. Alexander (1969) has pointed out that the significance of the complex three-dimensional arrangement of the myotome may be that fibers of the deep axial musculature shorten or stretch to a similar extent during a tail beat cycle, in spite of having different radial positions on the body.

2.2 Propulsion and Thrust Are Achieved by Corporeal Wave Form

A fish swims by successfully coupling muscular contraction into forward thrust. Contractions of the body musculature generate a series of undulating waves that, at least in the subcarangiform mode of swimming that we consider here, increase in amplitude near the tail. When the speed with which the wave is propagated along the body is greater than the water speed of the fish, the body elements that are at an oblique angle with respect to the long axis effectively "push" against the water, generating forward thrust. Several recent reviews deal with hydrodynamics of fish locomotion, and the reader is referred to these for more comprehensive accounts (Webb 1975, 1978, 1984; Wardle and Videler 1980; Blake 1983).

2.3 Power Output of the Musculature Determines Speed of Swimming

The amount of power developed by any muscle is described by the arithmetic product of (force generation)×(contraction velocity). For fish, the former term can be modified by the mass of fibers activated while the latter ultimately will relate to frequency of tail beat cycling. Both are influenced by the physiological and biochemical characteristics of the tissue and their optimal combination will be described by the force-velocity relationships of the muscle under study. The power necessary to overcome drag in the aqueous environment increases as an approximately cubic function of swimming speed (Webb 1975). Accordingly, a unit increase in swimming speed may require up to an approximately cubic increase in the volume of activated muscle.

The discrete anatomical arrangement of fiber types in the locomotory muscle mass of fish (see Sect. 2.1.2) provides an excellent opportunity to correlate muscular activity with specific, identifiable phases of locomotion. In general, electromyographic studies indicate that the superficial layer of red muscle is used during low speed swimming and white muscle is progressively recruited as swimming speeds increase (see Sect. 5.1.1). There are, however, apparent differences between species with regard to the point at which the white muscle mass is recruited. In some species, including rainbow trout, skipjack tuna, coalfish, and mackerel, it appears that some white muscle may be active during swimming speeds that can be sustained for indefinite periods (Webb 1971; Greer-Walker and Pull 1973; Brill and Dizon 1979a; Roberts and Graham 1979). In other species, including herring

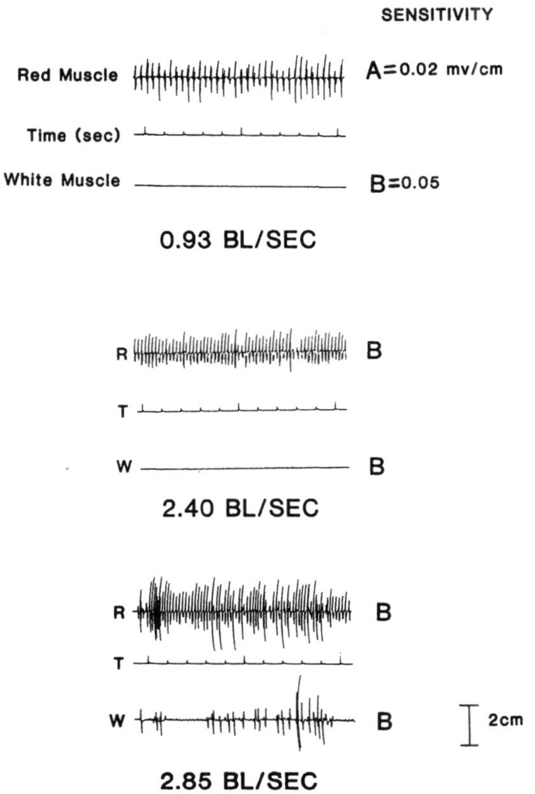

Fig. 3. Electromyographic recordings from slow oxidative (red) and fast glycolytic (white) muscle zones of striped bass while swimming at increasing velocities. Note that white muscle remains inactive (electrically quiescent) throughout the low-speed range and is activated only at higher swimming velocities (Sisson and Sidell 1987)

(*Clupea harengus pallasi*), bluefish (*Pomatomus saltatrix*), and striped bass (*Morone saxatilis*), there is a definite division of labor between red and white muscle with red muscle used exclusively at swimming speeds that can be maintained indefinitely (sustained swimming) and white muscle recruited only for high-speed and burst swimming (Bone et al. 1978; Freadman 1979a). This division of labor between red and white fiber types is particularly well illustrated in striped bass (Fig. 3).

Calculations based upon the mechanical performance of chemically skinned red type fibers from striped bass (see Sect. 4.1.1) indicate that power output of the red muscle mass peaks at the swimming speed of initial white muscle recruitment (Moerland and Sidell 1986b; Fig. 4). Recruitment of the white muscle mass, of course, is under neural control but, given that power output from red muscle peaks at a modest speed, further increases in swimming velocity must be derived from activation of the large mass of white muscle (Moerland and Sidell 1986b).

Because white fibers are thought to shorten by only 2%–3% of their initial length during contraction, as opposed to approximately 10% for red fibers, those white fibers initially recruited should therefore be well down on the ascending limb of their force-velocity and power curves (right side) with a large reserve for power demands of higher swimming speeds (Fig. 4). Progressive recruitment of greater proportions of the white muscle mass will reduce the mean load experi-

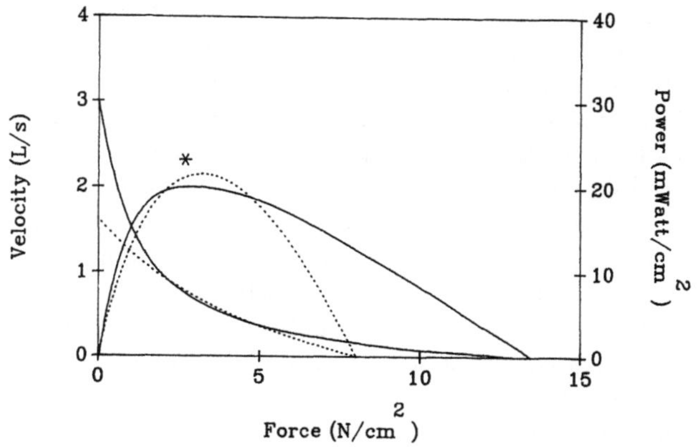

Fig. 4. Force-velocity and power output curves at 15 °C for fibers of white (*solid lines*) and red (*dotted lines*) muscles of striped bass. Power output was calculated as the product of force and velocity for fibers of 1 cm resting length. *Asterisk* (*) marks the point on the power curve of red muscle at which whtie muscle is initially recruited at 15 °C (0.78 $L_0\ s^{-1}$ and 2.72 N cm^{-2}). Data are from Moerland and Sidell (1986b). See text for additional information

enced by white fibers, shifting their position to the left (higher velocity and increased power) on their intrinsic force-velocity curve.

In a fashion similar to that described above, recruitment of the relatively large mass of faster contracting white muscle will effectively relieve the mean load borne by red fibers. As the burden of load is assumed increasingly by white muscle, velocity of contraction of red fibers will approach the unloaded speed of shortening. If, as in mammals, the unloaded shortening velocity of red fibers was greatly less than that of white, the untenable situation might arise where the white musculature must work actively to compress the red muscle mass. However, the difference in unloaded shortening velocities between red and white fibers of teleost fish is substantially less than that seen in mammals. It seems reasonable to conclude that even at high swimming speeds red muscle fibers are activated and contract at velocities consistent with high tail beat frequencies. Under these conditions, however, they contribute very little to overall power development of the trunk because of their small mass and highly unloaded position on the force-velocity curve.

3 Performance Characteristics at the Molecular Level: Contractile Proteins

The contractile proteins of teleost skeletal muscle are similar in structure, function, and distribution among different fiber types to those from striated muscles of higher vertebrates (Johnston 1983a). In both fish and higher vertebrates, red (slow) and white (fast) muscle types possess different isoforms of myosin light chains and at least some of the calcium regulatory proteins (Moerland and Sidell

1986a; Johnston 1982a, 1983b). Myosin from fish is very unstable compared to that from mammals, however, and conventional techniques of isolation yield aggregated preparations of very low enzymatic activity (Connell 1961; Johnston 1982a). As a consequence, nearly all studies of the catalytic properties of myosin from fish are carried out with relatively stable preparations in which myosin is complexed with actin, most often in the form of actomyosin or intact myofibrils.

Temperature directly influences reaction rates and successful mechanisms of compensation must provide a means to overcome the impact of temperature upon catalytic function. As pointed out by Somero (1978), adaptation of catalytic function to temperature must be accompanied by concomitant adaptations of regulatory sensitivity and structural integrity of the enzyme protein. As we shall see, adaptation to temperature in the contractile protein complex of teleost fish combines the elements of compensation of reaction rate, structural stability, and the regulation of catalytic activity.

Two general types of mechanism can be employed at the molecular level to ameliorate thermal sensitivity of biochemical reaction rates. First, the effects of temperature may be offset by alterations in the quantity of enzyme present within the cell. In poikilotherms, for example, activities of potentially rate-limiting enzymes from energy metabolism frequently increase after long-term exposure to cold (Hazel and Prosser 1974; see Sect. 4.3). The second possible mechanism entails a qualitative change in enzyme complement, meaning that a thermal variant of the protein is expressed which is suited specifically for catalysis at the prevailing temperature. Mechanisms of the first type generally are not applicable to contractile proteins of muscle. Myofibrils typically compose 40%–70% of cell volume in teleost muscle and most of the remainder is occupied by mitochondria and sarcoplasmic reticulum; there simply is little intracellular space for compensation by means of quantitative adjustments in the volume density of contractile proteins (Sidell and Johnston 1985). Note, however, that increases in total tissue mass will achieve the same compensatory effect in power development as increasing the quantity of contractile units within a cell, but at a higher level of biological organization (Sect. 4.2).

3.1 Differences in Protein Structure/Function Have Developed over Evolutionary Time

Natural selection over evolutionary periods of time has yielded contractile proteins that are specifically adapted for function at a given temperature. For example, when assayed at a common cold temperature, the activities of myofibrillar ATPases from polar teleosts are considerably higher than those from tropical species (Johnston and Walesby 1977). This increase in catalytic activity at cold temperatures is achieved by an alteration in reaction thermodynamic parameters that serves to lower the thermal sensitivity of catalysis (Fig. 5). Although free energies of activation ($\Delta G^{\neq}$) are relatively similar among species, enthalpies of activation ($\Delta H^{\neq}$) show regular decrement with decreases in normal cell temperature. This is accompanied by a concomitant increase in the entropic contribution ($\Delta S^{\neq}$) to the free energy of activation (Johnston and Goldspink 1975; Johnston

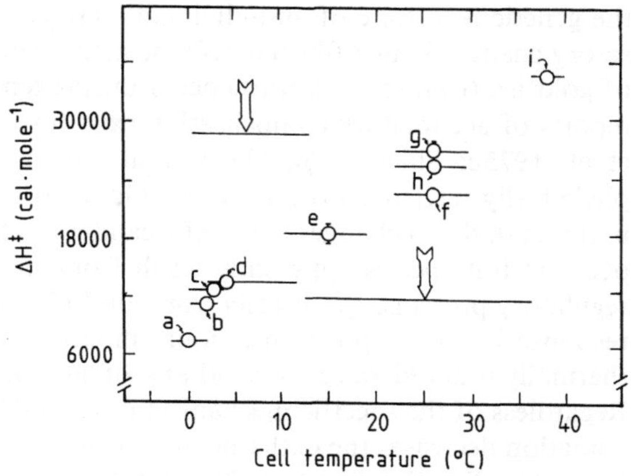

Fig. 5. Activation enthalpies ($\Delta H^{\neq}$) of fish fast muscle myofibrillar ATPases. The normal range of habitat (cell) temperatures experienced by each species is indicated by a *horizontal line*. *Points* represent mean ±SEM of 6–8 fish. The exact position of the point refers to the environmental temperature of each species prior to capture. *a Champsocephalus gunnari* (Antarctic); *b Notothenia neglecta* (Antarctic); *c Notothenia rossii* (Antarctic); *d Cottus bubalis* (North Sea); *e Salmo trutta* (northern Europe); *f Abudefduf oxydon* (Indian Ocean); *g Pomatocentrus uniocellatus* (Indo-Pacific); *h Dassyllus carneus* (Indo-Pacific); *i Tilapia grahami* (Hotsprings equatorial soda lake, East Africa). *Fundulus heteroclitus* is indicated by an *arrow*. Compared to the other species studied, the mummichog *F. heteroclitus* experiences greater fluctuations of habitat temperatures and its ATPase has a significantly lower temperature dependence (i.e., lower $\Delta H^{\neq}$). See text, Sections 3.1, 3.3 (Sidell et al. 1983)

and Walesby 1977). The site of this adaptation is apparently the actin-myosin complex and does not involve calcium regulatory proteins of the thin filament (Johnston and Walesby 1979; see Sect. 3.2). These genetically fixed adaptations to cold permit the needed increases in catalytic activities at low temperatures, but apparently modifications in protein structures necessary to achieve the effect result in lower thermal stabilities for the contractile protein complex. When incubated at 37 °C, half life for inactivation of catalytic function is only minutes for preparations from the Antarctic species *Champsocephalus gunnari, Notothenia neglecta,* and *N. rossii*, but in the order of 9–10 h for *Tilapia grahami* endemic to East African hot springs (Johnston and Walesby 1977). Increases in catalytic activity at cold temperatures are thought to be attained via more flexible and open tertiary protein structure and this same property results in decreased thermal stability for the myofibrillar proteins of cold-adapted species (Somero 1978).

3.2 Thermal Acclimation Induces Changes in the Myofibrillar Proteins of Some Fish

Individuals of many teleost species in temperate latitudes are subjected to considerable seasonal variations in temperature. In contrast to adaptations that result from the action of natural selection over evolutionary periods of time (Sect. 3.1), acclimatory compensation must be accomplished by mechanisms that lie within the genetic repertoire of an individual. Several studies have shown that compensatory changes in myofibrillar ATPase activity in both red and white musculature of goldfish (*Carassius auratus*) occur during temperature adaptation and similar reports of acclimatory compensation exist for carp, tench, and roach (Johnston et al. 1975c; Sidell 1980; Heap et al. 1985). Note that all of these fish are phyletically related cyprinid species. Goldfish are by far the best-studied system in this regard and changes in specific catalytic activity of myofibrillar ATPase that occur in this species apparently result from structural modifications of calcium regulatory proteins, specifically troponin I (Johnston 1979, 1983b). Whether this response is due to posttranslational modification of extant proteins or due to thermally-induced selective synthesis of new variants is not known at present. Regardless of the specific mechanism of expression, the net result is that cold acclimation decreases the enthalpy of activation for ATP hydrolysis, making the effects of acclimation functionally similar to the trends noted for evolutionary cold adaptation (Sect. 3.1). A difference between adaptation in the evolutionary and acclimatory frames of time is that the former involves the actomyosin fraction of the contractile protein complex, whereas the latter involves only the calcium regulatory proteins.

Capacity to adapt to long-term perturbations in temperature by altering specific catalytic activity of the myofibrillar ATPase described for cyprinid species is not widespread among eurythermal teleost fish. No acclimatory changes in myofibrillar ATPase activity are found in at least three phyletically divergent species whose natural histories include marked seasonal changes in temperature (chain pickerel, *Esox niger* – Sidell and Johnston 1985; killifish, *Fundulus heteroclitus* – Sidell et al. 1983; striped bass, *Morone saxatilis* – Moerland and Sidell 1986a). Further, a survey of the available literature indicates that most other poikilotherms also are unable to alter the catalytic or contractile properties of muscle in response to long-term changes in temperature (Bennett 1984). Thus, acclimatory responses of the type described for cyprinid species illustrate an interesting mechanism of adaptation to temperature, but one that appears to be unusual among eurythermal vertebrates. It has been suggested that capacity for acclimatory modification of myofibrillar ATPase is related to ploidy of the species under study (Sidell and Johnston 1985). Polyploid fish have the unusual ability to respond to thermal changes by expressing temperature-specific isoforms of enzymes of energy metabolism (Somero 1975). All examples published to date of fish that show acclimatory modifications of the contractile protein complex belong to polyploid taxa. A notable exception to the pattern is the relatively stenothermal *Salmo gairdneri*, which is a member of the polyploid family Salmonidae, but which does not show an acclimatory change in myofibrillar ATPase activity (Penney and Goldspink 1981).

3.3 Environments of Acutely Unstable Temperature Correlate with Unique Contractile Protein Characteristics

Large and rapid changes in water temperature are rare in nature due to the high heat capacity and attendant thermal inertia of the aquatic medium. Situations that do approach this extreme of thermal variability are the rocky intertidal and saltmarsh areas of temperate zones. For example, the temperature of tidal pools along the Atlantic coast of Maine (USA) can drop 10–15 °C in less than 1 h as water that has been warmed by solar radiation is replaced by cold ocean water of the incoming tide. The killifish *Fundulus heteroclitus* is a common fish of this Atlantic coast intertidal zone and observations indicate that the locomotory capability of this species is not substantially impaired during the rapid changes of temperature that accompany tidal cycles (Sidell et al. 1983). The contractile protein complex of the white musculature of *F. heteroclitus* is adapted specifically to maintain catalytic function and regulatory sensitivity to Ca^{2+} over a range of naturally experienced temperatures; similar changes in temperature would impair severely the function of the contractile system from other more stenothermal fish. Activity of myofibrillar ATPase from this killifish is relatively insensitive to changes in temperature across the range 13–35 °C, as evinced by a low enthalpy of activation ($\Delta H^{\neq}$; Fig. 5). These temperatures coincide with those naturally experienced during tidal cycles in the warm summer months. In contrast, the activation enthalpy at temperatures below 12 °C is comparatively high, indicating that contractile function may be impaired substantially during the winter months when this species is relatively torpid (Sidell et al. 1983). As discussed in Section 3.1, low thermal sensitivity is often associated with an increased susceptibility to thermal denaturation, presumably because mechanisms leading to a decreased thermal sensitivity of catalysis also require a more open and flexible tertiary protein structure. The myofibrillar complex from *F. heteroclitus* is an exception to this pattern; although the enthalpy of activation over the 13–35 °C range is similar to that of polar species (Fig. 5), the thermal denaturation curve for catalytic function and regulatory sensitivity to Ca^{2+} is more similar to that of tropical and equatorial species (Sidell et al. 1983). The exact mechanism that allows these minnows to pull off this molecular sleight-of-hand is still unknown.

4 Adaptations to Temperature at the Tissue and Cellular Levels

The global aim of muscular adaptations to different body temperatures in fish must be to maintain rates at which chemical energy in the form of metabolic fuels can be converted into mechanical work performed against the environment, i.e., locomotion. This overall sequence consists of two major phases: (1) energy metabolism, in which a significant proportion of the chemical bond energy released during catabolism of foodstuffs is conserved in the "high energy" terminal phosphate ester of ATP, and (2) contraction, in which energy derived from hydrolysis of ATP is converted to mechanical work by the contractile protein lattice of the muscle cell. This section of our essay will document mechanisms observed at the cellular and tissue levels that contribute to temperature compensation in each of these functions in fish muscle. Because adaptations in metabolic production of ATP are required only if the ATP-utilizing machinery maintains its energetic demand at different body temperatures, it is logical that we first examine thermal relations of the latter process.

4.1 Biochemical Data Suggest Testable Hypotheses About Cellular Contractile Performance

Results of molecular-level studies of piscine contractile systems described in Section 3 lead to several testable hypotheses about mechanical performance at the cellular/tissue level:

1. Evolutionary modifications revealed by biochemical comparisons of contractile proteins from polar, temperate, and tropical fish imply that mechanical performance of muscle at different temperatures should be well correlated with the species' normal thermal habitat.
2. Maximum unloaded contraction velocities (V_{max}) of fish muscle fibers should be sensitive to acute changes in temperature based upon apparent correlation between maximal myofibrillar ATPase activities and contraction velocities of homologous muscles established by Barany (1967). (Note, however, that little can be said about the other determinant of power development "force" based upon biochemical data.)
3. Eurythermal fish capable of biochemically modifying their contractile proteins during temperature acclimation (cyprinid species) should show corresponding compensations in mechanical performance of their muscles after thermal acclimation. Mechanical performance of muscle cells from those species incapable of modifying contractile elements, however, should be refractory to thermal acclimation.

During recent years, data has been accumulated that support at least the general validity of each of these hypotheses.

4.1.1 Experimental Preparations

Two primary types of experimental preparation have been utilized to examine the mechanical performance of fish muscle in vitro. The first is the isolated muscle block preparation developed by Wardle (1975) and variously modified by other investigators (Brill and Dizon 1979b; Yamawaki and Tsukuda 1986; McVean and Montgomery 1987). This preparation consists of a 1–2 g block of excised muscle mounted in a saline bath and affixed between two securing points, one of which is attached to a force transducer. The muscle is electrically stimulated to contract via pulses through mass electrodes either mounted in the saline bath or in direct contact with the tissue. Bath temperature is regulated by an external refrigerated circulator.

The second type of in vitro preparation which has been employed extensively with fish muscle is isolated "skinned" muscle fibers. Application of this technique to muscle fibers from fish was led by I. Johnston and his group at the University of St. Andrews, Scotland. The preparation consists of carefully dissected single white muscle fibers (or, in the case of much smaller red muscle cells, a small bundle of four to five fibers) attached between two small glass hooks and immersed in a physiological salt solution. One of these hooks is mounted to an extremely sensitive force transducer and specific modifications of the apparatus permit recording of isometric (Johnston and Sidell 1984) or isotonic contractions and length changes (Altringham and Johnston 1982). Fibers are chemically "skinned" by treatment with a mild nonionic detergent which disrupts the integrity of the sarcolemma permitting internal ionic activities to be set by concentrations in subsequent bathing solutions but leaving the entire myofibrillar apparatus intact. The muscle cell can be induced to contract by immersing it in a salt solution containing both an enzymatically based ATP-regeneration system and excitatory concentrations of Ca^{2+}. Relaxation of the fiber is accomplished by immersion in a solution containing the selective Ca^{2+} chelator, EGTA [ethylene glycol-bis-(β-aminoethyl ether) N,N'-tetraacetic acid].

The muscle block preparation offers advantages in rapidity and ease of setup and can be executed with widely available equipment. Among its disadvantages is that a significant percentage of the fibers contained in the block may be damaged by the dissection procedure. Actual excitation of the muscle also is probably not via direct electrical stimulation of the muscle cells which possess relatively inexcitable membranes (Johnston 1980b), but rather is likely via more easily excitable motor neurons distributed throughout the intact block. The latter feature does not necessarily diminish the biological significance of results because the preparation does represent the neuromuscular organization of the tissue in vivo. However, attribution of responses to a particular cellular mechanism is extremely difficult.

Attractiveness of the skinned fiber preparation is that it permits more discrete experimental examination of performance by the undisrupted contractile protein lattice. Nervous input is absent and plasma membrane and cellular metabolic influences are removed. Another feature adding to uniformity of experimental conditions is that the resting sarcomere length of each fiber examined is adjusted carefully to 2.3 μm by monitoring diffraction of laser light passed through the fiber (Altringham and Johnston 1982). Disadvantages of the preparation are its requirement for highly specialized equipment and that handling of the delicate single cells also can be extremely tedious.

4.1.2 Evolutionary Adaptations in Contractility Are Discernible

Certainly the differences in both protein structure and activation thermodynamics observed between myofibrillar proteins from species found in polar, temperate zone, and tropical waters (Sect. 3) suggest that the contractile machinery of polar fish should be most competent at cold temperature while that of tropical fish should perform better than other species' muscles at very warm temperatures. Although data on mechanical performance of homologous muscles is not as clear in pattern as, for example, that for thermostability of protein structure, they do tend to support the concept of evolutionary thermal adaptation.

McVean and Montgomery (1987) compared contraction times of muscle block preparations from an Antarctic notothenioid and a temperate zone marine species at measurement temperatures of 0–16 °C. At very cold temperatures, muscle from the Antarctic species indeed was capable of twofold faster contraction than that from the temperate zone representative. When measured at their respective normal habitat temperatures, however, muscle of the temperate species (14 °C) contracted twice as fast as that from the notothenioid (−1.9 °C). The results imply that evolutionary compensation in contractile performance has developed but that this compensation is not complete (i.e., in their natural habitats the warmer-bodied species should still show better locomotory performance than the polar species). Although the general concept is consistent with more extensive data available from isolated muscle fiber experiments, results from the latter studies produce a modified view of the roles of contraction velocity and force generation in evolutionary adaptations for muscular power development.

Comparison of maximal unloaded contraction velocities (V_{max}) and maximal isometric force production (P_0) of isolated "skinned" muscle fibers of fish from widely divergent thermal habitats reinforces the correlation of contraction velocity and normal body temperature (Table 2). Equally striking, however, is the lack of correlation between body temperature and force development. When measured at normal body temperatures, muscle fibers from polar species appear capable of greater force generation that those from warmer thermal habitats (Table 2). These results have prompted Johnston and Altringham (1985) to suggest that the primary evolutionary adaptation to temperature in power development by fish muscles resides in the ability of crossbridges to generate force, not in contraction speed of the muscles per se. Indeed, in a subsequent report, Altringham and Johnston (1986) present data that indicate muscles from polar fish are capable of generating more force per ATP hydrolyzed at cold temperatures than tissues from temperate zone species. Unlike the behavior of muscle block preparations, single

Table 2. Maximum unloaded contraction velocities and isometric tension developments by chemically skinned white muscle fibers from polar, temperate zone, and tropical fish species

Species	V_{max} ($L_0\,s^{-1}$)	P_0 ($kN\,m^{-2}$)	Measurement temperature (°C)	Reference[a]
Antarctic (−2° – +2 °C)				
Chaenocephalus aceratus	1.1 ± 0.09	214 ± 14	0	a
Notothenia rossii	0.9 ± 0.09	260 ± 22	0	a
Trematomus hansoni	0.9 ± 0.05	261 ± 21	0	a
Notothenia neglecta	–	225 ± 22	0	b
Temperate zone (4° – 25 °C)				
Myoxocephalus scorpius	1.54	150 ± 20	20	c
Gadus morhua	1.01	187 ± 18	8	d
Esox niger	1.7	209 ± 21	25	e
Morone saxatilis	2.90	135 ± 14	15	f
Tropical (15° – 30 °C)				
Tilapia mossambica	2.57	–	18	g
Mugil cephalus	4.6 ± 0.3	210 ± 28	20	b
Carangus melampygus	4.0 ± 0.3	160 ± 20	20	b
Katsuwanus pelamis	8.1 ± 0.7	180 ± 10	30	b
Euthynnus affinis	–	188 ± 0.8	30	b

[a] *References: a* Johnston and Harrison (1985); *b* Johnston and Brill (1984); *c* Sidell and Johnston (1985); *d* Altringham and Johnston (1982); *e* Sidell and Johnston (1985); *f* Moerland and Sidell (1986b); *g* Flitney and Johnston (1979).

fibers from polar, temperate zone, and tropical species display very similar unloaded contraction velocities at 0 °C (Johnston and Altringham 1985). The apparent discrepancy in behavior between muscle blocks and single fibers may be attributable to enhanced neuromuscular transmission at cold temperatures in blocks from polar fish (MacDonald 1981), and may not be due to adaptations in contractile elements.

Regardless of confusion about the specific responses of contraction velocity and force generation, the combined data are in agreement that power development is most competent at cold temperature in muscles from polar fish while the tissue from tropical animals performs best at warm temperatures (i.e., >25 °C), clearly reinforcing the concept of evolutionary adaptation in locomotory performance.

4.1.3 Thermal Sensitivity of ATPase Does Not Accurately Predict Contraction Velocity

Although interspecific comparisons of myosin ATPase-specific activity do yield accurate predictions of relative contraction velocities from the corresponding muscles (Barany 1967), extrapolation of this relationship to effects of temperature intraspecifically appears valid only in qualitative terms. Only three studies have compared thermal sensitivity of myofibrillar ATPase and contraction velocity within the same species (Johnston and Sidell 1984; Sidell and Johnston 1985; Moerland and Sidell 1986a, b). Results from experiments with both marine (*Myxocephalus scorpius*), fresh water (*Esox niger*), and anadromous (*Morone saxatilis*) fish consistently showed that myofibrillar ATPase activity is more temperature-dependent than unloaded contraction velocity of the same muscle tissue. Such a result is not particularly surprising given the differences in mechanical restrictions of the two preparations. Supercontraction of isolated myofibrils to very short sarcomere lengths and differences in mechanical constraints for crossbridge cycling between the preparations probably account for the lack of proportionality between these two parameters.

4.1.4 Not All Eurythermal Fish Display Acclimatory Adaptations in Cellular Contractility

In a fashion similar to that seen for biochemical parameters of the myofibrillar system, acclimation to different temperatures affects mechanical performance at the cellular level quite differently in cyprinid and noncyprinid species.

Cyprinids. Although compensation of myofibrillar ATPase activity during thermal acclimation has now been demonstrated in a number of cyprinid species (Johnston et al. 1975c; Sidell 1980; Heap et al. 1985), contractile performance at the tissue or cellular level has been examined to date in only two, goldfish (*Carassius auratus*) and mirror crap (*Cyprinus carpio*). Yamawaki and Tsukuda (1986) have measured the work output of muscle blocks from goldfish acclimated

Table 3. Mechanical performance at 7 °C of skinned muscle fibers isolated from warm- and cold-acclimated carp (*Cyprinus carpio*)[a]

Parameter (units)	Acclimation temperature	
	7 °C	23 °C
Oxidative (red) fibers	(9)	(10)
Isometric force ($kN\,m^{-2}$)	76 ± 9	47 ± 6
Unloaded contraction velocity, V_{max} ($L_0\,s^{-1}$)[b]	0.95 ± 0.08	0.42 ± 0.05
Maximum power output ($W\,kg^{-1}$)	8.5 ± 0.8	2.4 ± 0.5
Glycolytic (white) fibers	(16)	(10)
Isometric force ($kN\,m^{-2}$)	178 ± 15	76 ± 10
Unloaded contraction velocity, V_{max} ($L_0\,s^{-1}$)	1.8 ± 0.5	0.62 ± 0.04
Maximum power output ($W\,kg^{-1}$)	19.3 ± 2.2	6.9 ± 0.6

[a] Data are from Johnston et al. (1985). All values shown are mean ± SEM for the number of fibers indicated in parentheses.
[b] L_0 = Initial muscle length.

to 8 ° and 25 °C over a range of experimental temperatures. Their results show that muscle from cold-acclimated fish is capable of performing significantly greater mechanical work than that of warm-acclimated counterparts between 5 ° and 25 °C. These results are in complete accord with predictions based upon thermally induced changes in biochemical characteristics of the myofibrillar complex of this species (Sect. 3.2).

Contractile performances of both slow and fast muscle fibers from thermally acclimated carp also indicate temperature compensation in contraction velocity, ability to generate force, and maximum power output of the muscle cells (Johnston et al. 1985; Table 3). When these parameters are measured for fibers from 7 ° and 23 °C-acclimated fish at an experimental temperature of 7 °C, each fiber type from cold-acclimated fish shows two- to threefold greater maximal contraction velocity, maximal isometric tension, and calculated maximal power output than homologous fibers from warm-acclimated animals. When measurements are attempted at 23 °C, however, performance of fibers from warm-acclimated fish is normal while those from cold-acclimated animals consistently fail to relax completely following activation, display highly significant residual (Ca^{2+}-insensitive) force development, and very slow contraction velocities (Johnston et al. 1985). These temperature-specific differences in performance of fibers are presumably attributable to molecular variants of Ca^{2+}-regulatory proteins of the myofibrils which are expressed at different acclimation temperatures in cyprinid species (Johnston 1979; Sect. 3.2). Regardless of the proximate cause, the outcome is clear; during the acclimation process, contractile properties of cyprinid muscles are modified to achieve temperature compensation in their mechanical performance. Although the adaptive significance of these changes appears compelling, currently available information suggests that the responses are restricted to only the cyprinid group and are not widespread among eurythermal fish.

Noncyprinids. In every noncyprinid fish species studied to date, regardless of other measures of eurythermality, temperature acclimation does not affect contractile function of muscle when measured at biochemical or cellular levels. For example, chain pickerel (*Esox niger*) are powerful piscivorous fresh water fish native to northeastern North America where they seasonally experience habitat temperatures ranging from 4 ° to >25 °C. Yet, both contraction velocity and force development by muscle fibers from pickerel acclimated to 5 ° or 25 °C are identical (Sidell and Johnston 1985). Despite these results, the species obviously maintains activity over this temperature range as evinced by its successful winter capture by anglers using live bait through holes in ice-covered ponds. Recent studies with both anadromous and marine species have yielded similar seemingly paradoxical results.

We have examined thermal relationships of muscle fiber mechanics in striped bass (*Morone saxatilis*), a strong-swimming, anadromous, and eurythermal species that experience large seasonal temperature fluctuations and undertake extensive spawning and alimentary migrations. Both slow and fast muscle fibers from this species show sensitivity to acute temperature change, but acclimation to warm or cold temperatures has no significant effect on its maximal unloaded contraction velocities or maximal isometric force development (Moerland and Sidell 1986b). Johnston and Wokoma (1986) also have reported similar results with North Sea flounder (*Platichthys flesus*). They were unable to detect any differences in contractile properties of muscle fibers from this species after acclimation of fish to 5 ° or 23 °C for up to 2 months.

At least superficially, the absence of alteration in contractile performance by muscle cells from temperature-acclimated noncyprinid species suggests that these fish are incapable of locomotory adaptations to changes in body temperature. But such a conclusion is premature and serves to remind us of the caution which must be exercised by an increasingly reductionist community of biologists in extrapolating cellular/biochemical data to the whole organismal level. After all, swimming activity of the intact fish is powered by more than a single isolated muscle fiber. As we will see in Section 5, actual measurements of swimming performance of whole animals drives home this cautionary note.

4.2 Adaptations in Tissue Anatomy – Thermal Acclimation Induces Shifts in Proportions of Different Muscle Fiber Types

It has long been recognized that many eurythermal fish species have the capacity to compensate their overall metabolic rate over a period of days, weeks, or months to overcome the initial effect of changes in body temperature. Early studies of metabolic temperature compensation focused primarily on whole organismal measures of routine metabolism, such as oxygen-utilization rates (reviewed by Prosser 1973). Skeletal muscle represents a large fraction of the organism's body mass and locomotion is a major component of routine metabolism. It therefore is not surprising to find that significant adaptive changes occur in muscle during thermal acclimation that correlate meaningfully with observations of whole animal metabolism. Although many studies have been conducted at the biochemical level (reviewed by Hazel and Prosser 1974, Sidell 1983), most changes in enzyme activities that have been recorded now seem attributable to thermally induced alterations in the anatomical organization of skeletal muscle at both tissue and ultrastructural levels. Given the emphasis placed upon differential responses of cyprinid and noncyprinid fish in earlier sections, it is at once surprising and somewhat reassuring that these anatomical adaptations appear to be very widespread among all eurythermal species.

Slow oxidative (red) muscle fibers have a much greater capacity for aerobic metabolism than that of fast glycolytic muscle cells (Bone 1978; Johnston 1980a). It follows that one possible mechanism underlying increases in aerobic metabolism of locomotory musculature that accompany cold acclimation is an elevation in the relative proportion of red muscle fibers within the tissue. This response now has been confirmed in phyletically divergent species. The first observation to suggest that thermal acclimation causes a shift in proportions of muscle fiber types in fish was that of Smit et al. (1974). On the basis of careful dissections and weighings, they reported that the proportion of red muscle tissue in myotomes of goldfish decreases with increasing acclimation temperature. Subsequently, Johnston and Lucking (1978) used histochemical techniques to establish that combined proportions of red and aerobic intermediate fiber types in trunk musculature of the same species increase by approximately 1.6-fold during acclimation from 31 ° to 3 °C. Cold-induced red fiber proliferation of goldfish also has been corroborated by independent biochemical methods (Sidell 1980). Red fiber proportion in striped bass similarly increases from 9% to 15% of the locomotory muscle mass after acclimation from 25 ° to 5 °C (Jones and Sidell 1982; Sidell and Morse unpublished).

Metabolic consequences of increasing red fiber mass during cold acclimation are obvious; in toto aerobic capacity of the trunk musculature increases accordingly. On face value, however, the adaptive significance of increased aerobic metabolic capacity per se is initially obscure. How can a greater red fiber mass contribute to enhanced locomotory ability at cold temperature? Obviously, if enhancement occurs, it must be manifest as improved sustained swimming capability given that this behavior is powered exclusively by red fibers in most fish (see Sect. 2).

At the simplest approximation, an increase in red muscle mass of the trunk should result in a proportional increase in potential for force generation when red fibers are maximally recruited. Force is one of the multiplicands (along with contraction velocity) in determining power development. However, the maximal sustainable speed that an animal achieves at any given temperature is related linearly to its tail beat frequency (TBF) which, in turn, must be predicated upon contraction speed of the red fibers. Any satisfactory explanation of the adaptive significance of red muscle proliferation should incorporate an elevation of contraction velocity of the fibers. We recently have proposed a model that encompasses these considerations (Sisson and Sidell 1987).

Mass of red muscle in the trunk increases during the process of cold acclimation. Thus at any swimming speed up to and including maximal sustainable performance, more fibers are available for recruitment in the cold-acclimated animal; i.e., the load on the tail may be distributed over a greater number of fibers. This

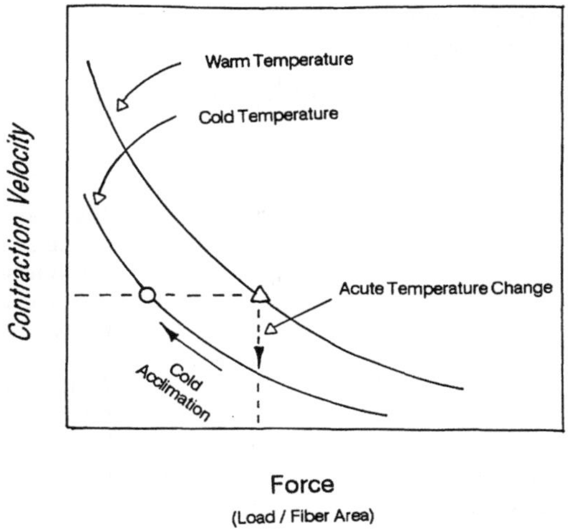

Fig. 6. How proliferation of red muscle mass at cold temperature may contribute to maintenance of sustained swimming speed. Theoretical force-velocity curves for individual red muscle fibers at warm and cold temperatures are illustrated. An acute decrease in temperature moves the force-velocity curve downward. The *upward, solid arrow* indicates acclimation to the lower temperature where red muscle proliferation decreases the load on each individual fiber, moving it to the left (higher velocity) on its intrinsic curve. (After Sisson and Sidell 1987)

will lead to reduction in the actual load borne by any individual red muscle fiber, shifting its position on the fiber's intrinsic force-velocity curve to the left (i.e., higher contraction velocity). The result should be an expansion of the limits of TBF/swimming speed within the sustainable mode during the process of cold acclimation (Fig. 6). If this model is accurate, elevation in fiber contraction velocities during cold acclimation must be associated with greater rates of myosin crossbridge cycling and, therefore, elevation in the intact tissue's demand for ATP integrated over time. Consequently, it seems only reasonable that adaptive compensations in ATP-generating machinery of energy metabolism must occur concurrently during thermal acclimation.

4.3 Metabolic Adaptations Ensure ATP Supply at Different Body Temperatures

Measurements of oxygen utilization in vitro clearly demonstrate that cold acclimation is accompanied by marked elevation in rate of aerobic metabolism by both muscle fiber types in fish (Hochachka and Hayes 1962; Dean 1969; Smit et al. 1974). Enhanced aerobic character of locomotory muscle thus is due to intracellular changes within fiber types in addition to shifts in their proportions. The effect is most patent in red muscle because of the tissue's high mitochondrial content (Jones and Sidell 1982).

4.3.1 Mitochondrial Densities of Muscle Cells Are Altered by Thermal Acclimation

When the vectors of change in muscular enzyme activities that occur as a result of thermal acclimation of fish are examined carefully, an interesting pattern emerges. Those enzymes from pathways of energy metabolism that increase in ac-

Table 4. Patterns of activity or concentration response to temperature acclimation by enzymes from fish muscle

Enzyme/protein	Species	Tissue	Acclimation temperatures (°C)	Activity ratio[a] (C/W)	Reference[b]
Cytochrome oxidase	Goldfish	Whole epaxial	5, 25	1.66	1
	Green sunfish	Whole epaxial	5, 25	1.94	2
	Striped bass	Red fibers	5, 25	1.97	3
	Carp	Red fibers	7, 23	1.58	4
		White fibers		2.03	
	Chain pickerel	Red fibers	5, 25	1.63	5
		White fibers		1.15	
		Heart		1.94	
	Flounder	Red fibers	5, 23	2.1	6
Cytochrome c	Green sunfish	Whole epaxial	5, 25	1.51	7
Succinic dehydrogenase	Goldfish	Whole epaxial	5, 25	1.84	8
	Green sunfish	Whole epaxial	5, 25	1.35	2
Citrate synthase	Striped bass	Red fibers	5, 25	1.57	3
	Carp	Red fibers	8, 23	2.25	4
		White fibers		2.83	
	Chain pickerel	Red fibers	5, 25	1.59	5
		White fibers		1.42	
		Heart		1.42	
Carnitine palmitoyl-transferase	Carp	Red fibers	7, 23	3.15	4
	Flounder	Red fibers	5, 23	2.6	6
	Chain pickerel	Red fibers	5, 25	1.36	5
		White fibers		1.67	
		Heart		1.56	
β-Hydroxyacyl CoA dehydrogenase	Carp	Red fibers	7, 23	1.92	4
		White fibers		1.29	
Hexokinase	Carp	Red fibers	7, 23	1.78	4
		White fibers		4.67	
	Chain pickerel	Red fibers	5, 25	1.75	5
		White fibers		NS	
		Heart		1.97	
	Flounder	Red fibers	5, 23	1.6	6
Glucosephosphate isomerase	Green sunfish	Whole epaxial	5, 25	0.083	2
6-Phosphofructo-kinase	Striped bass	Red fibers	5, 25	NS	3
		White fibers		NS	
	Flounder	Red fibers	5, 23	NS	6
		White fibers		NS	
Pyruvate kinase	Carp	Red fibers	7, 23	NS	4
		White fibers	7, 23	NS	4
		white fibers		NS	
	Striped bass	Red fibers	5, 25	NS	3
		White fibers		NS	
	Flounder	Red fibers	5, 23	NS	6
		White fibers		NS	
Lactate dehydrogenase	Carp	Red fibers	7, 23	NS	4
		White fibers		NS	
	Goldfish	Whole epaxial	5, 25	0.74	8

Table 4 (continued)

Enzyme/protein	Species	Tissue	Acclimation temperatures (°C)	Activity ratio[a] (C/W)	Reference[b]
	Green sunfish	Whole epaxial	5, 25	0.90	2
	Striped bass	Red fibers	5, 25	NS	3
		White fibers		NS	
	Chain pickerel	Red fibers	5, 25	0.82	5
		White fibers		0.70	
		Heart		0.90	
	Flounder	Red fibers	5, 23	NS	6
		White fibers		NS	

[a] Ratio of activities between acclimation groups (cold/warm) when assays are performed at a common (usually intermediate) temperature.
[b] References: *1* Wilson (1973); *2* Shaklee et al. (1977); *3* Jones and Sidell (1982); *4* Johnston et al. (1985); *5* Kleckner and Sidell (1985); *6* Johnston and Wokoma (1986); *7* Sidell (1977); *8* Sidell (1980).

tivity during cold acclimation and decrease in titer during acclimation to warm temperature are consistently associated with the mitochondrial compartment of the cell (Table 4). Enzymes from pathways located in the cytoplasmic compartment are equally consistent in showing a lack of compensatory change. Although these biochemical data have been accumulating for many years, their ultrastructural correlates have come to light only recently.

Johnston and Maitland (1980) were the first to report that the fraction of cell volume occupied by mitochondria, V_v (mit, f) in fish muscle shows a striking inverse correlation with acclimation temperature. Homologous muscles from crucian carp acclimated to 2° and 31 °C show V_v (mit, f) that are greater at the cold temperature by 81%, 92%, and 231% in slow, intermediate, and fast muscle fiber types, respectively. Similar responses in mitochondrial density now have been reported during temperature acclimation in several phyletically divergent fish species (Table 5) and in both muscular and nonmuscular somatic tissues from other ectotherms (frog muscle: Ballantyne and George 1978; fish liver: Campbell and Davies 1978). Apparently, the response is due to true proliferation, not simply enlargement of extant mitochondria, since geometries of the individual organelles are unaltered by thermal acclimation (Tyler and Sidell 1984; Egginton and Sidell 1988) and cristae densities within the mitochondria remain unchanged (Egginton and Sidell 1989). Mounting evidence suggests that maintenance of aerobic metabolic potential as body temperature cools is accomplished by proliferation of cellular mitochondrial populations. Two additional lines of evidence further indicate that high mitochondrial densities may be a fundamental requirement for maintenance of aerobic metabolism at cold cell temperatures: (1) Activities of mitochondrial enzymes increase during cold acclimation in virtually all tissues of eurythermal ectotherms (Hazel and Prosser 1974). (2) Activities of mitochondrial enzymes are very high in aerobic muscles of polar fish (Johnston and Harrison 1985; Sidell et al. 1987; Torres and Somero 1988) and mitochondrial densities of

Table 5. The effect of thermal acclimation upon fraction of cell volume occupied by mitochondria ($V_v$mit,f) in fish muscles

Species	Fiber type	Acclimation temperature (°C)	Mitochondrial volume density ($V_v$mit,f)
Crucian carp[a]	Red	28	0.147 ± 0.007
(*Carassius carassius*)		2	0.314 ± 0.013
	White	28	0.016 ± 0.002
		2	0.061 ± 0.007
European eel elver[b]	Red	29	0.188 ± 0.052
(*Anguilla anguilla*)		10	0.264 ± 0.055
Goldfish[c]	Red	25	0.056 ± 0.005
(*Carassius auratus*)		5	0.201 ± 0.013
	White	25	0.008 ± 0.001
		5	0.014 ± 0.001
Striped bass[d]	Red	25	0.286 ± 0.018
(*Morone saxatilis*)		5	0.448 ± 0.024
	White	25	0.027 ± 0.003
		5	0.040 ± 0.004

[a] Johnston (1982a).
[b] Egginton and Johnston (1984).
[c] Tyler and Sidell (1984).
[d] Egginton and Sidell (1989).

these tissues are among the highest reported for any fish (Fitch et al. 1984; Johnston and Harrison 1987).

Quantitative increases in mitochondrial density at cold temperature may contribute to conservation in rates of cellular ATP production in two ways. First, increases in the intracellular concentration of mitochondrial enzymes may help offset reduced catalytic turnover rates at cold temperature (Hazel and Prosser 1974). In other words, although catalysis by each individual enzyme molecule proceeds more slowly, the aggregate rate of product formation is conserved. Secondly, mitochondrial proliferation will reduce mean diffusion path lengths for compounds traversing between cytosolic and mitochondrial compartments of the cell and increase the area of exchange surface between these two compartments. Both effects will serve to compensate for lowered cytosolic diffusion coefficients of compounds at cold temperature, conserving metabolic fluxes (Sidell 1983, 1988). Calculations based upon quantitative stereological analyses of electron micrographs from muscle of thermally acclimated fish (Tyler and Sidell 1984; Egginton and Sidell 1989) suggest that these ultrastructural adaptations indeed can compensate for measured decreases in diffusion of small molecules through cytosol of fish muscle (Sidell and Hazel 1987).

4.3.2 Does Fuel Preference Shift at Different Body Temperatures?

Insufficient data are currently available to evaluate whether changes in body temperature cause predictable and wholesale shifts in reliance of fish muscle upon alternate metabolic fuels. The limited extant information does tend to suggest that lipids play an increasingly important role in energy metabolism of oxidative muscle as body temperatures decrease.

In eurythermal fish, cold acclimation results in significant increases in activities of enzymes intimately involved in fatty acid oxidation, such as 3-hydroxyacyl CoA dehydrogenase (3-HOAD) and carnitine palmitoyltransferase (CPT) (Kleckner and Sidell 1985; Johnston et al. 1985). More direct measurements of $^{14}CO_2$ evolution from labeled fatty acids by muscle also are consistent with the idea of enhanced lipid catabolism at cold temperature. Oxidative muscle from striped bass acclimated to 5 °C catabolizes palmitic acid at twice the rate of that from 25 °C-acclimated fish, even when measurements are made at the respective acclimation temperatures (Jones and Sidell 1982). Finally, muscle tissues from chronically cold-bodied Antarctic fish display activities of 3-HOAD and CPT at 0 °C which are comparable to those from temperate zone species measured at much warmer assay temperatures (Sidell et al. 1987; Johnston and Harrison 1985; Sidell and Crockett 1987).

Recent studies with mammalian muscles suggest that the high mitochondrial densities characteristic of slow muscle from cold-bodied fish may be intimately involved with preferred catabolism of fatty acids. Dudley et al. (1987) have shown conclusively that increases in mitochondrial densities of rat skeletal muscles after endurance training cause a much greater sensitivity of mitochondria to cytosolic respiratory control. Tissues with high mitochondrial content maintain a higher adenylate energy charge (i.e., lower ADP, AMP, and P_i concentrations) during active contraction than those with lower mitochondrial densities. These intracellular conditions ensure low concentrations of potent positive modulators of flux-generating steps in glycolysis and glycogenolysis, thus favoring oxidation of non-carbohydrate substrates (Holloszy and Coyle 1984).

At this point, we have described numerous physiological and biochemical responses of fish muscle-to-temperature change occurring over widely different time courses. We further have offered interpretations of the adaptive significance of these responses in many cases. For this process to become more than an exercise in arm-waving, we must turn to measurements of whole animal performance in testing the validity of our hypothesis.

5 Whole Animal Swimming Performance

Substantial literature exists documenting various measures of swimming performance of intact fish at different temperatures. For a detailed compendium of interspecific comparisons, interested readers should refer to Beamish's (1978) excellent review on swimming capacity of fish. Perusal of these data indicates that regardless of the performance criterion employed (sustained, prolonged, or "burst" speed), swimming performance among fish species appear to be dominated by features of lifestyle and body form rather than temperature per se. Relatively few experiments have focused on those intraspecific effects of temperature upon performance that may be correlated directly with physiological characteristics of fish muscle. During the last several years, however, a few studies have measured parameters of swimming performance in fish that do shed light on the adaptive value of thermally induced changes in muscle tissue described in the preceding sections.

5.1 The Sustained Swimming Mode

Quantifying locomotory capacity of fish within the sustained swimming mode has proved to be most amenable to experimentation. Sustained swimming is the scope of speeds which can be maintained indefinitely without fatigue. The process is supported exclusively by aerobic metabolism and, therefore, relies upon highly oxidative slow/red muscle tissue activity and cannot involve significant excitation of the much larger mass of rapidly fatiguing glycolytic (white) muscle (see Sect. 2.3). Sustained swimming in fish thus presents the ideal situation to the muscle physiologist: a quantifiable behavior which can be attributed to a known and anatomically discrete muscle mass.

5.1.1 Methodologies

Two major experimental approaches may be utilized to measure sustained swimming performance in fish. The first is observational and entails subjecting animals to stepwise increases in swimming speed for standard time intervals until the animal fatigues. Experiments generally are conducted by placing fish in a water tunnel where they must swim to maintain station against controlled velocities of water flow (see Beamish 1978 for description of various apparatus designs). Fatigue is defined operationally as the point where the fish no longer maintains stations and is swept back by water flow against a retaining screen. A "critical swimming speed", or U_{crit} is computed by the formula originally developed by Brett (1964):

$$U_{crit} = u_1 + [(t_1/t_2) \times u_2] \ ,$$

where

u_1 = the highest velocity maintained for a complete time interval;
u_2 = the magnitude of experimentally employed stepwise velocity increment;
t_1 = the time swam at speed increment where "fatigue" occurred, and
t_2 = the interval employed in stepwise tests.

Although experimentally tedious, measurement of U_{crit} is a noninvasive technique and yields reliable data with species that adapt well to swimming in performance tunnels.

The second means for assessing maximal sustainable swimming speed of fish is an invasive method that assumes any speed requiring recruitment or activation of glycolytic muscle is, by definition, nonsustainable. Fine hook electrodes are surgically implanted in trunk muscle at zones of homogeneous fiber type composition. After recovery from surgery, the animal is forced to swim in a performance tunnel at increasing speeds while electrical activity is recorded from each muscle type through the implanted electrodes (for technical details, see Johnston et al. 1977, Freadman 1979a, Rome et al. 1984, Sisson and Sidell 1987). The highest swimming velocity where white muscle remains electrically quiescent (is not activated) is considered equivalent to maximal sustainable swimming speed. Although several authors have speculated that some activation of fast muscle fibers may occur during sustained swimming (reviewed by Bone 1978; see Sect. 2.3), data from species where both U_{crit} and initial white fiber recruitment speeds have been carefully compared argue against this idea.

Freadman (1979a, b) measured both U_{crit} and speed of initial white fiber recruitment in striped bass (*Morone saxatilis*) and bluefish (*Pomatomus saltatrix*). Swimming speeds where electrical activity first was recorded from white muscle of these species were 3.2 and 4.5 body lengths per second (bl s^{-1}) in striped bass and bluefish, respectively. U_{crit}s (30-min interval) measured in the same apparatus were 3.28 ± 0.13 bl s^{-1} for striped bass and 4.61 ± 0.17 bl s^{-1} for bluefish, indicating excellent agreement between the two experimental approaches.

5.1.2 Sustained Swimming is Sensitive to Acute Temperature Exposure

Certainly all in vitro measures of both contractile and metabolic capacity of oxidative muscle tissue suggest that performance of red muscle should deteriorate with acute decreases in temperature. This expectation largely has been confirmed in studies of swimming animals.

Rome et al. (1984) electromyographically measured speeds of initial white muscle recruitment in mirror carp acclimated to 15 °C (Fig. 7). When tests were performed at different test temperatures, they found that the same fish were able to attain swimming velocities of 46 ± 6 cm s^{-1} at 20 °C before recruiting white muscle, while a speed of only 26 ± 3 Cm s^{-1} could be attained at 10 °C ($Q_{10} = 1.8$). Significant thermal sensitivity in U_{crit} of mirror carp acclimated to 28 °C (but not those acclimated to 10 °C) also has been reported by Heap and Goldspink (1986). Striped bass also show clear thermal sensitivity to acute temperature exposure in swimming speeds powered by red muscle alone. In this species, the Q_{10} for initial white muscle recruitment speed is approximately 1.5, regardless of the animal's previous thermal history (Sisson and Sidell 1987).

All of these data are consistent with the interpretation presented by Rome et al. (1984) which they refer to as "compression of recruitment order". Because muscle performance is impaired by acutely decreasing temperature, colder bodied animals must activate more muscle fibers to attain any given swimming speed, therefore necessitating recruitment of anaerobic fast fibers at lower absolute swimming velocity. That is, the order of the recruitment pattern is retained at all temperatures, but recruitment of motor units is numerically "compressed" over a smaller speed range. The obvious outcome of this situation is that acute decreases in body temperature must lead to decrements in both maximal sustained and burst swimming capabilities of the animal. To be adaptive, any changes in

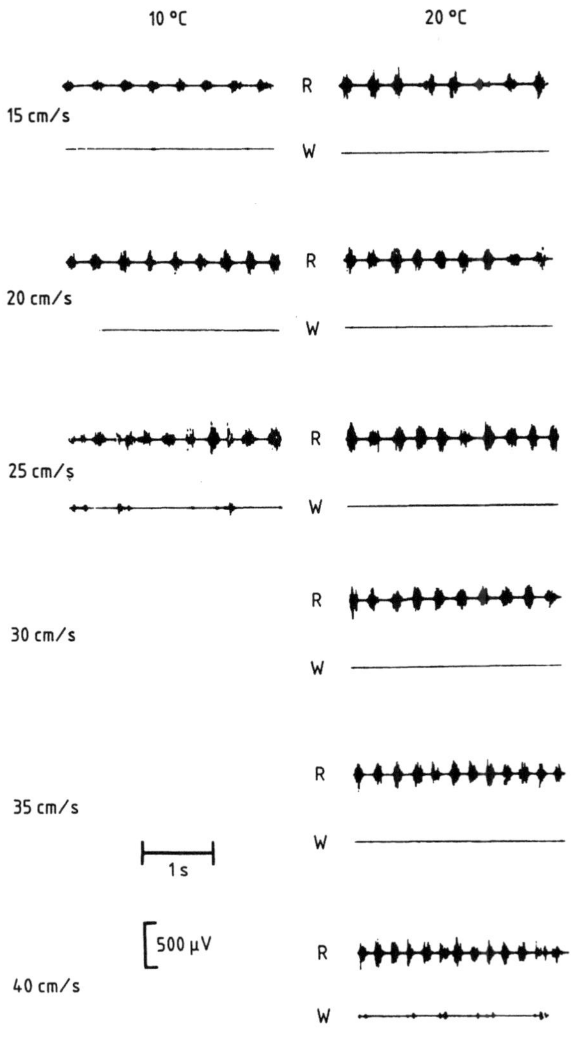

Fig. 7. Electromyographic recordings from red and white muscle of a single carp swimming at 10° and 20°C after relatively acute (48 h) exposure. Note that activation of white muscle fibers is required to power much slower swimming speeds at cold temperature, demonstrating "compression of recruitment order" (see text) (Rome et al. 1984)

the physiology of muscle which accompany longer-term thermal acclimation must compensate for this acute thermal sensitivity.

5.1.3 Temperature Acclimation Modifies Sustained Swimming Performance

The many alterations in both contractile and metabolic machineries of muscle that occur during thermal acclimation apparently do contribute to enhancement of sustained swimming performance in fish. Rome et al. (1985) were the first to examine systematically the effect of thermal acclimation on sustained locomotory ability. They employed electromyography to compare speeds of initial white muscle recruitment by 8°- and 26°C-acclimated mirror carp at both warm and cold

Table 6. Speed of initial white muscle recruitment by mirror carp as a function of muscle temperature and acclimation temperature[a]

Fish	Swimming velocity (cm s^{-1}) 10 °C	20 °C
	Cold-acclimated (8 °C)	
1	28	43
2	33	48
3	33	43
4	32	48
5	28	38
Mean ± SEM	30.8 ± 1.2	44 ± 1.9
	Warm-acclimated (26 °C)	
6	23	48
7	21	38
8	21	38
9	22	48
10	22	43
Mean ± SEM	21.8 ± 0.4	43 ± 2.2

[a] Data from Rome et al. (1985).

test temperatures (Table 6). Their results clearly show that at 10 °C test temperature, cold-acclimated animals are capable of reaching a significantly higher swimming velocity than their warm-acclimated counterparts before having to activate the white muscle mass.

If the interpretations of in vitro data presented earlier are accurate, improved cold temperature performance in cyprinid species, such as carp, may be ascribed to the combined effects of: (1) molecular alterations in Ca^{2+}-regulatory proteins of the myofibrils which leads to elevated ATPase activity and fiber contraction velocity; (2) proliferation of red muscle fibers which enlarges the mass of aerobic muscle available to power low-speed swimming; and (3) elevations in mitochondrial density of slow fibers, resulting in adaptive increases in aerobic capacity to supply ATP to the contractile elements. The data, however, do not permit a clear definition of the relative roles of each of these concurrent responses to adaptation of locomotory performance. It is particularly interesting to question whether changes in myofibrillar architecture, which are apparently unique to cyprinids, are critical to enhanced performance. A partial answer has come from recent experiments with striped bass conducted in our laboratory.

As we have described earlier (Sects. 2, 3), thermal acclimation of striped bass has no effect on any measured parameter of muscle contractile performance at molecular or isolated fiber levels of organization. This species, however, does show enlargement of red muscle mass and metabolic and ultrastructural responses to thermal acclimation that are analogous to those seen in cyprinid species. In a comparative sense, striped bass may be viewed as a naturally occurring experimental control that permit us to assess whether changes in myofibrillar characteristics are necessary for enhanced performance. The answer to this question appears to be "no".

Table 7. Swimming speeds of initial white muscle recruitment by thermally acclimated striped bass[a,b]

Test temperature (°C)	SS ($bl\ s^{-1}$)	
	Acclimation temperature 9 °C	25 °C
9	1.95 ± 0.33 (7)	–
15	2.46 ± 0.13 (5)	1.84 ± 0.10 (8)
25	–	2.82 ± 0.26 (5)

[a] Data from Sisson and Sidell (1987).
[b] All data are expressed as mean ± 95% confidence intervals. All entries are significantly different from each other at $\alpha = 0.05$. Number of animals is given in parentheses.

We recently have used EMG techniques to measure swimming speeds of initial white muscle recruitment by thermally acclimated striped bass at different test temperatures (Sisson and Sidell 1987). When results were compared between 9 °- and 25 °C-acclimated fish, swimming at a common test temperature of 15 °C, cold-acclimated animals were able to swim 34% faster than warm-acclimated fish while relying exclusively upon power developed by red muscle tissue (Table 7). In fact, the sustainable speed achieved by cold-acclimated animals swimming at 9 °C was equivalent to performance of warm-acclimated fish at 15 °C. Improved sustained swimming performance thus can be attained through combined metabolic compensation and quantitative adjustment of red fiber mass in the absence of changes in myofibrillar contractile elements. At present, the best explanation of this behavior appears to be the load-distribution mechanism described in Section 4.2.

5.2 Burst Speed Swimming

The relationship between temperature and burst speed swimming abilities of fish is far more difficult to resolve than characteristics of low speed swimming. This is due partially to inherent difficulties encountered in inducing true maximal burst performance of the animal under constrained conditions of the laboratory. The issue is clouded further by significant semantic differences in defining the exact locomotory phase corresponding to "burst speed performance". By convention, fish biologists operationally have defined burst speed swimming as the maximal speed attained by the animal for a period of less than 15 s (Hoar and Randall 1978). Using this definition, a prevailing consensus has developed that burst speed swimming of fishes is relatively independent of temperature intraspecifically (Beamish 1978). However, this notion is somewhat misleading in light of our current understanding of physiological mechanisms that temporally come into play upon maximal activation of the fast glycolytic musculature in fish.

It is now clear that the earliest phases of fast muscle activation are supported energetically by discharge of the tissue's substantial creatine phosphate pool (Driedzic et al. 1981; Dobson et al. 1987). During a very brief interval (probably lasting only a few seconds) after the onset of maximal activation, rephosphorylation of ADP to ATP occurs at enormously high rates by the transfer of the high energy phosphate linkage from creatine phosphate (CrP) to ADP. This reaction is catalyzed by creatine phosphokinase (CPK) whose catalytic potential is extremely high in fast muscle fibers. For the duration of this very brief initial phase of activation, ATP supply is functionally limitless and power development by the muscle is limited by maximal activity of the contractile elements. Immediately following this initial phase, but still within the period classically referred to as "burst" swimming, glycogenolysis is rapidly activated and the bulk of ATP supply to contractile elements is derived from anaerobic glycolysis of glucosyl residues to lactic acid (Driedzic et al. 1981; Dobson et al. 1987). Although the glycolytic potential of fish fast muscle is very high, it is probably exceeded by maximum capacity of the tissue's actomyosin ATPase to hydrolyze ATP. This latter phase of burst swimming thus is probably limited by maximal rates of glycolytic ATP generation, not contractile elements per se.

5.2.1 Burst Speed Swimming and Acute Temperature Change

We now can evaluate whether maximal burst speed swimming is sensitive to acute temperature change. Certainly data presented earlier on thermal sensitivity of molecular and cellular contractile performance combined with the apparently sound proposal of "compression of recruitment order" suggest that maximal generation of mechanical power by the locomotory muscle should be thermally sensitive. These conditions correspond to the CrP discharge phase of "burst" swimming as described above. In fact, limited empirical evidence does exist to support this contention.

Among the most interesting studies of maximal swimming speeds in fish are those of Wardle and co-workers who videotaped high-speed swimming of unrestrained fish (Wardle 1975, 1979). Wardle argues that fastest twitch contraction times measured with muscle blocks from several species correlate well with their maximal tail beat frequencies (Wardle 1975). Unfortunately, temporal resolution of the taping system employed appears inadequate to define accurately maximal tailbeat frequencies of the animals and the correlation must be considered as suggestive rather than definitive. If Wardle's interpretations prove correct, however, they would indicate that muscle contraction times limit maximal swimming speed. These experiments do show that both contraction times and maximal swimming speeds of these species are similarly temperature-dependent.

How then can we reconcile the prevailing view of thermal independence of "burst" speed swimming with this idea? The logical explanation is that glycogenolytic ATP production limits performance over most of the conventional "burst" speed range at all temperatures and that this process shows a relatively low thermal sensitivity. Unfortunately, no experimental information is available at present to assess temperature dependence of maximal glycolytic flux rates in fish muscle. It is interesting to note, however, that such behavior would also explain why adjustments in maximal catalytic potentials of glycolytic enzymes are not required during thermal acclimation (Table 4).

5.2.2 Burst Speed Swimming and Thermal Acclimation

In light of temperature effects on contractile performance already described, initially it appears that only cyprinid species have the ability to compensate maximal swimming speeds after thermal acclimation. Acclimatory adjustments in myofibrillar proteins which lead to enhanced contraction times and power development of fast muscle (Sects. 3, 4) are apparently unique to this animal group. Does this mean that all other eurythermal fish suffer a decrement in swimming ability at any/all speeds and that this impairment can only be compensated for in a sustained swimming mode? Recent data suggest that such a conclusion based solely upon muscle physiology is a misleading oversimplification. Other factors must be considered including effects of temperature upon physical characteristics of the medium (water) and several behavioral adjustments available to the animal which may affect its hydrodynamic characteristics and swimming capacity. We will now turn to these final topics of our essay.

6 The Interaction Between Organism and Environment: Propulsion at Different Temperatures

It should be possible to describe quantitatively propulsion of a swimming fish by measuring kinetic parameters and applying them to hydrodynamic models. A practical limitation of this approach lies in the difficulty of accurately determining the requisite kinetic parameters and in the choice of a suitable model (reviewed by Webb 1975). Fortunately, simplified models have been developed in recent years that permit reasonable estimates of thrust and power output to be made by considering principally the kinematics of the trailing edge (tail) of a swimming fish (Lighthill 1969; Webb 1975, 1978; Wu 1977).

6.1 Does Temperature Influence the Relationship Between Swimming Speed and Tail Beat Frequency?

Relatively few studies specifically have addressed the effects of temperature on the kinematics of swimming fish. Most available data relate swimming speed to tail beat frequency, a parameter which is readily obtained from cinematographic or electromyographic records. In general, swimming speed can be described as a linear function of tail beat frequency. Temperature has no discernible effect on this relationship for carp and goldfish (Smit et al. 1971; Rome et al. 1984). The slope of the line relating swimming speed and tail beat frequency increases with decreasing temperature for rainbow trout (Stevens 1979), but decreases with decreasing temperature in largemouth bass and striped bass (Stevens 1979; Sisson and Sidell 1987). The most comprehensive set of data is available for striped bass (Fig. 8) and we will use this information to examine further the effects of temperature on the kinematics of swimming fish.

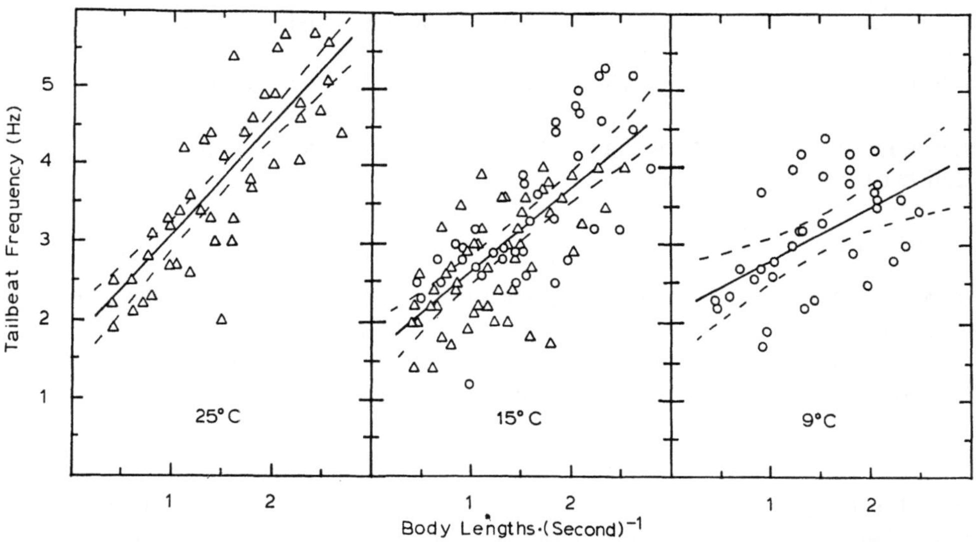

Fig. 8. Tail beat frequency (TBF) versus swimming speed (SS) at test temperatures of 25 °, 15 °, and 9 °C for striped bass acclimated to 25 °C (*triangles*) and 9 °C (*circles*). *Dotted lines* indicate 95% confidence limits for best-fit linear regression. When data were curve-fitted to linear and several nonlinear equation forms, the linear model was most favorable. For 25 °C: TBF = [1.49±0.10 SS]+[1.61±0.18]. No significant difference between acclimation groups was found at 15 °C; best-fit regression line for pooled data: TBF = [1.05±0.11 SS]+[1.64±0.16]. For 9 °C: TBF = [0.69±0.19 SS]+[2.14±0.29] (Sisson and Sidell 1987)

Data in Fig. 8 clearly indicate that the relationship between tail beat frequency and forward swimming speed is not fixed in striped bass. Rather, the tail beat frequency necessary to achieve a given swimming speed decreases with temperature. Two explanations are possible: either the forward thrust delivered per tail beat is greater at cold temperatures, or the power necessary to achieve a given swimming speed is less at cold temperatures. The former possibility could result from changes of several types in trailing edge kinematics (Sect. 6.3) and, as such, this explanation focuses on behavioral responses of the animal to temperature. The latter possibility is extrinsic to the organism and entails the physical properties of water; we will consider this factor next because it is applicable to all fish experiencing temperature change.

6.2 Does Drag Vary with Temperature?

As mentioned in the introduction, density and viscosity are two physical properties of the locomotory medium that are determinants of drag experienced by a swimming fish. Both density and viscosity of water change with temperature. The impact of temperature change on density is relatively small, causing differences of only approximately 0.2% between 5 ° and 25 °C. Kinematic viscosity of water, however, increases markedly with decreasing temperature from 0.89 centistokes at 25 ° to 1.52 centistokes at 5 °C.

The total drag on a swimming fish is the sum of frictional (skin) and pressure (form) drag. Frictional drag results from the intermolecular cohesion of water molecules, and is quantified as viscosity. Pressure drag is the result of unfavorable pressure gradients created by the distortion of flow about an object. Frictional drag can be estimated by applying the standard Newtonian equation (Hoerner unpublished):

$$D_f = \frac{1}{2} \varrho_e \cdot S \cdot U^2 \cdot C_f \ , \tag{1}$$

where D_f is frictional drag, ϱ_e is the density of the environment, S is the total wetted surface area, U is velocity, and C_f is the frictional drag coefficient. The frictional drag coefficient, in turn, is a function of the conditions of flow about the body and of its Reynolds number, R_L. The Reynolds number is given by:

$$R_L = [(L\,U)/\nu] \ , \tag{2}$$

where L is the length of the body and ν is the kinematic viscosity of the medium. When flow in the boundary layer about the body is laminar, the frictional drag coefficient is given as:

$$C_f = 1.33\, R_L^{-0.5} \ , \tag{3}$$

and when boundary layer flow is turbulent (Webb 1975):

$$C_f = 0.072\, R_L^{-0.2} \ . \tag{4}$$

Pressure drag is usually calculated as a multiple of frictional drag. This quantity is highly dependent upon body form and on the conditions of flow in the fluid layer immediately adjacent to the body. In general, pressure drag is higher for objects with higher R_L and is far greater when flow at the boundary layer is separated than when attached.

It is apparent that R_L for an individual fish will vary with temperature because the viscosity of the medium changes with temperature. For our example of striped bass (body length, 17 cm) swimming at 2 bl s^{-1}, R_L will be approximately 3.8×10^4 and 6.5×10^4 at 9° and 25 °C, respectively. Substitution of these values for R_L into Eqs. (1), (3), and (4) indicates that the frictional drag experienced by our model fish at 9 °C will be approximately 30% greater than at 25 °C when flow in the boundary layer is laminar; the differential reduces to 10% when flow is turbulent. Obviously, this effect cannot explain the observation that in striped bass, the tail beat frequency necessary to achieve a given swimming speed decreases in cold. In fact, as pointed out by Stevens (1979), this type of analysis indicates that more power per stroke will be necessary to achieve a given swimming speed at cold temperatures. However, it is possible that there are positive consequences to the increased viscosity of cold water. Increasing the viscosity of the medium could serve more effectively to damp disturbances of flow at the boundary layer (Webb 1975). This could have the net effect of reducing pressure drag by delaying the point along the body at which the boundary layer separates.

Unfortunately, we have little or no direct information to evaluate whether or not drag experienced by a swimming fish can vary substantially with temperature. The pressure field about an actively oscillating fish is obviously very complex, but our analysis is based upon models developed for relatively simple nonflexing objects. Therefore, the conditions for boundary layer flow (and therefore drag) of a swimming fish may be very different from those assumed in our calculations. We look forward to future contributions from our colleagues in biomechanics/hydrodynamics to resolve the net impact of changes in water viscosity. Rudimentary analyses of caudal fin kinematics, however, yield somewhat more satisfying estimates of possible behavioral contributions to swimming ability.

6.3 The Interaction of Temperature and Trailing Edge Kinematics

The mean thrust (T) of a swimming fish can be approximated by the following equation (Wu 1977):

$$T = K \cdot \varrho_e \cdot b^2 \cdot \omega^2 \cdot a^2 \cdot [1-(U/c)] \ , \qquad (5)$$

where b = the characteristic depth of the trailing edge (tail), ω = the tail beat frequency, a = the tail beat amplitude, c = the velocity of propagation of the propulsive wave along the body, and K = a constant of proportionality. The quantity c, in turn, is the product of ω and λ, where λ is the wavelength of the body propulsive wave. The expression U/c thus represents the ratio of the water speed of the animal (U) to the velocity of the body propulsive wave (c) and is related to the "propellor efficiency" (Webb 1975).

It is apparent from Eq. (5) that increasing ω will increase thrust if the other terms of the equation are constant. In spite of the uncertainties presented in Section 6.2, we will make the conservative assumption that the power necessary to achieve a given swimming speed is not altered substantially by temperature. Net thrust at 9 °C therefore equals that at 25 °C, meaning that changes in ω must be accompanied by offsetting changes in other parameters of Eq. (5): tail beat amplitude (a), depth of the trailing edge (b), or the propulsive wavelength (λ). We can estimate the magnitude of change necessary in these parameters to accomplish this balance. When thrust at two divergent temperatures is equivalent, Eq. (5) simplifies to:

$$(b_H/b_L) \cdot (a_H/a_L) = (\omega_L/\omega_H) \sqrt{[1-U/(\omega_L \lambda_L)]/[1-U/(\omega_H \lambda_H)]} \ , \qquad (6)$$

where the subscripts H and L denote high and low temperatures, respectively. This treatment assumes that the constant of proportionality (K) does not vary with temperature and ignores the small effect of temperature upon the density of the medium. From our example of striped bass (bl = 17 cm), now swimming at 2.5 bl s^{-1} (42 cm s^{-1}), ω is 5.3 and 3.9 Hz at 25 ° and 9 °C, respectively.

Initially we will assume that $\lambda = (0.8) \cdot bl = 13.6$ cm, and that λ is the same at our two temperatures. Substituting these values back into Eq. (6) yields:

$$(b_H/b_L)\cdot(a_H/a_L) = (3.9/5.3)\sqrt{(1-0.8)/(1-0.6)} = 0.5 \ .$$

Thus, with constant λ equivalent thrust can be provided at the lower temperature by doubling the tail beat amplitude, doubling the depth of trailing edge, or by any combination of the two that yields doubling of their product. If, however, $\lambda = 0.9$ at 9° and 0.7 at 25 °C (15.3 and 11.9 cm, respectively) the quantity U/c at the two temperatures converges and Eq. (5) further simplifies to:

$$(b_H/b_L)\cdot(a_H/a_L) = (\omega_L/\omega_H) = (3.9/5.3) = 0.7 \ . \tag{7}$$

Under these conditions, equivalent thrust at 9 °C can be achieved by a 30% increase in λ combined with a net increase of approximately 30% in amplitude, depth, or their product. This magnitude of change in depth or amplitude is modest and may be biologically feasible. Wardle (1979) has suggested that changes in the wavelength of the body propulsive wave might be a compensatory mechanism to offset the effects of cold on contraction velocity of muscle and λ is known to vary among species (Webb 1975). It is not clear, however, whether changes in λ for an individual are relevant or biologically possible. These parameters are accessible by high-speed cinematographic or videotaping techniques, so answers may be forthcoming.

Why changes in temperature should necessitate a reallocation among each of the above determinants of thrust is not immediately clear. We do know from studies of muscle fiber mechanics (Sect. 4.1) that of the two components of power output, force development by muscle of fish is less thermally dependent than contraction velocity. We also know that additional force may be generated by the trunk at cold temperature by numeric compression of fiber recruitment (Sect. 5.1). Both force-related factors may combine to enhance power development at cold temperature, despite slower contraction velocity, which would result in lower tail beat frequency.

7 Recapitulation and Conclusions

It is not difficult to argue that, for vertebrate ectotherms, temperature is the most pervasive of all environmental variables and locomotion is among the most critical of all behavioral traits. We have attempted in this essay to outline our current understanding of the interrelationship between these two factors in teleost fish. Not surprisingly, no single mechanism can be identified that explains the locomotory success of fish in their wide range of thermal niches around the globe.

We have seen, at the molecular level, that processes of natural selection have resulted in evolutionary modifications of contractile proteins for optimization of function in extremes of thermal habitat. In some species groups (e.g., cyprinids), these processes are mimicked over much shorter temporal spans of seasonal temperature changes. But, most eurythermal fish appear unable to modify the contractile characteristics of muscle cells after they are fully differentiated. This does not mean that the latter animals are at the mercy of vagaries in their thermal environment. All truly eurythermal fish apparently are capable of exploiting the remarkable plasticity of skeletal muscle tissue to help overcome thermal change. Varying the proportion of red skeletal muscle of the trunk inversely with adaptation temperature seems a particularly commonplace response for maintenance of sustained swimming ability. Also at the systemic level, eurythermal fish have the ability to activate progressively muscle fibers of the trunk in a temperature-dependent fashion. Although the sequence or pattern of fiber activation itself is independent of temperature, the number of fibers activated at any level of locomotory performance increases with decreasing body temperature. Viewed in this light, expansion of the mass of red muscle available for activation also has obvious adaptive implications with respect to sustained swimming ability.

Regardless of specific mechanism employed, enhanced mechanical work at cold temperature must be fueled by enhanced supply of chemical energy in the form of ATP. Thus, we have encountered very similar and widespread adaptations to temperature in the machineries of aerobic cellular energy metabolism. In all cases studied to date, mitochondrial densities of homologous muscles appear to vary inversely with adaptation temperature. This is true both intra- and interspecifically between animals of similar body form and lifestyle. There are reasons to believe that high mitochondrial density may be a fundamental requirement for maintaining a robust and tightly regulated aerobic metabolism at cold cellular temperature. High mitochondrial densities further may influence regulatory status of the cellular milieu to favor a shift in metabolic preference toward greater reliance upon fatty fuels. Although present data on this last point are enticing, more experimental effort will be required to resolve definitively this part of the metabolic story.

The relationship between body temperature and high-speed swimming abilities of fish is less clear-cut at present. The most reliable current physiological information seems to suggest that maximal swimming speed capabilities of fish should be impaired at cold temperatures unless specific adaptive variants of contractile elements are expressed (as is the case in evolutionarily adapted species and some polyploid groups). The issue is clouded by nomenclature inconsistencies in reference to exactly what constitutes burst activity. We have attempted to identify these sources of confusion between whole animal and cellular physiology viewpoints.

Finally, we have utilized recent data on the relationship between tail beat frequency and swimming speed of fish at different temperatures to identify parameters of the extrinsic medium and within the behavioral repertoire of fish which may influence swimming performance. Water viscosity, which changes inversely to temperature, may impact on the power requirements of swimming locomotion but current hydrodynamic models do not permit a definitive evaluation of the effect. Hydrodynamic analyses, however, do allow us to estimate that modest increases in tail beat amplitude or height of the trailing edge of the caudal fin (or their combination) may explain how some fish are capable of attaining higher swimming speeds with lower tail beat frequencies in cold water than at warmer temperatures.

Underlying each of the deductions presented in this chapter are the uniquely attractive aspects of skeletal muscle as an experimental system. No other tissue provides such clear and inviolate linkages between structure and function. In addition, no other vertebrate group provides the anatomical segregation of fiber types and associated discrete locomotory behaviors that are realized in fish. Without interposing too many "black boxes" it is possible to trace characteristics of whole animal locomotory capacity through biological organization to the molecular level. We anticipate that future studies of fish skeletal muscle will contribute even more toward identifying determinants of muscular performance by animals.

Acknowledgments. BDS' work is supported by NSF grants DPP 85-16137 and DCB 85-18442. TSM currently holds an NIH Postdoctoral Fellowship in the Department of Radiology, Harvard Medical School. The authors wish to express their thanks to Dr. Paul W. Webb for critically assessing the manuscript.

References

Alexander RM (1969) The orientation of muscle fibres in the myomeres of fishes. J Mar Biol Assoc UK 49:263–290

Altringham JD, Johnston IA (1981) Quantitative histochemical studies of the peripheral innervation of cod (*Gadus morhua*) fast muscle fibres. J Comp Physiol 143:123–127

Altringham JD, Johnston IA (1982) The pCa-tension and force-velocity characteristics of skinned fibres isolated from fish fast and slow muscle. J Physiol (London) 333:421–449

Altringham JD, Johnston IA (1986) Evolutionary adaptation to temperature in fish muscle cross bridge mechanisms, tension and ATP turnover. J Comp Physiol 156:819–822

Ballantyne JS, George JC (1978) An ultrastructural and histological analysis of the effects of cold acclimation on vertebrate skeletal muscle. J Thermal Biol 3:109–116

Barany M (1967) ATPase activity of myosin correlated with speed of muscle shortening. J Gen Physiol 52:197–216

Beamish FWH (1978) Swimming capacity. In: Hoar WS, Randall DJ (eds) Fish physiology, vol 7: Locomotion. Academic Press, New York London, pp 101–187

Bennett AF (1984) Thermal dependence of muscle function. Am J Physiol 247:R217–R229

Blake RW (1983) Fish locomotion. Cambridge Univ Press, New York, 203 pp

Bone Q (1978) Locomotr muscle. In: Hoar WS, Randall DJ (eds) Fish physiology, vol 7: Locomotion. Academic Press, New York London, pp 361–424

Bone Q, Kicenuik J, Jones DR (1978) On the role of the different fibre types in fish myotomes at intermediate speeds. Fish Bull 76:691–699

Brett JR (1964) The respiratory metabolism and swimming performance of young sockeye salmon. J Fish Res Board Can 21:1183–1226

Brill RW, Dizon AE (1979a) Red and white muscle fibre activity in swimming skipjack tuna, *Katsuwonus pelamis* (L.). J Fish Biol 15:679–685

Brill RW, Dizon AE (1979b) Effect of temperature on isotonic twitch of white muscle and predicted maximum swimming speeds of skipjack tauna, *Katsuwonus pelamis*. Environ Biol Fish 4:199–205

Campbell CM, Davies PS (1978) Temperature acclimation in the teleost, *Blennius pholis*: changes in enzyme activity and cell structure. Comp Biochem Physiol 61B:165–167

Carey FG, Teal JM, Kanwisher JW, Lawson KD, Beckett JS (1971) Warm-bodied fishes. Am Zool 11:135–143

Connell JJ (1961) The relative stabilities of skeletal muscle myosins of some animals. Biochem J 80:503–509

Dean JM (1969) The metabolism of tissues of thermally acclimated trout (*Salmo gairdneri*). Comp Biochem Physiol 29:185–196

Dobson GP, Parkhouse WS, Hochachka PW (1987) Regulation of anaerobic ATP-generating pathways in trout fast-twitch skeletal muscle. Am J Physiol 253:R186–R194

Driedzic WR, McGuire G, Hatheway M (1981) Metabolic alterations associated with increased energy demand in fish white muscle. J Comp Physiol 141:425–432

Dudley GA, Tullson PC, Terjung RL (1987) Influence of mitochondrial content on the sensitivity of respiratory control. J Biol Chem 262:9109–9114

Egginton S, Johnston IA (1984) Effects of acclimation temperature on routine metabolism, muscle mitochondrial volume density and capillary supply in the elver (*Anguilla anguilla* L.). J Thermal Biol 9:165–170

Egginton S, Sidell BD (1989) Temperature acclimation induces changes in subcellular structure of locomotory muscle from striped bass. Am J Physiol 256:R1–R9

Fitch NA, Johnston IA, Wood RE (1984) Skeletal muscle capillary supply in a fish that lacks respiratory pigments. Respirat Physiol 57:201–211

Flitney FW, Johnston IA (1979) Mechanical properties of isolated fish red and white muscle fibres. J Physiol (London) 295:49P–50P

Freadman MA (1979a) Role partitioning of swimming musculature of striped bass, *Morone saxatilis* Walbaum and bluefish, *Pomatomus saltatrix* L. J Fish Biol 15:417–423

Freadman MA (1979b) Swimming energetics of striped bass (*Morone saxatilis*) and bluefish (*Pomatomus saltatrix*): gill ventilation and swimming metabolism. J Exp Biol 83:217–230

Greer-Walker M, Pull G (1973) Skeletal muscle function and sustained swimming speeds in the coalfish, *Gadus virens* L. Comp Biochem Physiol 44A:495–501

Hagiwara S, Takahashi K (1967) Resting and spike potentials of skeletal muscle fibers in saltwater elasmobranch and teleost fish. J Physiol (London) 190:499–518

Hazel JR, Prosser CL (1974) Molecular mechanisms of temperature compensation in poikilotherms. Physiol Rev 54:620–677

Heap SP, Goldspink G (1986) Alterations to the swimming performance of carp, *Cyprinus carpio*, as a result of temperature acclimation. J Fish Biol 29:747–753

Heap SP, Watt PW, Goldspink G (1985) Consequences of thermal change on the myofibrillar ATPase of five freshwater teleosts. J Fish Biol 26:733–738

Hoar WS, Randall DJ (eds) (1978) Fish physiology, vol 7: Locomotion. Academic Press, New York London, p 13

Hochachka PW, Hayes FR (1962) The effect of temperature acclimation on pathways of glucose metabolism in the trout. Can J Zool 40:261–270

Holloszy JF, Coyle EF (1984) Adaptations of skeletal muscle to endurance training. J Appl Physiol 56:831–838

Hudson RCL (1969) Polyneuronal innervation of the fast muscles of the marine teleost *Cottus scorpius* L. J Exp Zool 50:47–67

Johnston IA (1979) Calcium regulatory proteins and temperature acclimation of actomysin ATPase from an eurythermal teleost (*Carassius auratus* L.). J Comp Physiol 129:163–167

Johnston IA (1980a) Specialization of fish muscle. In: Goldspink DF (ed) Development and specialization of skeletal muscle. Cambridge Univ Press, London, pp 123–148

Johnston IA (1980b) Contractile properties of fish fast muscle fibres. Mar Biol Lett 1:323–328

Johnston IA (1982a) Biochemistry of myosins and contractile properties of fish skeletal muscle. Mol Physiol 2:15–29

Johnston IA (1982b) Capillarisation, oxygen diffusion distances and mitochondrial content of carp muscles following acclimation to summer and winter temperatures. Cell Tissue Res 222:325–337

Johnston IA (1983a) Dynamic properties of fish muscle. In: Webb PW, Weihs D (eds) Fish biomechanics. Praeger, New York, pp 36–67

Johnston IA (1983b) Cellular responses to altered body temperature: the role of alterations in the expression of protein isoforms. In: Cossins AR, Sheterline P (eds) Cellular acclimatisation to environmental change. Cambridge Univ Press, London, pp 121–143

Johnston IA, Altringham JD (1985) Evolutionary adaptation of muscle power output to environmental temperature: force-velocity characteristics of skinned fibres isolated from antarctic, temperate and tropical marine fish. Pflügers Arch 405:136–140

Johnston IA, Brill R (1984) Thermal dependence of contractile properties of single skinned muscle fibres from Antarctic and various warm water marine fishes including skipjack tuna (*Katsuwonus pelamis*) and Kawakawa (*Euthynnus affinis*). J Comp Physiol 155:63–70

Johnston IA, Goldspink G (1975) Thermodynamic activation parameters of fish myofibrillar ATPase enzyme and evolutionary adaptations to temperature. Nature (London) 257:620–622

Johnston IA, Harrison P (1985) Contractile and metabolic characteristics of muscle fibres from antarctic fish. J Exp Biol 116:223–236

Johnston IA, Harrison P (1987) Morphometrics and ultrastructure of myocardial tissue in notothenioid fishes. Fish Physiol Biochem 3:1–6

Johnston IA, Lucking M (1978) Temperature induced variation in the distribution of different types of muscle fibre in the goldfish (*Carassius auratus*). J Comp Physiol 124:111–116

Johnston IA, Maitland B (1980) Temperature acclimation in crucian carp, *Carassius carassius* L., morphometric analyses of muscle fibre ultrastructure. J Fish Biol 17:113–123

Johnston IA, Sidell BD (1984) Differences in temperature dependence of muscle contractile properties and myofibrillar ATPase in a cold-temperate fish. J Exp Biol 111:179–189

Johnston IA, Tota B (1974) Myofibrillar ATPase in the various red and white trunk muscles of the tunny (*Thunnus thynnus* L.) and the tub gurnard (*Trigla licerna* L.). Comp Biochem Physiol 49B:367–373

Johnston IA, Walesby NJ (1977) Molecular mechanisms of temperature adaptation in fish myofibrillar adenosine triphosphatases. J Comp Physiol 119:195–206

Johnston IA, Walesby NJ (1979) Evolutionary temperature adaptation and the calcium regulation of fish actomyosin ATPases. J Comp Physiol 129:169–177

Johnston IA, Wokoma A (1986) Effects of temperature and thermal acclimation on contractile properties and metabolism of skeletal muscle in the flounder (*Platichthys flesus* L.). J Exp Biol 120:119–130

Johnston IA, Walesby NJ, Davison W, Goldspink G (1975a) Temperature adaptation of myosin in Antarctic fish. Nature (London) 254:74–75

Johnston IA, Ward PS, Goldspink G (1975b) Studies on the swimming musculature of the rainbow trout. I. Fibre types. J Fish Biol 7:451–458

Johnston IA, Davison W, Goldspink G (1975c) Adaptations in the Mg^{++}-activated myofibrillar ATPase activity induced by temperature acclimation. FEBS Lett 50:293–295

Johnston IA, Davison W, Goldspink G (1977) Energy metabolism of carp swimming muscles. J Comp Physiol 144:203–216

Johnston IA, Sidell BD, Driedzic WR (1985) Force-velocity characteristics and metabolism of carp muscle fibres following temperature acclimation. J Exp Biol 119:239–249

Jones PL, Sidell BD (1982) Metabolic responses of striped bass to temperature acclimation. II. Alterations in metabolic carbon sources and distributions of fiber types in locomotory muscle. J Exp Zool 219:163–171

Kleckner NW, Sidell BD (1985) Comparison of maximal activities of enzymes from tissues of thermally acclimated and naturally acclimatized chain pickerel (*Esox niger*). Physiol Zool 58:18–28

Korneliussen H, Dahl HA, Paulsen JE (1978) Histochemical definition of muscle fibre types in the trunk musculature of a teleost fish (cod, *Gadus morhue* L.). Histochemistry 55:1–16

Lighthill MJ (1969) Hydrodynamics of aquatic animal propulsion. Annu Rev Fluid Mech 1:413–446

MacDonald JM (1981) Temperature compensation in the peripheral nervous system: Antarctic vs temperate poikilotherms. J Comp Physiol 142:411–418

McMahon TA (1984) Muscles, reflexes and locomotion. Univ Press, Princeton, 323 pp

McVean AR, Montgomery JC (1987) Temperature compensation in myotomal muscle: Antarctic versus temperate fish. Environ Biol Fish 19:27–33

Moerland TS, Sidell BD (1986a) Biochemical responses to temperature in the contractile protein complex of striped bass, *Morone saxatilis*. J Exp Zool 238:287–295

Moerland TS, Sidell BD (1986b) Contractile responses to temperature in the locomotory musculature of striped bass, *Morone saxatilis*. J Exp Zool 240:25–33; also see Erratum J Exp Zool 240:410

Nemeth P, Hofer HW, Pette D (1979) Metabolic heterogeneity of muscle fibers classified by myosin ATPase. Histochemistry 55:1–6

Penney RK, Goldspink G (1981) Short term temperature acclimation in myofibrillar ATPase of a stenontherm *Salmo gairdneri* Richardson and an eurytherm *Carassius auratus*. J Fish Biol 18:715–721
Prosser CL (1973) Comparative animal physiology. Saunders, Philadelpia, pp 362–428
Roberts JL, Graham JB (1979) Effect of swimming speed on the excess temperatures and activities of heart and red and white muscle in the mackerel *Scomber japonicus*. Fish Bull 76:861–867
Rome LC, Loughna PT, Goldspink G (1984) Muscle fiber activity in carp as a function of swimming speed and muscle temperature. Am J Physiol 247:R272–R279
Rome LC, Loughna PT, Goldspink G (1985) Temperature acclimation improved sustained swimming performance in carp at low temperatures. Science 228:194–196
Schmidt-Nielsen K (1972) Locomotion: energy cost of swimming, flying and running. Science 177:222–228
Shaklee JB, Christiansen JA, Sidell BD, Prosser CL, Whitt GS (1977) Molecular aspects of temperature acclimation in fish: contributions of changes in enzyme activities and isozyme patterns to metabolic reorganization in the green sunfish. J Exp Zool 201:1–20
Sidell BD (1977) Turnover of cytochrome c in skeletal muscle of green sunfish (*Lepomis cyanellus* R.) during thermal acclimation. J Exp Zool 199:233–250
Sidell BD (1980) Responses of goldfish (*Carassius auratus* L.) muscle to acclimation temperature: alterations in biochemistry and proportions of different fiber types. Physiol Zool 53:98–107
Sidell BD (1983) Cellular acclimatization to environmental change by quantitative alterations in enzymes and organelles. In: Cossins AR, Sheterline P (eds) Cellular acclimatisation to environmental change. Cambridge Univ Press, London, pp 103–120
Sidell BD (1988) Diffusion and ultrastructural adaptations of ectotherms. In: Jones DP (ed) Microcompartmentation. CRC, Boca Raton pp 71–92
Sidell BD, Crockett EL (1987) Characterization of energy metabolism in Antarctic fishes. Ant J US 22:213–214
Sidell BD, Hazel JR (1987) Temperature affects the diffusion of small molecules through cytosol of fish muscle. J Exp Biol 129:191–203
Sidell BD, Johnston IA (1985) Thermal sensitivity of contractile function in chain pickerel, *Esox niger*. Can J Zool 63:811–816
Sidell BD, Moerland TS, Johnston IA, Goldspink G (1983) The eurythermal myofibrillar protein complex of the mummichog (*Fundulus heteroclitus*): adptation to a fluctuating thermal environment. J Comp Physiol 153:167–173
Sidell BD, Driedzic WR, Stowe BD, Johnston IA (1987) Biochemical correlations of power development and metabolic fuel preferenda in fish hearts. Physiol Zool 60:221–232
Sisson JE III, Sidell BD (1987) Effect of thermal acclimation on muscle fiber recruitment of swimming striped bass (*Morone saxatilis*). Physiol Zool 60:310–320
Smit H, Amelink-Koutstaal JM, Vijverberg J, von Vaupel-Klein JC (1971) Oxygen comsumption and efficiency of swimming goldfish. Comp Biochem Physiol 39A:1–28
Smit H, van den Berg RJ, Kijne-den Hartog I (1974) Some experiments on thermal acclimation in goldfish (*Carassius auratus*). Neth J Zool 24:32–49
Somero GN (1975) The roles of isozymes in adaptation to varying temperatures. In: Market CL (ed) Isozymes II. Physiological function. Academic Press, New York London, pp 221–234
Somero GN (1978) Temperature adaptation of enzymes: biological optimization through structure-function compromises. Annu Rev Ecol Syst 9:1–29
Stanfield PR (1972) Electrical properties of white and red muscle fibres of the elasmobranch fish *Scyliorhinus canicula*. J Physiol (London) 222:161–186
Stevens ED (1979) The effect of temperature on tail beat frequency of fish swimming at constant velocity. Can J Zool 57:1628–1635
Torres JJ, Somero GN (1988) Metabolism, enzyme activities and cold adaptation in Antarctic mesopelagic fishes. Mar Biol 98:169–180
Tyler S, Sidell BD (1984) Changes in mitochondrial distribution and diffusion distances in muscle of goldfish upon acclimation to warm and cold temperatures. J Exp Zool 232:1–9
Vogel S (1981) Life in moving fluids. Univ Press, Princeton, 352 pp
Wardle CS (1975) Limit of fish swimming speed. Nature (London) 255:725–727

Wardle CS (1979) Effects of temperature on the maximum swimming speed of fishes. In: Ali MA (ed) Environmental physiology of fishes. Plenum, New York, pp 519–531
Wardle CS, Videler JJ (1980) Fish swimming. In: Elder HY, Trueman ER (eds) Aspects of animal movement. Cambridge Univ Press, London, pp 125–150
Webb PW (1971) The swimming energetics of trout. I. Thrust and power output at cruising speeds. J Exp Biol 55:521–540
Webb PW (1975) Hydrodynamics and energetics of fish propulsion. Bull Fish Res Board Can 190:159 pp
Webb PW (1978) Hydrodynamics: non-scombrid fish. In: Hoar WS, Randall DJ (eds) Fish physiology, vol 7: Locomotion. Academic Press, New York London, pp 190–237
Webb PW (1984) Form and function in fish swimming. Sci Am 251:72–82
Wilson FR (1973) Enzyme changes in goldfish (*Carassius auratus* L.) in response to temperature acclimation. I. An immunochemical approach. II. Isozymes. PhD Thesis, Univ Illinois, Urbana
Wu T (1977) Introduction to the scaling of aquatic animal locomotion. In: Pedley TJ (ed) Scale effects in animal locomotion. Academic Press, New York London, pp 203–232
Yamawaki H, Tsukuda H (1986) Enhanced contractile activity of goldfish skeletal muscle related to cold acclimation. Comp Biochem Physiol 83A:549–553

Chapter 6

Prostaglandins (Eicosanoids) and Their Role in Ectothermic Organisms

T. Mustafa[1] and *K. C. Srivastava*[2]

Contents

[1] Institute of Biology, Odense University, Campusvej 55, 5230 Odense M, Denmark
[2] Department of Environmental Medicine, Institute of Community Health, Odense University, J.B. Winsløws Vej 19, 5000 Odense C, Denmark

Advances in Comparative and Environmental Physiology, Vol. 5

1 Introduction

Prostaglandins (PGs) are often mentioned in the context of mammalian physiology and pathophysiology. It is now known that PGs are present and play important roles in almost all mammalian tissues and fluids (Horrobin 1978). The term eicosanoids was introduced by Corey et al. (1980) to encompass various biologically active derivatives of eicosapolyenoic acids, especially arachidonic acid. The eicosanoid family includes PGs, thromboxanes (TXs), hydroperoxy- and hydroxyeicosatetraenoic acids (HPETEs and HETEs), the leukotrienes (LTs), and lipoxins (LXs). The term eicosanoids has not been used uniformally in this review. Terms like prostaglandins, thromboxanes, etc. have been preferred, and they appear at appropriate places in keeping with the tradition followed by various researchers who have contributed to the development of this area of research in biology and medicine. Although interest in the biological significance of these compounds was essentially based on the physiological studies in mammals traceable to the early part of this century, the PGs were not studied from nonmammalian organisms until the first report appeared of PGs in an invertebrate animal, the gorgonian coral *Plexaura homomalla* (Weinheimer and Spraggins 1969). Since then, occurrence of PGs in over 100 invertebrate species, several fish, and amphibian species has been reported. Apart from detection of PGs in a large number of invertebrates, fish, and amphibians, here termed as ectotherms, certain reports suggest that eicosanoids play fundamental physiological roles in representatives of many of these organisms. Such findings in the future may prove to be of great significance as they provide insights into the details of regulatory physiology. This area is now well established and it is noteworthy that the general field of eicosanoid research was honored by the 1982 Nobel Prize in Physiology or Medicine (Oates 1982).

Though over the years there has been a steady increase of literature describing the occurrence and physiological roles of eicosanoids in mammalian systems, comparatively little attention was paid to their distribution and physiological role(s) in nonmammalian species. The aim of this review is to provide a cognition of eicosanoid biochemistry and their physiological significance in major groups of nonmammalian organisms wherever data are available. Attempt has been made to guide the initiated and uninitiated through the eicosapolyenoic acid especially arachidonic acid metabolism. Since the appropriate nomenclature, detail knowledge of techniques and methods employed, distribution, and the physiological background come from decades of work on various mammalian systems, it is unavoidable and moreover useful to begin with a background from mammal studies.

1.1 Historical Background

In the 1920s at Berkeley, California research work on ovulation in relation to vitamin E produced variable data which was thought to be due to dietary lipid containing varying amounts of vitamin E. To sort out this problem, rats were administered a purified fat-free diet. This resulted in the appearance of a deficiency syndrome which was characteristically different from that of vitamin E deficiency. In order to find an explanation, rats were raised on different fats. The mystery was finally solved when fats were saponified and it was shown that the active substance which prevented the syndrome was located in the fatty acid fraction. This led Burr and Burr (1930) to conclude that the active principle was linoleic acid which the researchers called an "essential fatty acid" (EFA). Experiments performed with α-linolenic acid and arachidonic acid (AA) produced similar results. Thus, the term "essential fatty acids" came into existence.

Two New York gynecologists (Kurzrok and Lieb 1930) in their studies with artificial insemination reported that fresh human semen could react with human uterus causing strong contractions. This activity could be readily shown in vitro.

A few years later Goldblatt in England and von Euler in Sweden independently discovered and made studies that led to partial characterization of the activity demonstrated chemically as a lipid soluble compound which was different from other known endogenous substances, such as acetylcholine and histamine, called collectively biogenic amines, which are related to amino acids and are soluble in water. This lipid soluble acid could be extracted from semen and accessory glands of several animal species. The active prinicple, besides producing fall in blood pressure when administered intravenously by causing peripheral vasodilation, also produced contraction of several isolated smooth muscle preparations. Von Euler named this activity prostaglandin (PG) because of its high potency and solubility properties, and that this substance present in human semen was thought to originate in the prostate gland. Further, it was found to be a mixture of many closely related analogs which were responsible for the biological activity of the semen. Since then a large number of naturally occurring PGs have been characterized; of these one or two have been found to be present in almost every mammalian tissue examined. Although PGs are now known to be produced in many tissues and cells, the name prostaglandin has been firmly established by use.

Surprisingly, research work on these new compounds did not proceed further apparently due to World War II. Several years later at the instance of von Euler, Bergström in 1947 took up the problem of isolation, purification, and characterization of the active principles. For this purpose, Bergström and his associates at the Karolinska Institute, Stockholm prepared concentrates from thousands of Icelandic sheep vesicular glands. By employing the Craig-Counter-current Distribution method they observed that the biological activity was mainly present in a fraction containing a new type of highly active, lipid soluble, unsaturated hydroxy acids. They were thus able to isolate milligram amounts of the first two prostaglandins: prostaglandin E_1 (PGE_1) and prostaglandin E_2 (PGE_2). Due to the fact that PGs are lipid soluble acids, they are readily extracted from body tissues and fluids and equally easily separted from most other biologically active materials.

In a separate research development, more information on the EFA accumulated. Deficiency symptoms associated with a fat deficiency including testicular atrophy and increased permeability in the skin and cell membranes were observed by physiologists. Two dietary fatty acids, linoleic and α-linolenic acids, promoted growth in fat-deficient rats and thus were attributed EFA activity.

Though α-linolenic acid was shown to function similarly to linoleic acid but was definitely less active. Therefore, linoleic acid attracted much greater attention in research. EFA deficiency in humans is not easily seen under normal circumstances. This is due to two factors: a small amount (2–3% energy) of EFA is sufficient to inhibit EFA deficiency and small quantities are evidently available from all types of foods including the cheap ones from plant sources which are rather a rich source of linoleic acid. In humans EFA deficiency in the form of severe skin symptoms were demonstrated in infants who were fed milk in which EFAs were virtually absent (Hansen et al. 1962). Administration of EFA cured the skin symptoms. In cases of EFA deficiency in rapidly developing infants with protein malnutrition, who start to gain weight during treatment with diets high in protein and low in fat contents, the onset of several dermal lesions and an increased susceptibility to infections have been described (Paulsrud et al. 1972).

Until the 1960s, structural identification of EFAs or polyunsaturated fatty acids was achieved only through a tedious and time-consuming technique which involved fractional distillation and treatment of fatty acids with alkali to produce conjugated double bonds so that they could absorb in the ultraviolet region and thus measured spectrophotometrically. The entire concept of methodology changed with the introduction of gas-liquid chromatography as an analytical tool which enabled researchers to undertake a much more accurate investigation of the EFAs and their metabolism. It became clear that the EFA, linoleic and α-linolenic acids, belonged to two separate types in the sense that they were not interconvertible. However, one thing was common to them, that is, their *cis* double bonds were interrupted by methylene group(s). It was later demonstrated by in vitro techniques that linoleic acid was convertible to AA, and α-linolenic acid to docosahexaenoic acid (DHA), thus AA and DHA being the end products.

2 Precursors of Prostaglandins

With the application of new physicochemical techniques, especially a combination of gas-liquid chromatography and mass spectrometry, structural characterization of the primary prostaglandins, E_1, E_2, E_3, $F_{1\alpha}$, $F_{2\alpha}$, and $F_{3\alpha}$ was rendered possible. When it was realized that all PGs contained *cis* double bounds, suspicion arose that PGs could have arisen from essential fatty acids. Van Dorp in Holland and Bergström in Sweden showed independently that arachidonic acid was convertible into prostaglandins (van Dorp et al. 1964; Bergström et al. 1964).

In attempts to define the substrate specificity for the enzyme that produces prostaglandins, it was suggested that a minimum requirement for a fatty acid to serve as a precursor in prostaglandin production is to possess at least three *cis* double bonds in a methylene uninterrupted chain. It appears that besides naturally occurring ω6 acids, ω5 and ω7 fatty acids can also be converted respectively into ω-nor and ω-homo prostaglandins which show high biological activity. Interestingly, a correlation is found between the EFA activity of the fatty acids serving as precursors and the rate of formation of the biologically active prostaglandins derived from them (van Dorp 1971). Van Dorp has proposed two re-

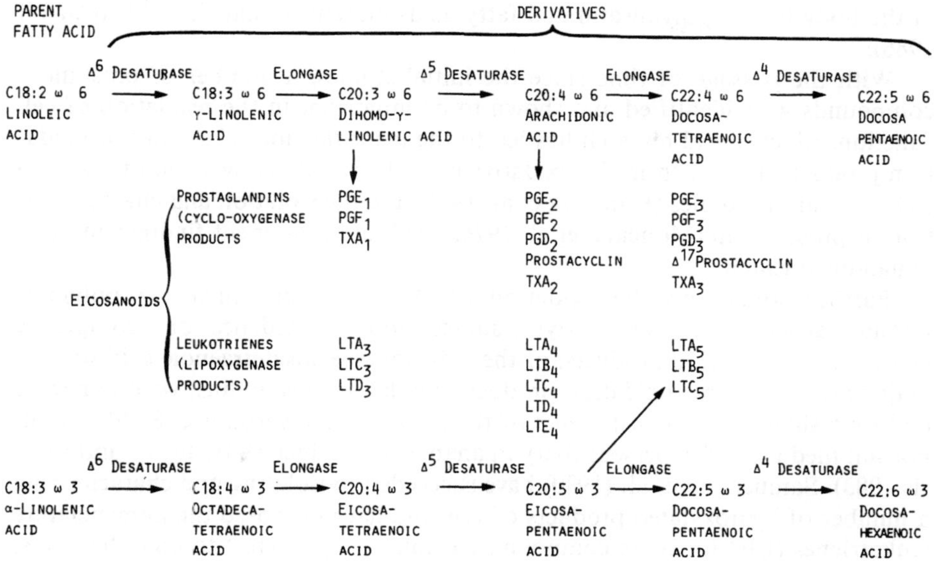

Fig. 1. Production of prostaglandins, thromboxanes, and leukotrienes from essential fatty acids (after Crawford 1983). The ω or η nomenclature is based on the system of numbering of the carbon chain from the methyl end; the number following these letters refers to the carbon atom on which the first double bond is located

quirements for a fatty acid to show EFA activity: (1) the fatty acid must be convertible into prostaglandins biologically, and (2) the prostaglandins thus produced must show biological activity. From this it would imply that the precursor acid must have a fixed structure of the carboxyl end and *cis* double bonds in the positions 8, 11, 14 or 5, 8, 11, 14 must be present. Although this proposition is logical and virtually complete, there are still a number of points that need clarification. α-Linolenic acid, an $\omega 3$ fatty acid is converted via eicosapentaenoic acid (timnodonic acid) (EPA, C20:5 $\omega 3$) into docosahexaenoic acid (clupanodonic acid) (DHA, C22:6 $\omega 3$). Eicosapentaenoic acid is convertible into prostaglandins of the three series (Fig. 1), though it does not serve as good a substrate as arachidonic and dihomo-γ-linolenic acids are for the prostaglandin synthase. Inspite of the fact that EPA appears to be a poor substrate compared to arachidonic acid for prostaglandin synthase, it is gaining a wider interest in the role of dietary polyunsaturated fatty acids in human health, especially in the amelioration of certain cardiovascular diseases. Most fish oils and other marine lipids contain very little linoleic acid or other $\omega 6$ fatty acids. The main polyunsaturated fatty acids in such oils are eicosapentaenoic and docosahexaenoic acids. EPA has the same structure as AA except for one additional double bond at the 3-position relative to the methyl end of the chain. The consumption of marine fish oils and other fishery products is related to the low incidence of cardiovascular diseases in Greenland Eskimos whose diet may contain an average of 4 to 5 g EPA a day (Bang 1987). The latter effect has been related to change in the balance of prostacyclin and thromboxane as a result of replacing $\omega 6$ polyunsaturated fatty acids in the body by $\omega 3$ polyunsaturated fatty acids present in marine lipids (Carroll 1986).

With the passage of time since the initial studies, a number of prostanoid compounds were identified and shown to be important in the regulation of cell function. Biochemical research has led to the identification of a series of oxidation products derived from the oxidative cascade of AA; conversion of AA into PGs and thromboxanes (Hamberg et al. 1975) and the isolation and characterization of prostacyclin (Moncada et al. 1976, 1977) are prominent findings in prostaglandin research.

Further studies on the oxidation of AA, this time involving polymorphonuclear leukocytes as the enzyme source produced evidence for a completely new series of oxidation products; of these 5-hydroxyeicosatetraenoic acid was the major product. These oxidation products, unlike prostaglandins, were noncyclic and were shown to be constituents of the slow reacting substance (SRS), an important mediator of hypersensitivity in anaphylaxis (Piper 1983; Taylor and Morris 1983). Samuelsson et al. (1980) have succeeded in isolating and characterizing a number of hydroxylated products of AA and have proposed the nomenclature leukotrienes (LTs) for these compounds. It may be of interest to note that EFA, PG, and LT (SRS) activities were described at about the same time in the 1920–30s. However, it took more than 30–40 years before it was realized that PGs and LTs were producible from EFAs. Finally, the types of eicosanoids produced depend on which unsaturated fatty acids are available as substrates. Specific unsaturated fatty acid availability is dependent on a number of integrated types of control. The type of dietary fat will determine which acids are potentially available for microsomal desaturation and chain elongation reactions which generate arachidonate and other precursors for the production of prostaglandins and other oxygenated derivatives. Once unsaturated acids are produced, the factors regulating the subsequent incorporation into and release from specific phospholipids are important determinants in dictating which fatty acids are available for cyclooxygenase and lipoxygenase enzymes. The production of PGs, TXs, and LTs from respective precursors and the derivation of the latter from parent fatty acids are shown in Fig. 1.

3 Release of the Substrate for Eicosanoid Synthesis

A major rate-limiting factor in the generation of eicosanoids is the availability of free AA. This concept finds support in the observation that by adding nonesterified arachidonate one observes enhanced formation of eicosanoids. Besides, steroids which diminish the release of arachidonic acid are established antiinflammatory agents (Flower and Blackwell 1979; Hirata et al. 1980). Cells generate eicosanoids in response to several stimuli which may be of physiological, pharmacological, or pathological origin. Physiological concentrations of certain hormones and autacoids and, more or less "unspecific" stimuli, e.g., mechanical trauma, cell injury, and nonphysiological (pharmacological) levels of several hormones are known to induce prostaglandin synthesis and release. It is desirable to make a distinction between the specific and unspecific stimuli because the pools for the two types of stimuli may be different.

Eicosanoids are not stored within cells and are released soon after they are produced. As the level of eicosanoid precursors in a free state is extremely low, it follows that the first step in eicosanoid generation would involve availability of the precursor acid at or close to the site of enzymatic conversion. Theoretically, precursor fatty acids could come from several stores, i.e., cholesterol esters, mono-, di-, and triglycerides, or phosphatides. In some tissues, particularly adipose tissue, a considerable amount of the substrate is present in the triglycerides. However, in most cells and tissues eicosanoid precursors are released from phosphatides, and this is the reason why considerable attention has been focused on these lipids as suppliers of precursors in eicosanoid synthesis.

3.1 Structure of Phosphatides

The relevant lipids which are involved in providing precursor fatty acids for the eicosanoid synthesis are members of the class of complex lipids. They are glycerophospholipids (phosphatides) which contain a three-carbon glycerol unit to which various functional groups are attached by ester linkages. Conventionally, the three-carbon atoms of the glycerol unit are referred to as α, β and α'. Hydroxyl groups in glycerol moiety at the α- and β-positions are esterified by fatty acids; hydroxyl group at the α'-position is esterified either by phosphoric acid to give phosphatidic acid, or more commonly to one of four phosphorylated bases giving phosphatidylcholine (lecithin), phosphatidylethanolamine, phosphatidylserine, or phosphatidylinositol (Fig. 2).

Enzymes which cleave fatty acids esterified to positions α and β are called phospholipase A_1 (PLA_1) and phosphalipase A_2 (PLA_2), respectively. The β-position is usually occupied by unsaturated fatty acids which include all eicosanoid precursors. Therefore, PLA_2 is considered as a potentially important regulatory enzyme in eicosanoid biosynthesis. The residual products of PLA_1 and PLA_2 are called respectively, α- and β-lysophosphatides. Cleavage of the glyceryl-phosphate bond is brought about by another enzyme, phospholipase C (PLC), which yields a diglyceride moiety and a phosphorylated base (inositol,

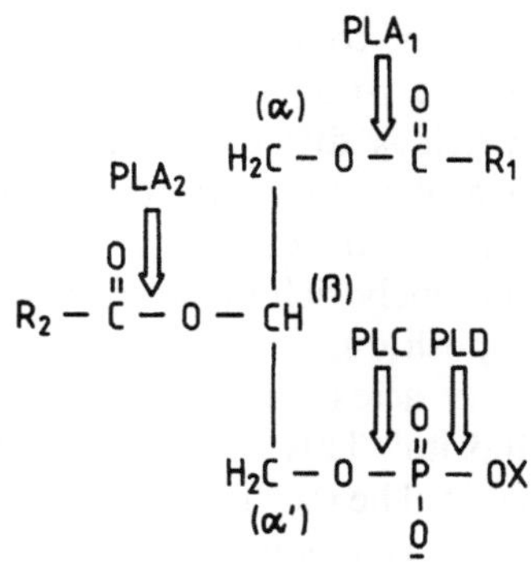

Fig. 2. Schematic representation of a phosphatide molecule showing the site of action of the phospholipases. *R_1, R_2*: Fatty acid residues in ester linkage. *X*: Polar group; the most common are choline, ethanolamine, inositol, or serine

choline, ethanolamine, or serine). The base moiety of the phosphatide can be cleaved by another enzyme, phospholipase D (PLD) (an enzyme present in plants) giving phosphatidic acid. Thus, AA can be obtained directly from the phosphatides by the action of PLA_2 or it may be liberated also from the products of PLC or PLD if the enzyme is functioning with a diglyceride lipase (in the case of PLC) or a phosphatidic acid-specific PLA_2 (in the case of PLD; Fig. 2).

3.1.1 Distribution and Characteristics of Phospholipases

Phospholipase A_1 and A_2 are a heterologous group of enzymes that are ubiquitously distributed throughout eukaryotic cells. They are both membrane-associated or in soluble forms. Those of the former type occur bound to the membranes in plasma, Golgi, or mitochondria. They function maximally at neutral pH and depend on the presence of calcium ions for their activity. Some instances of substrate specificity have been reported and this is explained by the specific locations within the cell membranes (van den Bosch 1980). For example, proteins with high molecular weight when present in the medium containing the cell are known to block release of arachidonic acid from membranes. Some PLA_2 enzymes become active when receptors localized on the cell surface are occupied (Blackwell et al. 1978; Humes et al. 1982). Fluidity changes occur in the membrane following receptor activation, and the biological effects are activation of adenyl cyclase and calcium-dependent adenosine triphosphatase or increase in influx of calcium and subsequent effects on cell behavior and release of cellular mediators and enzymes. Arachidonic acid is released following these reactions by a membrane-associated PLA_2 (Wolfe 1982). This may explain why substances like bradykinin and antigen-antibody complexes induce cells to produce prostaglandins in large quantities.

Phospholipase C, like PLA enzymes, is also more or less ubiquitously distributed. Unlike PLA, it does not occur as membrane-bound. It is present in lysosomes and is in soluble form. It shows optimal activity at acidic pH, is not dependent on calcium ions for its activity, and does not show substrate specificity. However, phosphatidylinositol-specific PLC enzymes have been reported (Shukla 1982). These enzymes require calcium ions for activity and possess neutral pH optima. Hydrolysis of arachidonate-rich phospholipid, phosphatidylinositol (PtdIns, containing myo-inositol linked to phosphate through the hydroxyl group on the 1-position) takes place in all organisms other than fungi by a mechanism which is different from that for other phospholipids. Phosphatidylinositol is hydrolyzed specifically by phospholipase C to diacylglycerol and inositol phosphate (inositol-1,2-cyclic phosphate and inositol-1-phosphate). The diacylglycerol (uniquely rich in stearic and arachidonic acids respectively occupying the α- and β-positions in the glycerol molecule) is then hydrolyzed completely to glycerol and fatty acids, or is recycled to phosphatidate and subsequently to phosphatidylinositol. This chain of reactions is characterized as phosphatidylinositol turnover cycle. The mechanisms involved in the control and function of this cycle have been extensively studied (Irvine 1982; Berridge 1983, 1984).

In addition to phosphatidylinositol, other mammalian inositides, disphosphatidylinositol (PtdIns4P) and triphosphatidylinositol (PtdIns4,5P_2) which are also rich in arachidonate may serve as substrates for phosphodiesterase or phosphomonoesterase. Agonists (hormones, neurotransmitters, etc.) act by stimulating the hydrolysis of triphosphatidylinositol by a phosphodiesterase enzyme to give diacylglycerol and inositol-1,4,5-triphosphate (Ins-1,4,5P_3). Diacylglycerol is converted to either monoacylglycerol or phosphatidic acid, whereas Ins-1,4,5P_3 is converted to inositol-1,4-diphosphate (Ins-1,4P_2) by triphosphatase. Ins-1,4,5P_3 acts to mobilize intracellular calcium, and diacylglycerol stimulates the same protein kinase C that can be activated by tumor-promoting phorbol esters. Thus, the main reaction of this transducing mechanism is a receptor-mediated hydrolysis of the phosphoinositides to give two compounds, diacylglycerol and inositoltriphosphate, both functioning as second messengers in a number of biological processes including cellular proliferation (Berridge 1984). Moreover, as phosphatidylinositol uniquely contains stearic acid at the α-position and arachidonic acid at the β-position, a coupled system of phospholipase C and a diglyceride lipase has assumed greater significance in eicosanoid biosynthesis. It is likely that specific stimuli induce prostaglandin production by activation of phospholipase C, whereas in pathophysiological situations unspecific stimuli induce the formation of these compounds in which phospholipase A_2 activation is involved.

Despite the lively debate about which of these enzymes is normally responsible for liberating fatty acid for eicosanoid synthesis, it now seems that it is probably phospholipase A_2. In some cells, e.g., platelets, the concerted action of PLC and PLA_2 is necessary for releasing arachidonic acid (Rittenhouse-Simmons 1979). PLA_2 is a membrane-bound enzyme found in virtually every cell type and organ in the body. The likely critical factor regulating the PLA_2 activity in cells is the availability of the calcium ions. Selective calcium ionophore agents such as A23187 stimulate phospholipase activity, although calcium chelation (with EDTA or EGTA) does not inhibit phospholipase activity. This strongly suggests that the necessary calcium ions come from an intracellular store. One thing which is clear is that PLA_2 is activated with release of arachidonic acid following a wide variety of mechanical, chemical, hormonal, immunological, and most likely environmental stimuli. Recently, it has been realized that the very high concentrations of prostaglandins previously found in homogenates of organs such as lung, kidney, and gut did not reflect the "basal" content as was first envisaged, but rather prostaglandins synthesized during homogenization (mechanical stimuli) due to activation of PLA_2. Thus, as a general rule, any stimulus, which disrupts or deranges membranes, releases intracellular calcium ions and activates phospholipase activity.

3.1.2 Inhibitors of Phospholipase Activity

Due to involvement of eicosanoids and hydrolysis products of phosphatides by the action of phospholipases in various pathological states in man, it became imperative to find agents that inhibited phospholipase activity.

Glucocorticoids have been known to interfere with the formation of eicosanoids for several years. They do not affect the eicosanoid-generating enzymes but suppress generation of eicosanoids in intact cells. This has led to the conclusion that these drugs interfere with the release of AA from membrane phospholipids by inhibition of membrane PLA_2 activity (Gryglewski et al. 1975; Nijkamp et al. 1976; Blackwell et al. 1978).

The glucocorticoids do not show a direct effect on PLA_2 but their effects are mediated through the synthesis of proteins which possess an antiphospholipase property. The mechanism is supposed to involve interaction with specific glucocorticoid receptors, which eventually leads to the synthesis of the proteins (Flower and Blackwell 1979; Blackwell et al. 1980; Hirata et al. 1980). Such proteins have been isolated from perfused guinea-pig lungs, rat peritoneal macrophages, and peritoneal neutrophils of rabbit. The polypeptides isolated from lungs and macrophages appear to be similar with a molecular weight of 15000 (macrocortin). Macrocortin may be already present stored in some cells from which the glucocorticoids induce their release first within 1–2 h followed by its resynthesis (3–4 h). On the other hand, synthesis of the protein (lipomodulin) in rabbit peritoneal neutrophils needs more time (5–16 h) and is not stored preformed in the cell. The molecular weight of lipomodulin is 40000. It is suspected that both macrocortin and lipomodulin are derived from a bigger molecule, since proteins with molecular weight over 200000 have been shown to possess some degree of antiphospholipase activity (Blackwell et al. 1982). In vivo generation of the two proteins mentioned above has been achieved by treating rats with dexamethasone. Peritoneal proteins from such rats were isolated with molecular weights 15000 and 40000; they were found to possess antiphospholipase activity. These proteins inhibited PLA_2 activity, prevented eicosanoid synthesis by cells, inhibited neutrophil chemotaxis, and partly inhibited carrageenan-induced pleurisy in the rat (Hirata et al. 1981; Blackwell et al. 1982).

Several other inhibitors have been described, and amongst them the important ones besides glucocorticoids are mepacrine, chlorpromazine, procaine, and propranolol. Their mechanism of action has also been defined, and this may be summarized as: (1) by interacting directly with the enzyme (e.g., bromophenacyl bromide); (2) by interfering with the binding of the substrate (phosphatide) (most of the inhibitors belong to this, e.g., polyamines such as putrescine and spermidine, chlorpromazine, volatile anaesthetics such as halothane); (3) by interacting or interfering with the binding of calcium ions which maintain the structural conformation of the enzyme (Blackwell and Flower 1983).

4 Factors that Influence Eicosanoid Synthesis

As described earlier prostaglandin biosynthesis requires a substrate fatty acid in a nonesterified (free) form which is made available from membrane phospholipids and other stores by the action of phospholipases. Evidence has been accumulating over the years suggesting that substrate availability is not the only physiological constraint that determines the prostaglandin biosynthetic capability of a tissue. This aspect has been explored in detail by Lands and his group; the eicosanoid synthesis examined by them is confined to prostaglandins. Three additional factors known to influence the rate of prostaglandin synthesis by prostaglandin H synthase have been recognized: (1) stimulatory action of hydroperoxides (Hemler et al. 1979; Hemler and Lands 1980); (2) inhibitory action of glutathione and glutathione peroxidase (Hemler et al. 1979); and (3) inhibition of prostaglandin synthesis through self-catalyzed inactivation of fatty acid oxygenases by substrates that they metabolize (Marshall et al. 1987).

4.1 Regulation of Substrate Availability

In prostaglandin synthesis availability of the substrate fatty acid is a major factor. Prostaglandin synthase has a Km value of approximately 2 μM for arachidonic acid whose concentration in human blood plasma and in tissues (Cenedella et al. 1975) has been reported to be ca. 20 μM. Even with a tenfold higher concentration than the Km of the prostaglandin synthase for arachidonic acid, one is required to provide extra arachidonic acid either by exogenous addition or through local lipase activity to induce a significant rate of synthase activity (Flower et al. 1973). Certain cytosolic factors present in the cell (Spector 1975) might bind unesterified arachidonic acid, and of these one may be the cytosolic protein content. Even as little as 5 mg ml^{-1} cytosolic protein increases the apparent Km for arachidonic acid to ca. 50 μM making 96% of this substrate inaccessible to the synthase. On basis of the assumption that an intact cell has ca. 100 mg ml^{-1} protein concentration, to ensure a significant prostaglandin synthesis either the concentration of arachidonic acid must be quite high or the acid would have to be close to the catalytic site (Marshall et al. 1987).

There also are other metabolic processes, i.e., acid-CoA ligases, oxygenases, and β-oxidation, which utilize arachidonic acid. Cyclooxygenase inhibition by other fatty acids released by phospholipase activity may, in addition, result in reduced synthesis of PGs and related products (Pace-Asciak and Wolfe 1968; Lands et al. 1973; Srivastava and Awasthi 1983a; Srivastava 1985a, b). In order to overcome some of these limitations, physical separation by way of compartmentation of the synthase could ensure an increased accessibility of the fatty acid substrate to the enzyme.

4.1.1 Regulation of Hydroperoxide Availability

Once a fatty acid substrate is made available in the nonesterified form, its oxygenation by cyclooxygenase is the next step in the production of prostaglandins, thromboxanes, and prostacyclin. The biosynthesis of these biologically potent substances of physiological origin is strictly regulated since they occur in much lower amounts in vivo than would be predictable from in vitro determinations of total biosynthetic capacity. Initial studies showing inhibitory effect of glutathione peroxidase on cyclooxygenase provided clear evidence that for its catalytic action cyclooxygenase required some peroxide which could be produced by cycloox-

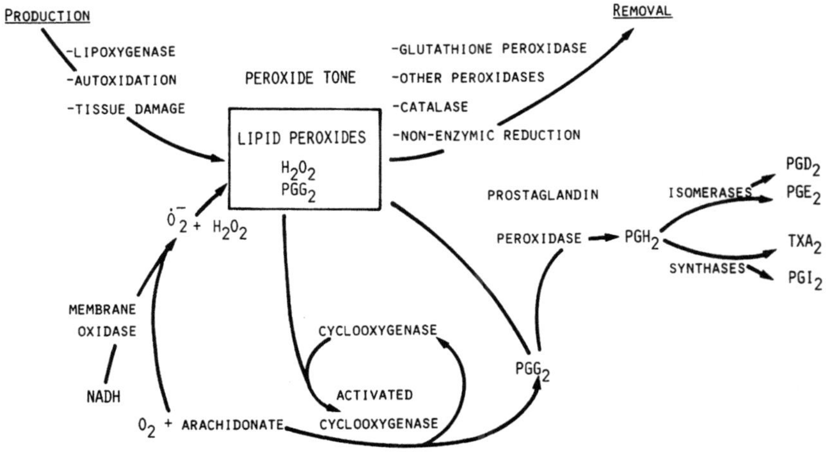

Fig. 3. Regulation of prostaglandin synthesis by factors that affect the peroxide tone of the cells. Balance between the formation and removal of lipid peroxides, $H_2O_2/\dot{O}_2^-$ system, and PGG_2 may regulate the rate of prostaglandin formation in cells (after Hemler et al. 1979)

ygenase itself. A wide variety of peroxides, including lipid hydroperoxides produced by lipoxygenase and autoxidation products from arachidonic acid, activate the cyclooxygenase (Hemler et al. 1979). Without the requisite levels of autoxidized fatty acids in vivo, peroxide-removing enzymes would have an inhibitory effect on the cyclooxygenase rendering it inactive (Fig. 3).

Processes, such as activation of membrane NADH oxidase, could generate stimulatory levels of hydrogen peroxide which indeed has been found to stimulate cyclooxygenase activity in mollusks (Morse et al. 1977). Moreover, as hydrogen peroxide and $\dot{O}_2^-$ have the capability of initiating membrane lipid oxidation, peroxidized lipids in sufficient amounts to stimulate cyclooxygenase may be produced (Fig. 3). A steady-state level of hydroperoxides has a major influence on cyclooxygenase activity.

For the activity of cyclooxygenase low concentrations ($10-100\,\mathrm{nmol\,l^{-1}}$) (Warso and Lands 1983) of hydroperoxides as activator are required. At higher levels (above $1\,\mu\mathrm{mol\,l^{-1}}$) they are inhibitory and toxic to many proteins. The hydroperoxide level is controlled by peroxidase. Conversion of PGG to PGH is brought about by PGH synthase, a peroxidase. This enzyme requires relatively much higher concentrations (micromolar) of hydroperoxides because of its low activity. Generally, most cell types have glutathione peroxidase and glutathione-S-transferase activity. Thus, glutathione peroxidase, if simultaneously present, would reduce the level of hydroperoxides which would lead to further inhibition of PGH production. For such an effect glutathione peroxidase requires factors which behave like phenol (Hanel and Lands 1982). Such factors are, of course, found in the cytosol of cells in ca. 1 mM concentrations. Lands and his group have attempted to assess the extent of suppression of PGH synthase activity by glutathione peroxidase in several tissues. Interestingly, PGH synthase is not so much dependent on peroxide-activator factors compared to glutathione peroxidase, and is probably an important enzyme producing lipid hydroperoxides in many tissues. It has been estimated that prostaglandin synthetic capacity of a tissue is a function of the ratio of glutathione peroxidase to cyclooxygenase activity (Marshall et al. 1987). This has been shown in several tissue preparations. However, on the basis of a high ratio of glutathione peroxidase to prostaglandin synthase in some tissues (e.g., kidney and spleen) low production of prostaglandins should take place, and this is not the case. This can be explained on the basis of the nature of localization of the peroxidases involved. PGH synthase is a membrane-bound enzyme localized in the endoplasmic reticulum and nuclear membrane. In contrast to this, glutathione peroxidase is a soluble enzyme located in the cytosol and is thus separated from micellar or membrane-bound peroxide. From this it would imply that cytosolic glutathione peroxidase would not be able to effectively suppress the reaction of membrane-bound PGH synthase with micellar or membrane-bound peroxides. This has actually been demonstrated in the macrophage cell line U937, bovine vesicular gland, and guinea pig spleen as the PGH synthase source with increasing additions of glutathione peroxidase. When PGH synthase is solubilized and mixed with increasing amounts of glutathione peroxidase, a complete inhibition of prostaglandin synthesis can be achieved (Marshall et al. 1987).

5 Cyclooxygenase Products

5.1 Prostaglandin Endoperoxide Synthase

Prostaglandin endoperoxides (PGG_2 and PGH_2) are formed from AA by the enzyme prostaglandin endoperoxide synthase (PES) (Samuelsson et al. 1975, 1978). PES possesses two activities: cyclooxygenase and a peroxidase activity. The cyclooxygenase activity of PES inserts two molecules of oxygen into AA to produce a 15-hydroperoxy-9,11-endoperoxide containing a cyclopentane ring (PGG_2). The peroxidase activity of PES reduces PGG_2 to its corresponding 15-hydroxy product (PGH_2). PES is a membrane-bound enzyme and has been detergent-solubilized and purified from sheep seminal vesicular gland (Roth et al. 1981; Pagels et al. 1983). It consists of two identical subunits with a molecular weight of about 72000. Each subunit requires one molecule of heme for maximal activity (Roth et al. 1981). It is probable that heme iron is at the active site of the enzyme and is critically involved in fatty acid oxidation. The specific activity of the enzyme falls between 2.4 and 23.4 $\mu mol\ min^{-1}\ mg^{-1}$ at 24 °C. The Km value for arachidonic acid and O_2 has been determined to be about 5 μM. Immunocytochemical studies have shown this enzyme to be localized in endoplasmic reticulum and nuclear membranes of 3T3 fibroblasts (Rollins and Smith 1980). In vesicular glands by such a technique the enzyme was found on the cytoplasmic side of the endoplasmic reticulum (Smith and Wilkin 1977).

Fatty acid oxidation catalyzed by cyclooxygenase occurs at first slowly and then accelerates (Hemler and Lands 1980). Exogenous hydroperoxides abolish this kinetic lag phase at 10^{-7} to 10^{-8} M concentrations, which is much below the

Km (10^{-5} M) of the peroxidase activity of PES (Marshall et al. 1987). The continuous presence of a hydroperoxide appears to be mandatory because the system (glutathione and glutathione peroxidase) that removes hydroperoxides terminates the PES-catalyzed fatty acid oxidation (Hemler and Lands 1980). Each pmol of PES is able to form about 1300 pmol of prostaglandin from arachidonic acid before it is inactivated (Kulmacz and Lands 1985). Deactivation of the cyclooxygenase seems to occur in intact cells as well as with purified enzyme preparations, and this may be an important factor in limiting in vivo prostaglandin biosynthesis (Brotherton and Hoak 1983).

The peroxidase activity of PES has a Km in the micromolar range (20 μM for PGG_1). It effectively catalyzes the reduction of lipid hydroperoxides and hydrogen peroxide, and undergoes self-deactivation (Egan et al. 1979). Various reducing cosubstrates promote the reduction reaction before the enzyme loses activity. In this process the cosubstrate may be covalently modified (Reed et al. 1984). Cooxidation by the hydroperoxide activity of PES has been implicated in the initiation of activation of certain procarcinogenic substances (Rice et al. 1985).

Inhibition of prostaglandin synthesis by antiinflammatory drugs has been demonstrated in a wide variety of cells and tissues. Inhibition has been demonstrated in whole animals. Nonsteroidal antiinflammatory drugs (NSAIDs) inhibit the cyclooxygenase but not the hydroperoxide activity of PES (Mizuno et al. 1982). Aspirin irreversibly inhibits the cyclooxygenase by covalently acetylating a lysine residue on the active site of the enzyme (Roth and Majerus 1975; Vane 1978). Aspirin-like drugs inhibit the abstraction of hydrogen from C-13 and therefore reduce the formation of cyclized prostaglandins. The same argument applies to explain their inhibitory role in the production of 11-HPETE and 15-HPETE (Siegel et al. 1980a). Some acetylenic fatty acids, such as eicosa-5,8,11,14-tetraynoic acid (ETYA), also inactivate the cyclooxygenase. Thus, chemicals that inhibit cyclooxygenase differ very much chemically and, depending on their chemical nature, bring about three types of inhibition: reversible competitive, irreversible, and reversible noncompetitive. The reversible competitive group includes fatty acids which are closely related to the substrate and have a comparable affinity for the enzyme but are not converted into oxygenated products. Aspirin serves as an example of an irreversible inhibitor. The reversible noncompetitive inhibitors possess antioxidation or radical scavanging properties. Paracetamol belongs to this class of inhibitor; it works by scavanging lipid peroxides which are needed in small concentrations to maintain the cyclooxygenase activity. The inhibitory action of indomethacin and fluorbiprofen is not explained by any of the mechanisms described above, although this is the most widely used cyclooxygenase inhibitor in prostaglandin biosynthesis studies. The time-dependent inactivation of the cyclooxygenase as observed with these inhibitors is related to the carboxylic acid groups and the aryl halogens (Rome and Lands 1975).

5.1.1 Bioconversion of Prostaglandin Endoperoxides

Thromboxanes. Prostaglandin endoperoxide H_2 (PGH_2) is further converted into the “two series” of prostaglandins and thromboxanes by separate enzymes. In blood platelets the principal product from PGH_2 is thromboxane A_2 (TXA_2) which contains an oxane ring. TXA_2 is readily hydrolyzed ($t\frac{1}{2} = 30$ s) to thromboxane B_2 (TXB_2), a compound with a hemiacetal structure (Hamberg et al. 1975). TXA_2 is a very potent agent which contracts vascular smooth muscle, produces platelet aggregation, and causes release of serotonin at concentrations below 7.4 ng ml^{-1} (20 nM). Essentially, maximal aggregation can be induced in fresh platelet-rich plasma (PRP) with 30 to 40 ng ml^{-1} (Lands et al. 1985).

Platelet thromboxane synthase is localized in the dense tubular membranes and produces TXA_2 and biologically inactive 12-hydroxy-heptadeca-5,8,10-trienoic acid (HHT) from PGH_2. A nonenzymatic formation of HHT from PGH_2 is not ruled out. Thus, from a molecule of PGH_2 half molecule each of TXA_2, HHT, and malondialdehyde (MDA) are produced (Diczfalusy et al. 1977). Endoperoxides that do not contain a Δ^5 double bond are degraded predominantly into C17 hydroxy acids. Thus, PGH_1 and Δ^4PGH_1 are converted only into hydroxy acids and malondialdehyde, whereas PGG_2 and PGH_2 are converted into a series of compounds which include 15-hydroperoxy-thromboxane A_2 and 12-hydroperoxy-heptadeca-5,8,10-trienoic acid (from PGG_2) (Hammarström 1980), thomboxane A_2, HHT, and MDA (from PGH_2).

Several inhibitors of thromboxane synthase have been described. Any substance that selectively inhibits thromboxane synthase would prove superior as an antithrombotic agent to a cyclooxygenase inhibitor such as aspirin. This is because by inhibiting the thromboxane synthase in platelets such an inhibitor would not hinder the formation of prostacyclin by blood vessels or other cells which would produce the latter either from their own endoperoxides or from those released from circulating platelets. Imidazole was the first such agent reported. It was found to inhibit thromboxane synthase at levels that had no effect on the cyclooxygenase (Needleman et al. 1977). Other inhibitors like certain substituted imidazole (dazoxiben), certain pyridine derivatives (OKY-1581) (Ito et al. 1983), prostaglandin analogs which contain a pyridine ring instead of the characteristic cyclopentane ring (Corey and Pyne 1983), and the PGH_2 analog 15-deoxy-9,11-azo-PGH_2 (Gorman and Lin 1983) have been described.

Prostacyclin. Prostaglandin endoperoxide PGH_2 is converted into prostacyclin (PGI_2), and enol-ether by blood vessel microsomes or fresh vascular tissue preparations (Johnson et al. 1976). Prostacyclin is unstable ($t\frac{1}{2} = 10$ min at pH 7.4, 24 °C) and readily incorporates a molecule of water to yield an inactive product 6-keto-$PGF_{1\alpha}$ (Johnson et al. 1976). Earlier, these compounds were shown to be produced in rat stomach (Pace-Asciak and Wolfe 1971). Vascular and nonvascular muscle cells also synthesize PGI_2 (Smith et al. 1983). Endothelial cell preparations from vascular tissue have been shown to possess higher synthetic capacity due to their higher cyclooxygenase content compared to monocytes (DeWitt et al. 1982). In large vessels PGI_2 synthetic capacity is highest at the intimal surface and gradually decreases from intima to adventia. However, all endothelia do not have the capacity to synthesize PGI_2 (Charo et al. 1984).

Homogeneous preparations of prostacyclin synthase have been obtained by affinity chromatography employing immobilized monoclonal antibody to prostacyclin synthase (DeWitt and Smith 1983). Prostacyclin synthase has a broad pH optimum. The enzyme has a molecular weight 52000 and a specific activity of 1–2 mmol PGI_2 min^{-1} mg^{-1} protein (DeWitt and Smith 1983). It belongs to a novel type of cytochrome P-450 like enzymes in which ferric heme iron does not change its redox state during enzymatic reactions (Ullrich and Graf 1984).

PGI_2 synthesis is a tightly regulated process which was initially shown to be inhibited by 15-HPETE ($IC_{50} = 0.48$ mg l^{-1}). But it was found later that this enzyme was inhibited by lipid peroxides. The inhibition by lipid peroxides is prevented by reducing agents that are reported to be radical scavangers. Vitamin C, a reducing agent, has been shown to induce increased synthesis of prostacyclin in rat aortic rings (Beetens and Herman 1983; Srivastava 1985c).

Prostacyclin activates adenyl cyclase via its receptor in cells which show response to prostacyclin. It elevates c-AMP in platelets (Gorman et al. 1977), vascular smooth muscle cells (Hajjar et al. 1982), and toad bladder epithelium (Pohlman et al. 1983). The antiaggregatory effect of PGI_2 on platelets and increase in short circuit current in toad bladder is mediated by increased cAMP levels. Membranes from cells that respond to prostacyclin have been shown to possess a specific protein receptor for it with a binding capacity of 350 fmol mg^{-1} membrane; the molecular weight being 82000 (Leigh et al. 1984).

Prostaglandin E_2. Prostaglandin E synthase isomerizes PGH_2 to PGE_2 by a microsomal enzyme from sheep vesicular gland. The enzyme has been partially purified with a Km value of 10 μM for PGH_1 and showed specific activity of 1–2 μmol PGE_2 min^{-1} mg^{-1} protein (Ogino et al. 1977). Several thiols, including glutathione, protect it from inactivation; this enzyme from kidney is stimulated by glutathione (Sheng et al. 1983). Glutathione is required as a cofactor in this reaction. Agents which react with and change sulfhydryl compounds inhibit this enzyme (Sheng et al. 1982).

Kidney from many animals produces PGE_2 as the major arachidonic acid product which counteracts ADH-induced water reabsorption (Sato and Dunn 1984). PGE_2 is also supposed to play a role in the modulation of renin release along with PGI_2 in mammals.

Prostaglandin F_2. The enzyme which catalyzes the conversion of PGH_2 to PGF_{2a} has been obtained from bovine lung in 540-fold purity. It is a monomeric protein with a molecular weight of 30500. This enzyme also converts prostaglandin D_2 (PGD_2) to PGF_{2a}, but it prefers PGH_2 as substrate (Watanabe et al. 1985).

Prostaglandin D_2. Prostaglandin D synthase converts PGH_2 to PGD_2. This conversion can be catalyzed by serum albumin with a Km value of 6 μM. The enzyme has been obtained in electropheric homogeneity from rat brain cytosol (Shimizu et al. 1979a) and spleen (Christ-Hazelhof and Nugteren 1979). The enzyme from brain has a molecular weight of about 80000–85000. It is stabilized by glutathione but does not require this thiol for activity (Shimizu et al. 1979a). The enzyme, isolated from spleen does not require glutathione and has a molecular weight of 26000–34000. Enzymes from both tissues are deactivated by compounds that react with sulfhydryl groups. This enzyme has been thoroughly

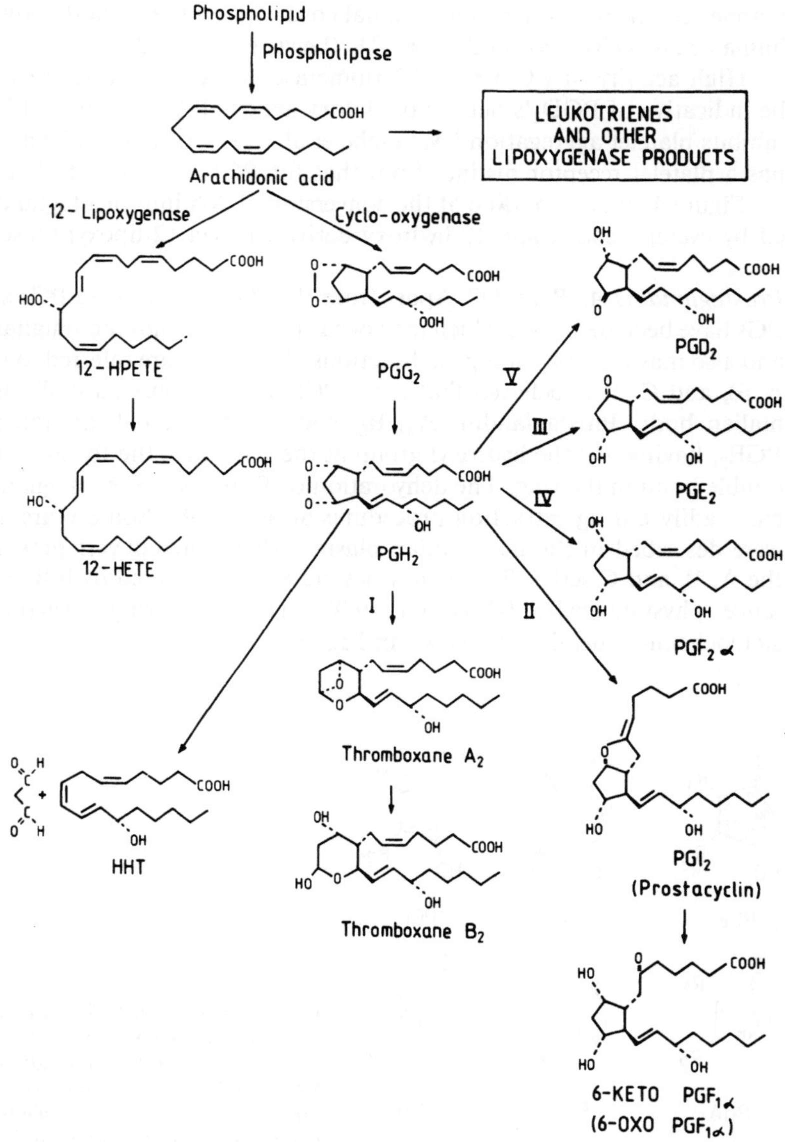

Fig. 4. An outline of the conversion of arachidonic acid into prostaglandins catalyzed by cyclooxygenase, and into 12-hydroxy derivatives via 12-lipoxygenase. *I* Thromboxane synthase; *II* prostacyclin synthase; *III* PG-endoperoxide E isomerase; *IV* prostaglandin F reductase; *V* PG-endoperoxide D isomerase. For further details of leukotrienes and other lipoxygenase products biosynthesis see Fig. 6

mapped in rat tissues. Brain and spinal cord contain the highest activity. Rat and human mast cells also produce PGD_2 (Lewis et al. 1982).

High activity of PGH to PGD isomerase in the central nervous system may be indicative of PGD_2's neuromodulatory actions (Shimizu et al. 1979b). PGD_2 inhibits platelet aggregation (Watanabe et al. 1982), increases platelt c-AMP, and has a platelet receptor distinct from that for PGI_2 (Schafer et al. 1979).

Figure 4 gives an outline of the conversion of AA into prostaglandins catalyzed by cycloxygenase, and 12-hydroxy derivatives via 12-lipoxygenase.

Prostaglandins A, B, and C. Apart from the above mentioned PGs several other PGs have been described which may occur naturally in nonmammalian organisms and also may have physiological functions. These PGs are referred to by the letters A, B, and C. It is believed that these PGs do not occur naturally in the mammalian body. Prostaglandins A_2, B_2, and C_2 are the dehydration products of PGE_2, having lost the hydroxyl group at the C-11 with the formation of another double bond in the ring. The dehydration of E to A series prostaglandin may occur readily during extraction procedures at acidic pH. Some years ago enzymes were described in the mammalian plasma which interconvert prostaglandins of the A, B, and C series. These enzymes are now considered of little or no importance physiologically (Horrobin 1978). The basic ring structures and the stereochemical details are shown in Fig. 5.

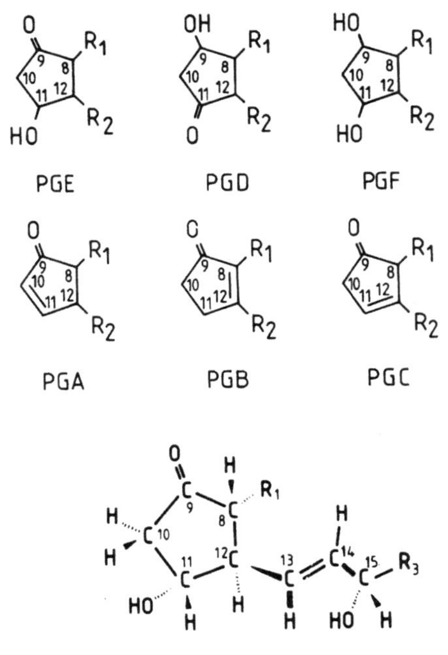

Fig. 5. Upper panel: The basic ring structures of the prostaglandins (*PGE, PGD, PGF, PGA, PGB,* and *PGC*). R_1 = Side chain (α-configuration) with 7 carbon atoms having either no double bond or one double bond; R_2 = side chain (β-configuration) with 8 carbon atoms having either one double bond or two double bonds. *Lower panel:* Natural E-type prostaglandin with stereochemical details. Side chains at positions 8 and 12 have α- and β-configurations respectively. Hydroxyl group at position 11 has an α-configuration while that at position 15 has a sinister (S) configuration. Side chains R_1 and R_3 contain 7 and 5 carbon atoms, respectively

6 Lipoxygenase Products

Lipoxygenase activity has been shown to produce six different hydroperoxy-eicosatetraenoic acids (HPETEs). The specificity of the lipoxygenase is determined by the position of the carbon atom of the fatty acid molecule on which the hydroperoxy group is located.

6.1 The 12-Lipoxygenase Enzyme and Its Products

This enzyme was first demonstrated in mammalian platelets in which it converts arachidonic acid to 12-S-hydroperoxy-eicosa-5,8,10,14-tetraenoic acid (12-S-HPETE). The catalytic reaction involves a stereospecifically controlled removal of the pro-S-hydrogen atom at C-10 followed by insertion of a molecule of oxygen at C-12 (Hamberg and Hamberg 1980).

The 12-lipoxygenase enzyme has been shown to be present in both cytosol and membrane of platelets (Nugteren 1982). The enzyme from the cytosolic fraction of human platelets has been shown to possess a molecular weight of about 100000 to 160000. The platelets continue to synthesize 12-HETE for a long time (about 2 h) after stimulation with collagen, although TXB_2 synthesis stops after few minutes (Hwang 1982). The Km value for arachidonate is 5 μM for the rat lung enzyme (Yokoyama et al. 1983), and 3.4 μM for the human platelet enzyme. In platelets for its activity this enzyme requires none of the cofactors and exhibits activity in the absence of exogenous calcium.

The 12-lipoxygenase activity in platelets is not inhibited by aspirin and other NSAIDs, but the activity is inhibited by ETYA, nordihydroguaiaretic acid (NDGA), BW755C, and several other compounds (Higgs and Vane 1983; Salari et al. 1984). 12-HPETE is reduced to the corresponding hydroxy analog (12-HETE) supposedly by a peroxidase activity associated with the 12-lipoxygenase enzyme. Selenium-dependent glutathione peroxidase can also bring about this conversion (Bryant et al. 1982).

Some biological activity has been attributed to 12-lipoxygenase products. The hydroperoxide inhibits in vitro collagen-induced platelet aggregation with IC_{50} of 2–6 μM, and at higher concentrations (IC_{50} = 15–25 μM) can even inhibit platelet cyclooxygenase (Bryant et al. 1983). The hydroxy analog (12-HETE) stimulates the migration of aortic smooth muscle cells in vitro at a concentration as low as femtomolar (Nakao et al. 1982).

6.2 The 15-Lipoxygenase Enyzme and Its Products

15-Lipoxygenase was obtained from the cytosol of rabbit neutrophils in a partially purified form (Narumiya et al. 1981). Its molecular weight has been determined to be 68000 by gel filtration chromatography. Its Km value for arachidonate as substrate is 28 μM. Agents which modify sulfhydryl groups such as ETYA (IC_{50} = 1.7 μM) and BW755C (IC_{50} = 75 μM) inhibit its activity which is slightly potentiated by indomethacin at higher concentrations (560 μM). 15-HPETE is decomposed nonenzymatically into several products, such as its hydroxy analog (15-HETE), its 15-keto-analog, 13-hydroxy-14,15-epoxy-eicosa-5,8,11-trienoic acid, and 11,14,15-trihydroxy-eicosa-5,8,12-trienoic acid (Narumiya et al. 1981).

15-HETE has been reported to inhibit the neutrophil 5-lipoxygenase and platelet 12-lipoxygenase activity ($IC_{50} = 1-5$ µM). This product does not affect the cyclooxygenase activity (Vanderhoek et al. 1982b). On the other hand, 15-HETE ($EC_{50} = 3$ µM) has been shown to stimulate the formation of 5-HETE and leukotriene B_4 (LTB_4) in a cultured mast/basophil cell line (Vanderhoek et al. 1982a). These observations have led to suggest that 15-HETE may have a regulatory role in several lipoxygenase pathways (Vanderhoek et al. 1982a).

6.3 The 5-Lipoxygenase Enzyme and Its Products

The 5-lipoxygenase produces several biologically active compounds called leukotrienes. These compounds have been found to be produced by elements of blood, such as neutrophils, eosinophils, monocytes, and in mast cells and keratinocytes (Borgeat and Samuelsson 1979; Rouzer et al. 1980; Jörg et al. 1982). Tissues from lung, spleen, brain, and heart also produce leukotrienes (Morris et al. 1980; Malik and Wong 1981). These compounds were known to be produced much earlier. Cobra venom stimulated production of a substance in the lung, which was called slow reacting substance (SRS) (Feldberg and Kellaway 1938). A similar substance was observed to be produced after immunological challenge; the name given was slow reacting substance of anaphylaxis (SRS-A). The chemical identification of SRS-A was made recently. It was found to be a mixture of peptidoleukotrienes LTC_4, LTD_4, and LTE_4 (Murphy et al. 1979). The name leukotrienes was given because they were originally discovered in leukocytes and they contain a conjugated triene structure in their molecules. The subscripts indicate the number of double bonds present. Thus, the leukotrienes originating from arachidonic acid ($\omega 6$) will belong to the 4-series, those from eicosapentaenoic acid ($\omega 3$) to the 5-series, and those from eicosatrienoic acid (Mead Acid) ($\omega 9$) to the 3-series. Due to the presence of the conjugated triene structure they absorb characteristically in the UV region of the spectrum. The sequence of reactions by which various leukotrienes are produced have been defined (Fig. 6). For more detailed information on cyclooxygenase products, products of the 12- and 15-lipoxygenases and on leukotrienes, the reader may refer to a recent review by Needleman et al. (1986).

The most recently discovered metabolites of arachidonic acid are called lipoxins (LXs) which possess a characteristic conjugated tetraene structure. Human leukocytes produce two main LXs: LXA, and LXB. Lipoxins appear to be produced enzymatically by the action of 5-lipoxygenase on 15-HPETE, the major product of 15-lipoxygenase. Lipoxin A has been shown to generate active oxygen species in human neutrophils and may represent a physiological mechanism of host defense other than that of LTB_4.

Metabolism of AA by 5-, 12- and 15-lipoxygenases, and conversion of 5-HPETE into various leukotrienes is shown in Fig. 6.

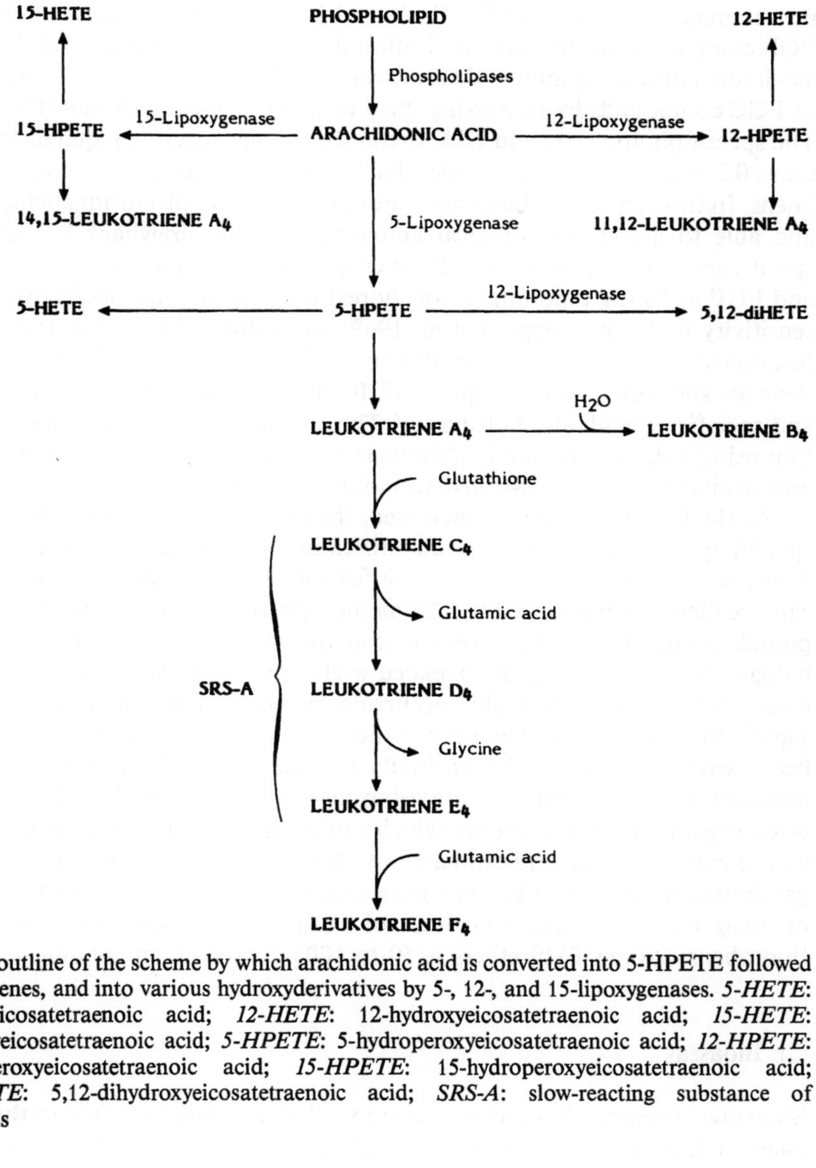

Fig. 6. An outline of the scheme by which arachidonic acid is converted into 5-HPETE followed by leukotrienes, and into various hydroxyderivatives by 5-, 12-, and 15-lipoxygenases. *5-HETE*: 5-hydroxyeicosatetraenoic acid; *12-HETE*: 12-hydroxyeicosatetraenoic acid; *15-HETE*: 15-hydroxyeicosatetraenoic acid; *5-HPETE*: 5-hydroperoxyeicosatetraenoic acid; *12-HPETE*: 12-hydroperoxyeicosatetraenoic acid; *15-HPETE*: 15-hydroperoxyeicosatetraenoic acid; *5,12-diHETE*: 5,12-dihydroxyeicosatetraenoic acid; *SRS-A*: slow-reacting substance of anaphylaxis

7 Quantitative Determination of Prostaglandins, Thromboxanes, and Leukotrienes

During the early part of the 1960s when prostaglandins were isolated and characterized, the main tool employed for their detection was bioassay which employed smooth muscle stimulating properties of these compounds. This was soon followed by other methods of detection such as the spectrophotometric method for the measurement of prostaglandins E, A, and B, and gas-liquid chromatography of PGF compounds using flame ionization detectors (Samuelsson et al. 1971). The gas-liquid chromatography (GLC) method was later applied to the determination of PGE compounds by converting them to stable O-methyl oximes (Gréen 1969). The spectrophotometric and GLC methods, though specific, required PGs in the range 0.2 to 20 μg, a level far higher than is usually encountered in biological specimens. In this respect the bioassay appeared to be more promising, being sensitive and able to detect physiological amounts, the only drawback is that it lacks specificity. With the discovery of the enzymatic technique for the assay of PGE and PGF in 500 pg amounts, it was hoped that the method would overcome the sensitivity problem (Änggård et al. 1969). Unfortunately, the method could not be applied to biological samples due to several factors. With the introduction of the mass spectrometric technique in 1970, the problems associated with sensitivity and specificity gradually faded away. This method employed multiple ion detection using a deuterated carrier (Samuelsson et al. 1970). Ultimately, a method is now available that is well-suited to biological studies.

As the list of eicosanoids increased, the two main problems – low level and specificity – encountered in the quantification of prostaglandins and related products, had to be solved. This became feasible with the development of electron capture detectors used in GLC. The method provides greatest sensitivity for compounds having intrinsically large electron affinities. Thus, compounds containing halogen atoms or having an extended π electron systems (PGB) would be most easily detected. As naturally occurring prostaglandins of known biological significance do not have halogens in their structure, chemical modifications have been used to make use of the sensitivity and selectivity offered by electron capture detectors. Prostaglandins and thromboxanes contain hydroxyl, carbonyl, and carboxyl organic functional groups which can be derivatized to give compounds that can be easily detected and estimated by this method. In general, electron capture gas chromatography can be used successfully to prostaglandin levels of 50 ng g^{-1} or 50 ng ml^{-1}. The detection limit of most prostaglandin methyl ester pentafluorobenzyloxime TMS ethers is 50 to 100 pg injected on-column.

7.1 Bioassay

Bioassay, in general, has contributed more than any other method to the development of the field of prostaglandins. Bioassays of the cascade and other types combined with antagonists of other biologically active compounds have proved useful. This technique has proved of immense help in understanding several physiological and pathological processes in which prostaglandins are involved. Thus, the role of lungs in the removal of prostaglandins during circulation, the role played by prostaglandins in the autoregulation and maintenance of blood flow to the kidney, the participation of prostaglandins in inflammation, mediation of pyrogen fever by prostaglandins, and release of rabbit aorta contracting substance (RCS), later identified as TXA_2, from lungs during anaphylactic reactions are some examples to substantiate the claim. It would be no exaggeration to say that without the bioassay technique researchers would have failed to recognize the biologically active and at the same time unstable products (PG-endoperoxides, TXA_2 and PGI_2) of AA metabolism.

Prostaglandins contract almost all smooth muscles from the intestinal tract and, therefore, preparations of gut from several species have been used for their detection and determination. Isolated preparations of arteries and veins are now being preferred because they enable the researcher to differentiate between prostaglandins, unstable PG-endoperoxides, TXA_2, and prostacyclin. For a detailed discussion of the principles, the reader may refer to a concise treatment by Moncada et al. (1978).

7.2 Radioimmunoassay (RIA)

Levine and van Vunakis (1970) were the first to use the RIA for the determination of PGs; they developed this method for PGE_1 and $PGF_{1\alpha}$. Since then this technique has been used at an accelerating pace covering several PGs, PG metabolites, PG analogs, and thromboxanes. Compounds such as HHT, HETE, etc. do not give rise to good antisera as their structures, unlike PGs and thromboxanes, are far from specific. Such compounds, therefore, cannot be quantified by RIA.

Radioimmunoassay utilizes the principle of competition between unlabeled and labeled molecules of a certain compound for the binding sites on an antibody produced against the same compound. A known amount of the labeled compound is added to all the tubes, whereas the amount of unlabeled compound is known but varied (for tubes used in making standard curve) or unknown (samples in which the quantity is to be determined). In each experimental procedure tubes for "zero binding" are included (no antibody added). Due to competition between the labeled and unlabeled molecules of a substance, in the presence of few unlabeled molecules, more of the labeled molecules will be bound to the antibodies. In contrast, when larger amounts of unlabeled molecules are present, the labeled molecules are displaced from the binding sites increasing the radioactivity of the unbound fraction. After separation of the unbound and antibody-bound fraction, the radioactivity is determined. The exact amount of the substance in an unknown sample is then determined with the help of a standard curve.

Radioimmunoassay offers certain advantages when compared with other quantitative methods: (1) a high sensitivity, the detection limit may be as low as 10 fmol of PG in a sample; (2) precision and accuracy is comparable to other methods; (3) the specificity under certain circumstances may be fairly good; (4) rapidity assay of 100 samples can be handled in a day or two including processing of data. However, there are also some drawbacks which one has to take into account. The major drawback is that the method is not a specific one in all cases. The other disadvantage is that the method is quite expensive contrary to the belief of many. A detailed discussion of principles for pitfalls of RIA and PGs and thromboxanes is described elsewhere (Granström and Kindahl 1978).

7.3 Gas-Liquid Chromatography – Mass Spectrometry

Apart from the high cost involved in such an apparatus, GC-mass spectrometry is quite suitable for the determination of PGs, both from the specificity and sensitivity point of view. In this method of analysis and determination, the compounds are extracted, purified, and derivatized. Extraction is achieved in organic solvents at acidic pH (3.0–4.0) after removing the nonpolar compounds, such as glycerides with petroleum ether at pH 8–8.5 which brings the PGs to the fully dissociated form (water soluble soaps).

To the sample is added a deuterated compound (e.g., $^2H_4$-$PGF_{2\alpha}$) followed by purification and derivatization. Purification is achieved by employing thin-layer and/or column chromatography. Derivatization is carried out by modifying various organic functional groups of the molecule; the carboxyl function is changed to methyl ester using diazomethane, the ketone group to the methyloxime, and the hydroxyl group to the acetate or trimethylester. The ratio of deuterated and non-deuterated molecules is measured by mass spectrometry. Since the amount of added deuterated compound is known, the amount of unlabeled compound can be easily calculated. As the deuterated molecules and the unlabeled ones are identical in physicochemical properties, the labeled molecules serve as carriers during purification and derivatization steps, in addition to serving as internal standards during the last stages of gas chromatographic-mass spectrometric analysis. To achieve satisfactory results, the compounds have to be extracted and purified using several steps. To assure a degree of confidence, one is required to label the deuterated carrier molecule with a radioactive isotope or alternatively to add a radioactive tracer of high specific activity. An important condition for increasing the sensitivity of this method is to obtain a low $H/^2H$ ratio in carrier molecules. A discussion of the principles and a detailed account of the analytical procedure are given by Gréen et al. (1978).

Gas chromatography-mass spectrometry is considered to be the most sensitive and reliable method for the determination of prostaglandins in biological samples in which they are present in extremely small amounts. However, this method is expensive and beyond the means of many laboratories. In most research centers, investigators still use radiolabeled AA and study the conversion of AA to PGs, thromboxanes, and related compounds in the biological systems. For the separation and determination of the radiolabeled AA metabolites, TLC is commonly used. Separation of PGs by TLC was first introduced by Gréen and Samuelsson (1964). The method is simple, fast, relatively less expensive, and well-suited for studying metabolic products of AA (Bailey et al. 1977; Srivastava et al. 1982; Srivastava and Awasthi 1983b; Tsunamoto et al. 1987).

7.4 High-Pressure Liquid Chromatography

In the last decade by combining the standard liquid chromatographic procedures for the separation of eicosanoids and the identification by spectrophotometric, fluorescent, or radioactive detection it became possible to quantitate various PGs and leukotrienes by HPLC. Terragno et al. (1981) were able to separate PGF_1, $PGF_{2\alpha}$, PGD_2, PGE_1, PGE_2, PGB_2, PGA_2, 6-keto $PGF_{1\alpha}$, 6-keto PGE_1, 15-keto-PGE_2, and TXB_2 on a reverse phase column with acetonitrile-water-phosphoric acid as the mobile phase. PGs were detected at 192 nm. Although this study was done using standard PGs with a sensitivity of 30 ng each, it has now become possible by employing more advanced UV detectors to quantitate as little as 1 ng each of the five diene PGs from animal tissues (Kinoshita et al. 1987; Palm and Levine 1987). For routine detection of PGs in animal tissues where PGs may occur in picogram levels UV detection may be unsatisfactory. The sensitivity of this method can be increased severalfold by coupling the prostaglandin with a suitable chromophore, e.g., p-(9-anthroloxy)phenacyl bromide. The lower limit of detection of the phenacyl bromide ester is 280 pg (ultraviolet 254 nm) and approximately 50 pg (fluorescence 249 nm emission, 413 nm cutoff). HPLC has played a much greater role in the separation and identification of lipoxygenase products, the leukotrienes. This is due to the fact that these compounds contain four double bonds and include a conjugated triene unit. Leukotrienes are easily separated by reverse phase HPLC with acetonitrile- or methanol-acidified water mobile phase with a UV detection between 235–280 nm (Hamilton and Karol 1982).

8 Occurrence, Distribution, and Release of Eicosanoids

Eicosanoids have a very short half-life and have been extracted from many animal tissues, plants, and even bacteria. The general consensus among researchers in the field is that the eicosanoid concentration of tissue homogenates reflects the biosynthesis from endogenous arachidonic acid during handling and homogenization of the samples. In order to evaluate the basal eicosanoid levels in a particular tissue ceratin precautions must be followed to prevent the biosynthesis of eicosanoids during sample preparations. Firstly, barbiturate anaesthetics should be avoided as they cause increases in PGs concentration. Secondly, biochemical reactions should be stopped in the first few moments of tissue removal by rapid freezing under liquid nitrogen or by homogenizing directly into organic solvents. Finally, these problems may be avoided by working on the body fluids carefully drawn without stressing the organism. Most determinations of arachidonic acid metabolite content of animal tissues have come from experiments where little or no effort has been made to prevent endogenous eicosanoid biosynthesis during sample preparation. In the few reports where such precautions have been taken, endogenous levels have been low (pg g^{-1} tissue range) and are likely insufficient to produce any significant biological action in that tissue. Thus, these basal or endogenous levels provide us with insufficient information about the amount and type of eicosanoid formed in a particular organ. A more valuable measure is to evaluate the eicosanoid-synthesizing ability of the tissue in question. In such experiments, tissue homogenates, microsomal fractions, or supernatants are incubated with labeled arachidonic acid or with one of the endoperoxides (PGG_2 or PGH_2) for a known period of time; the various metabolites synthesized then can be separated and identified using one or the other methods as described earlier. The fact that eicosanoids can be extracted from a particular tissue when incubated with substrate arachidonate may not necessarily indicate a physiological function for those PGs in that tissue in vivo. Almost all mammalian tissues (except erythrocytes) tested to this date produce some form of eicosanoids but how many of these do so naturally in the body is another question. Below, we provide a description of eicosanoid occurrence and distribution in nonmammalian species. Because of the various methods of determination used and units employed to report the results we have avoided reporting the actual concentrations or activities found in different organisms.

8.1 Eicosanoids in Invertebrates and Fish Species

Since the discovery of PGs in coral *P. homomalla* (Weinheimer and Spraggins 1969) more than 100 nonmammalian species have been reported to synthesize various eicosanoids. PG from coral, the 15-epi-PGA_2 is not pharmacologically active in the mammalian biological assay available but Bundy et al. (1972) demonstrated that these inactive compounds can be chemically converted into known biologically active prostaglandins. Christ and van Dorp (1972) compared the PG-synthesizing ability of various tissue homogenates from five invertebrate species by incubating the *cis*-(8,9,11,12,14,15-$^3H6$) eicosatrienoic acid. PGE_1 and B_1 were detected using a thin-layer chromatography system. Substrate conversion was less than 4% in all the marine invertebrate species examined, i.e., in *Anthoplexaura* (horny coral), *cyanea* (jelly fish), *Mytilus* (mussel), and *Homarus* (lobster). As the physiological significance of such a low specificity in these species is uncertain, this study suggested that the PG-synthesizing ability may be a tissue-specific occurrence rather a species-specific phenomenon. Also, Nomura and Ogata (1976), using a mammalian smooth muscle contraction bioassay technique, reported the presence of PGs in invertebrate species such as mussels, scallops, crawfish, and the bluecrab. Morse et al. (1978) described prostaglandin-endoperoxide synthase (PES) activity from the tissues of 41 coelentrate species. This PES assay spectro-photometrically monitors at 340 nm the obligatory cooxidation of the colorless aromatic cofactor L-epinephrine as it is converted to the intensely red adrenochrome product. We have attempted this assay in our laboratory and found it nonspecific for PES activity. When purified bovine serum albumin was added instead of tissue extract a considerable increase in optical density was recorded, indicating nonspecific autocatalytic activity in this method of determination. Thus, this assay must be looked upon with caution. A prostaglandin F-like compound has been isolated by Clare et al. (1982) from barnacle *Balanus balanoides*. Korotchenko et al. (1983) found PG-like compounds having smooth-muscle contracting activity from ten echinoderm species as well as prostaglandin activity in forty other invertebrates. Levine and Kobayashi (1983) by employing high-performance liquid chromatography and a radioimmunoassay technique, clearly demonstrated the presence of PGE_2, $PGF_{2\alpha}$, 6-keto-$PGF_{1\alpha}$, thromboxane B_2, and lipoxygenase product, 12-HETE in four invertebrate species including *Terpios zeteki* (Porifera), *Aplysia californica, Helix aspersa* (Mollusca), and *Arabica* (Echinodermata). PG metabolites from these species were also immunoreactive against antisera for prostaglandins, thromboxane, and 12-HETE derived from arachidonic acid. This means that immunoreactive compounds detected in the invertebrates possess some or all of the structures recognized by the appropriate antibodies irrespective of their precursors, that is, they may be derivatives of AA, or derived from other polyunsaturated fatty acids (Levine and Kobayashi 1983). Prostaglandin PGE_2 in *Modiolus demissus* (marine bivalve), and PGE_2 and PGF_{2a} in *Ligumia substrota* (freshwater bivalve) have been reported by Freas and Grollman (1980) and Saintsing et al. (1983), respectively. Srivastava and Mustafa (1984, 1985) showed that in *Mytilus edulis* (marine bivalve) arachidonic acid metabolism not only leads to the synthesis of PGE_2 and PGF_{2a} but also to PGD_2, thromboxane, and prostacyclin by gill, mantle, and adductor tissues. PGE_2 synthesis was in the range of 350–700 ng g^{-1} wet wt h^{-1} in all the above tissues tested, mantle being the most active tissue in this respect. In contrast, body wall tissues of *Arenicola marina* (polychaete) and *Ascaris suum* (nematode), and the digestive and reproductive tissues of *A. marina* metabolize AA into PGE_2, PGF_{1a}, PGD_2, TXB_2, and 6-keto-PGF_{1a} but at a much lower rate. At present, the starfish: *Astesias rubens, Marthoseterius glacialis*, and *Luida cilialic* oocytes are the only known source among aquatic invertebrates, where AA is metabolized into (8R)-HETE, a lipoxygenase product (Meijer et al. 1986a).

Christ and van Dorp (1972) were unable to show PG synthesizing ability from the tissues of insects. Destephano and his colleagues were first to report the presence of PGs in an insect, the house cricket, *Acheta domesticus* (Destephano et al. 1974). At present, the occurrence of prostaglandins is known in a wide variety of tissues from over a dozen insect species. Like the mammalian tissues all the insect tissues tested so far, indicated PG-synthesizing ability but highest PG activity was found in the reproductive organs. For example, Setty and Ramaiah (1979) reported in situ levels of 4193 ng PGE and 2527 ng PGF g^{-1} in the male reproductive tissues of *B. mori*. As the insect physiologists have concentrated their efforts on the role of PGs in insect reproductive function, the most common PGs found in insects have been of the E and F series. For further details regarding prostaglandins in insects the reader is referred to a review by Brady (1983).

Eicosanoid biosynthesis in fish due to multiple substrate availability is rather complex. In fish, along with AA other polyunsaturated fatty acids such as α-linolenic acid, eicosapentaenoic acid (EPA) (C20:5ω3), and docosahexaenoic acid (DHA) (C22:6ω3) are also found (Mai et al. 1981). The ratios of these fatty acids differ depending on the species and the tissue examined (Christ and van Dorp 1972), and dietary and seasonal variations may also alter the prostaglandin substrate availability in fish. Scanty information is available describing the synthesis of prostaglandins in fish tissues. (3_H)-Eicosatrienoic acid was converted up to 3–9% into PGE by the various tissues of carp, tench, and trout (Christ and van Dorp 1972). Also, Ogata and Nomura (1975) have demonstrated PGs from the intestinal tract of shark *Triakis scylla* and Nomura et al. (1973) reported the presence of PGs in testis and semen of teleosts. Synthesis of diene prostaglandins in freshwater fish has been reported by Bandyopadhyay et al. (1982). Trout gill tissue in vitro has been shown to produce PGE_1 from eicosatrienoic acid (Christ and van Dorp 1972), PGE_2 from arachidonic acid (personal observation), PGE_3 from eicosapentaenoic acid, and C22-PGF_4 from docosohexaenoic acid (Mai et al. 1981). Such evidence suggests that several substrates are available to fish tissues for prostaglandin synthesis. To date no single study has examined prostaglandin synthesis patterns utilizing all four substrates in a single tissue using a common detection technique. Like human skin, plaice skin (*Pleuronectes platessa*) converts AA into PGE_2 which has been implicated in inflammation reactions (Andersen et al. 1981). There is no direct evidence that fish blood thrombocytes and platelets produce prostaglandins. A study by Woodward et al. (1981) indicated that trout thrombocyte aggregation is dose-dependent and could be inhibited by preincubating the thrombocyte-rich plasma with adenosine, acetylsalicylic acid, or prostaglandin E_1. These properties are comparable to mammalian platelet aggregation where thromboxane A_2 is involved. Among mammalian tissues and cell types, the erythrocyte is the only tissue which is unable to convert AA into prostaglandin. In contrast, erythrocytes of toad fish (*Opsanus tau*) have a high capacity to convert exogenous radiolabeled arachidonate to prostaglandin E_2 and to prostaglandin E_{2a} (Cagen et al. 1983). Arachidonic acid was metabolized into PGF and PGI by gold fish heart (Herman et al. 1984), and into PGF_{2a}, PGE_2, PGD_2, and TXB_2 by flatfish (*Platichthys flesus*) and eel (*Anguilla anguilla*) hearts (Srivastava and Mustafa 1984). While the above mentioned examples indicate that tissues from fish species can synthesize prostaglandins, there is only one study available showing that trout gill is capable of converting AA and docosahexaenoic acid via lipoxygenase into 12-hydroxy-derivatives in response to calcium ionophore A23187 (German et al. 1986). Little is known about generation of the third group of eicosanoids, leukotrienes in fish. However, gill filaments, analogous to lung tissue, from a freshwater eel, *Anguilla rostrata*, produced LTC_4, D_4, and E_4 from exogenously labeled AA in response to A23187 (Piomelli 1985). Another study indicated that AA is the preferred substrate for leukotriene biosynthesis in plaice neutrophils despite the general abundance of EPA in marine teleost tissues (Tocher and Sargent 1987). There are several reports concerning PG production by reproductive organs of various fish species which will be discussed in a later section.

9 Physiological Significance

It is axiomatic that a variety of cells release and oxygenate polyunsaturated fatty acids into various eicosanoids. Mammalian tissues are able to synthesize up to 30 different eicosanoids, each with a differnt biological activity. Even a single organ can synthesize different forms of eicosanoids which may have a similar or opposite biological activity in that organ. Due to these tissue-specific variations in cyclooxygenase and lipoxygenase metabolites in the mammalian systems, one must not interpret the role of PGs from the mammalian literature as a set rule of prostaglandin effects in the tissues of invertebrates. As the information about the roles of PGs have been gleaned over several decades of work on mammalian systems, we will interpret the background as a loose set of guidelines for describing the role of PGs in ectotherms. In mammals PGs are of significance both under physiological and pathophysiological conditions. Their pharmacological activity enables them and their analogs to be put to clinical use in labor induction, in menstrual regulation, in relieving asthmatic condition, in reducing hypertension, in blood-clot inhibition, in nasal decongestation, and in the treatment of duodenal ulcers. Below, we describe a few of the physiological roles of eicosanoids attributed to the nonmammalian systems.

10 Modulation of Fluid and Electrolyte Fluxes

10.1 Bivalve Mollusks

The observation that prostaglandins regulate ion fluxes and water flow in kidney, gill, and across the membranes is derived from the response after treatment of organisms or tissues with appropriate cyclooxygenase and lipoxygenase inhibitors. Body fluids of freshwater bivalve ***Liguimia subrostrata*** are hyperosmotic to its aquatic medium. Injection of indomethacin, the cyclooxygenase inhibitor elicited an increase in sodium flux, while PGE_2 injections under comparable experimental conditions reduced the sodium influx five-fold from control values. In both experiments, chloride concentration and output of sodium remained unchanged. This prompted Graves and Dietz (1979) to suggest that prostaglandin E_2 regulated the ion fluxes in the freshwater bivalve. In another study Saintsing et al. (1983) identified PGE_2 and $PGF_{2\alpha}$ in blood and the gills (the primary site of Na transport) of the bivalve *L. subrostrata*. The concentration of both PGs was significantly reduced when the bivalves were injected with dexamethasone (PLA_2 inhibitor) or indomethacin, causing an increase in sodium flux. The stimulatory effect of the two inhibitors was nullified by coinjection of PGE_2, confirming the earlier view that prostaglandin E_2 is part of the sodium flux regulating system in the freshwater bivalve.

Other than prostaglandins, serotonin and cAMP have been implicated in positive regulation of Na transport in freshwater mussels (Dietz and Graves 1981). Injections of serotonin and cAMP in pond-water acclimated mussels caused a significant suppression up to 64 and 46% of circulating PGE_2 levels, respectively, but no significant effect on $PGF_{2\alpha}$ levels. From this observation Saintsing et al. (1983) inferred that there exists an antagonistic relationship between PG and serotonin/cAMP in regulation of Na transport. Hormonal regulation of PGs and cAMP concentration has been reported in the toad bladder (Zusman et al. 1978). Water permeability of the toad bladder is decreased by PGE_2 and increased by cAMP. Though a specific mechanism by which this function is accomplished is not known, however, it is probable that both PGE_2 and cAMP may be acting indirectly on Na transport as the effects of these compounds persist up to a period of 2–3 h (Saintsing et al. 1983).

Gill tissues of marine bivalve *Modiolus demisus* also synthesize PGE_2 and the synthesis is dependent on the ionic and osmotic state of the surrounding seawater (Freas and Grollman 1980). In addition to the osmotic effects of PG in marine bivalve, Freas and Grollman (1981) also reported ion-specific, PG-binding receptor sites in homogenates of gills, mantles, siphons, adductor, and upper and lower visceral masses. Receptor sites were specific for PGA_2, a dehydrated product of PGE_2. In *modiolus demisus* changes in ionic concentration altered both synthesis (Freas and Grollman 1980) and binding of prostaglandin. This may suggest that the binding site may play a role in the cellular mechanism involved in the actions of prostaglandins as a calcium ionophore (Carsten and Miller 1977). Another example of PG receptor sites from a nonmammalian source is of prostaglandin receptors on frog erythrocytes (Lefkowitz et al. 1977). These receptor sites exist in two different conformational states with each having a different affinity for PGE. The shifting of the equilibrium between these states is not dependent on ionic species but on the presence of a guanine nucleotide (Lefkowitz et al. 1977).

10.2 Fish and Amphibians

In man and other experimental animals PGE, PGA, and PGI series of protaglandins act as natriuretic, i.e., they increase urine volume and urinary output of Na^+ and Cl^- ions (Johnston et al. 1967; Bolger et al. 1978). The mechanism of this natriuretic effect has been the subject of intensive research. It is now widely believed that PGs are natriuretic by one or more of the following mechanisms: (1) by inhibiting the active reabsorption of Na^+ and Cl^- ions as PGE_2 decreases the transepithelial voltage, resulting in a decreased activity of the Na^+-K^+-ATPase in cells of the loop of Henle (Stokes 1979), (2) by antagonizing the fluid-retaining effect of vasopressin in the collecting tubules, and (3) by dilating renal medullary blood vessels thereby producing "washout" of the medullary osmotic gradient.

In fish and amphibians PGs also appear to regulate fluid and electrolyte excretion directly or indirectly via catecholamines, cortisol, prolactin, or vasopressin. In rainbow trout prostaglandin E_2 injections induced a marked antidiuresis initially, reflecting a reduced filtering population and thus reduced the overall glomerular filteration rate. Infusion of PGE_2 at 200 ng min^{-1} kg^{-1} rapidly decreased the glomerular filteration rate by reducing the populations of filtering nephrons and later by increasing the renal tubular water reabsorption (Brown and Bucknall 1986). The tubular antidiuretic action of PGE_2 may reflect a direct effect on renal tubular adenylate cyclase resulting in increased water permeability. Comparable effects were also reported in toad bladder but only with very high PGE_2 concentrations (Handler 1981). In rainbow trout injections of PGE_1 sharply increased the cortisol output in a dose-dependent manner while administration of dexamethasone and indomethacin, the latter a general inhibitor of prostaglandin synthesis, were shown to reduce circulating cortisol levels (Gupta et al. 1985). This indicates that water and electrolyte regulation in fish may be under joint regulation of cortisol, i.e., the mineralcorticoid hormone and prostaglandins. Horseman and Meier (1978) suggested that another hormone, the prolactin in fish, may act through prostaglandin synthesis in the process of osmoregulation. Reduced influx of water and efflux of Na^+ and Cl^- at the gill appear to be the most significant features of prolactin effects. In the fundulus, prolactin causes an increase in the plasma chloride concentration which could be prevented by the injections of inhibitors of prostaglandin synthesis prior to prolactin injection. Treatment of fish with PGE_1 alone increased the plasma chloride concentrations in PG inhibitor-treated fish. These observations led Horseman and Meier (1978) to theorize that some of prolactin's osmoregulatory actions are mediated in conjunction with prostaglandin. Results of Gupta et al. (1985) for trout and for fundulus (Horseman and Meier 1978) report a possible common mechanism of hormone and prostaglandin interaction in regulation of teleost osmoregulation. Yet another example of prostaglandin and hormone vasopressin interaction to regulate osmotic water flow was reported from toad bladder (Orlof et al. 1965). Recently, PGE_2 has been shown to play an important role as a negative-feedback modulator of the renal effects of vasopressin. However, the situation may be complicated by the fact that vasopressin also stimulates TXA_2 biosynthesis in some organs. Unlike PGF_2, TXB_2 the stable analog of thromboxane A_2 stimulates osmotic water flow in the toad bladder. Exactly how prostaglandin and thromboxane elicit their action via vasopressin in toad bladder has been recently reported by Marumo (1986) and is described as below: PGE_2 at 10^{-8} M concentration but not PGD_2, PGF_2, and PGI_2 even at 10^{-7} M inhibited the vasopressin-stimulated osmotic water flow in the toad bladder. Both the vasopressin- and cyclic AMP-stimulated water flow across the toad bladder is enhanced with the addition of indomethacin. In contrast, PGE_2 inhibited both vasopressin- and cAMP-stimulated water flow suggesting that PGE_2 inhibits the adenylate cyclase activity. Use of calmodulin inhibitor suppressed cyclic AMP-stimulated water flow in a dose-dependent manner indicating that adenylate cyclase is Ca^{2+} dependent. On the basis of these observations it may be suggested that PGE_2 inhibits vasopressin-stimulated adenylate cyclase activity by modulating the Ca^{2+} concentration. The synthesis of thromboxanes is known to be stimulated by vasopressin (Burch and Halushka 1982). In toad bladder TXB_2 elevated the vasopressin-stimulated water flow but showed no effect on cyclic AMP-stimulated water flow. Thus, PGE_2 and TXB_2 were suggested to be negative or positive modulators of vasopressin-stimulated water flow at the site of cyclic AMP generation. If a comparable mechanism exists in fish gill, gut, or in kidney remains to be verified.

11 Cardiovascular System

Prostaglandins and leukotrienes have significant roles in the mammalian cardiovascular system. PGE_2 and PGI_2 series are usually vasodilator in most vascular beds while $PGF_{2\alpha}$, TXA_2, and leukotrienes are normally vasoconstrictor. The local formation of PGs may be necessary for the regulation of blood flow within particular organs of the body. For nonmammalian sources information concerning the role of eicosanoids in the cardiovascular system is nonexistent. In invertebrates, especially bivalves and crustaceans, such an attempt may be very rewarding because of the simplicity of their circulatory systems. In fish cyclooxygenase activity has been indicated in goldfish, eel, and flounder heart (Cagen et al. 1983; Herman et al. 1984; Srivastava and Mustafa 1984) and toadfish erythrocytes (Cagen et al. 1983). In all of the tissues PGE_2 is formed. Like mammalian systems, prostaglandin E_2 has been shown to be hypotensive in the rainbow trout (Brown and Bucknall 1986). Hypotensive effects of PGE_2 have also been reported in the bullfrog (Leffler et al. 1980; Herman et al. 1982). In trout, a variable degree of bradycardia in response to intravenous injections of PGE_2 was also evident (Brown and Bucknall 1986). In the carp, bradycardia in response to PGE_2 injections was due to vagal inhibition (Peyraud-Waitzenegger et al. 1976). Injections of PGE_2 in both species also caused tachycardia which is usually seen in mammalian hearts following PGE administration. Roles of PGI_2 and TXA_2 have not been investigated in any of the fish or amphibian species. Because of their vasodilative and vasoconstrictive properties the roles of prostaglandins may be of great significance both in regulation of hypoxia and of body fluids in fish and amphibian species.

12 Reproductive Functions

Cyclooxygenase products of arachidonic acid influence essentially all aspects of female mammalian reproductive functions from the release of gonadotropic hormones from pituitary gland to parturition and lactation. The order of prostaglandin involvement in various mammalian reproductive functions appears to be $PGE_2 > PGF_1 > PGF_2 > PGI_2 > PGD_2$. These metabolites may act directly on different reproductive organs/processes or via interaction and modulation of various reproductive hormones and cAMP. Below we describe various examples of involvement of PGs in the reproductive functions of nonmammalian species.

12.1 Barnacle Egg-Hatching

Following a spring bloom barnacle *Balanus balanoides* mantle tissue synthesizes a compound which causes hatching and subsequent liberation of stage 1 nauplii and egg masses (Crisp 1956). The hatching substance was extracted from adult barnacles and was studied by gas chromatography and mass spectrometry (Clare et al. 1982, 1985). The hatching substance was found to have a structure comparable to prostaglandin F and showed the smooth muscle stimulating ability of PGF_1 and PGF_2 under comparable conditions (Clare et al. 1982). In a more recent study Holland et al. (1985) extracted 50 kg of barnacles, and the active compound was purified through four sequential systems of TLC and subjected to gas chromatography-mass spectrometry (GC-MS). Derivatized and hydrogenated derivatives of the hatching factor yielded a single major GC peak. Mass spectra of the hatching factor was not comparable with a PG but with another eicosanoid, 10,11,12-trihydroxy-5,8,11,17-eicosatetraenoic acid which probably is a lipoxygenase derivative of C20: 5 $\omega 3$. This observation of Holland et al. (1985) is not in agreement with that of Clare et al. (1985) where hatching extract prepared from aspirin-treated animals failed to induce hatching. This suggests that the barnacle hatching substance isolated by Clare et al. (1985) was a cyclooxygenase product and was either a PG or PG-like substance. This decrepency must still be resolved.

12.2 Egg Production and Spawning in Mollusca

Direct and indirect evidence of the involvement of PGs in the reproductive function of snails has recently been shown (Kunigelis and Saleuddin 1986). Direct injections of PGE_2 in the virgin snails *Helisoma durgi* caused no increase in egg mass production and even death followed in some isolated cases. But when 25, 50, and 100 ng PGE_2 doses were placed into the genital opening in a viscous form, cumulative egg production after 4 weeks was about 200, 425, and 650 eggs per animal, respectively. The ovotestis, bursa copulatrix, oothecal gland, and seminal vesicle of mated and virgin individuals of *H. durgi* exhibited ample cyclooxygenase activity. It was noticed that after mating, synthesis of PGF_{2a} was unchanged in ovotestis and bursa copulatrix tissues while PGE_2 synthesis was decreased in both tissues. Variations in the PG synthesis ability in virgin and mated snails and stimulation of egg mass by PGE_2 suggests a significant role of PG in reproduction in *H. durgi.* Radioactive arachidonic acid is converted into coeluting products with 6-keto-PGE_1, PGE_2, PGA_2/B_2, and TXB_2 by accessory sex gland homogenates of another snail, *Lymnaea stagnalis* (Clare et al. 1986). In this snail species the effect of PG administration and effects of mating on PG synthesizing ability were not studied; it is possible that PGs synthesized in the accessory glands may have some reproductive function.

Addition of H_2O_2 in alkaline seawater causes spawning in both male and female abalones and in *Mytilus californianus* (Morse et al. 1977). This induction of spawning was time-dependent (spawning begins after 3 ± 0.5 h of H_2O_2 addition) and could be blocked by aspirin, added to the seawater 15 min before H_2O_2. The inhibition of the H_2O_2 stimulating of spawning by aspirin suggests that cyclooxygenase activity is involved in this process. Form this observation it is inferred that a certain increase in the level of hydrogen peroxide has a stimulatory influence on cyclooxygenase activity resulting in higher levels of PGs which initiated spawning. Regulation of hydroperoxide availability and PG production have been described earlier in this review. Apart from the above observation, the presence of PGs in the reproductive organs from several marine invertebrates has been reported by Nomura and Ogata (1976). During the sexual maturation of ovaries in *Crassostrea gigas*, PGF_2 levels rise preceding spawning (Ono et al. 1982). Though the exact mechanism of PG actions in molluscan reproduction is unclear, it is believed that PGs may have a role in spawning.

12.3 Starfish Oocyte Maturation

In three species of starfish, *Astrias rubens, Marthosterius glacialis*, and *Luidia ciliaris*, micromolar quantities of AA and eicosapentaenoic acid or nanomolar concentrations of hydroxyeicosatetraenoic acid can induce oocyte maturation. In contrast, PGs did not stimulate maturation, and also known PG inhibitors did not impede oocyte maturation. On the other hand, three lipoxygenase inhibitors, quercetin, eicosatetraenoic acid, and butylated hydroxytoluene, effectively inhibited hormone-induced maturation. Instead of PGs, three lipoxygenase products, i.e., 12- and 15-hydroxyeicosatetraenoic acids (HETEs) and their corresponding hydroperoxyderivatives (HPETEs) activated maturation (Meijer et al. 1984). Maturating starfish oocytes convert the labeled AA into biologically active HETE-containing fractions (Meijer et al. 1986b). The lipoxygenase inhibitor ETYA is shown to inhibit conversion of AA into HETE and stimulation of oocyte maturation. From these lines of observations it is evident that in certain starfish species it is the lipoxygenase product HETE, not the PGs, which can stimulate oocyte maturation. Meijer et al. (1986a) showed that (8R)-HETE is the only active lipoxygenase product produced by the starfish oocytes. In nonmammalian species maturation of starfish oocytes by a lipoxygenase product presents a unique example and opens the possibility of further investigation for the role of lipoxygenase products in reproductive functions.

12.4 Egg-Laying in Insects

Like mammals, prostaglandins in insects have been studied in more detail with relation to reproduction than in other physiological functions in any of the nonmammalian species. The PGE and F series have been reported in reproductive tissues of house cricket *Acheta domesticus* (Destephano and Brady 1977; Murtaugh and Denlinger 1982), housefly *Musca domestica* (Wakayama et al. 1980), the Australian field cricket *Teleogryllus commodus* (Loher et al. 1981), and the cabbage looper *Trichoplusia ni* (Hagen and Brady 1982). PG synthase activity in the reproductive organs in vitro has been reported in *A. domesticus* (Destephano et al. 1976), *M. domestica* (Wakayama et al. 1980), *T. commodus* (Loher et al. 1981; Tobe and Loher 1983), *Lagusta migratoria* (Lange 1984), and *B. mori* (Bhagya-Lakshami and Ramaiah 1984).

The role of prostaglandins in egg laying has been best understood in two cricket species, *A. domesticus* and *T. commodus*. Evidence obtained with each species indicates that during mating a spermatophore is transferred from male to female which contains the prostaglandin synthase complex and converts the PG substrate into PGE_2, and oviposition behavior is initiated through a mechanism still unclear. The above role of PG in the cricket egg-laying process is suggested after the observation of localization of PG-biosynthetic activity in spermatophores and spermathecae from mated, but not from virgin crickets (Loher et al. 1981, 1985). Also, estimation of quantities of PGE_2 in spermathecae from virgin and mated crickets revealed ca. 500 pg in mated and none in virgin crickets (Loher et al. 1981; Stanley-Samuelson et al. 1983). Further, since PGE_2 levels in

T. commodus spermatophores were only in insignificant quantities compared with those in spermathecae, Stanley-Samuelson et al. (1983) concluded that it is the PG-synthesizing activity, not the PG itself, that is transferred during mating. Another example of PGs effect on egg-laying behavior is reported in the silkmoth *B. mori* which also transfers sperms during copulation via spermatophores (Setty and Ramaiah 1980). Levels of prostaglandin E and F increased significantly in female reproductive organs following mating. To test the possibility that the increase in the levels of PGE and PGF was due to direct transfer of PGs or the PG-synthase complex, Setty and Ramaiah (1980) treated *B. mori* males with PG-synthase inhibitors before mating which reduced the postmating egg-laying activity. Similar injections into females had no effect upon egg-laying. These data suggest a fundamental difference from the cricket egg-laying pattern, as it appears in *B. mori* PGs themselves, rather than PG-synthase activity, are transferred during mating. Instead of PG synthase or PG transfer during mating, Wakayama et al. (1986) reported that arachidonic acid injected into the male *M. domestica* prior to mating is transferred to females during the mating process. The higher amount of radioactivity associated with PGF_2 in mated females suggests that the majority of the radioactivity in the PG of females is formed after transfer of precursor, rather than direct transfer of PG. It is not known whether PG synthase is transferred during mating of *M. domestica*. However, the presence of PGE_2 and PGF_{2a} in ovaries of virgin females suggest that ovaries contain PG-synthase activity prior to mating. Though PGs have been detected and implicated in the release of egg-laying behavior in the above mentioned species of insects, this may not be considered as a general role of PGs in insects. PGE_2 and PGF_{2a} occur in virgin cabbage loopers *Trichoplusia ni* and are shown to increase in adult female reproductive tracts by three- and two-fold, respectively. However, if PGs are injected into adult females or PG inhibitors are injected, no effect on egg-laying behavior or reduction in postmating oviposition was noticed (Hagen and Brady 1982). From these observations, at the present moment, it may be concluded that PGs are important for egg-laying in some species, such as crickets, but less so or not at all in others. The exact role of PGs in the release of egg-laying behavior is unclear but increases in cricket hemolymph PG titer after mating suggest that in insects PGs may act as a circulating hormone at some site distant from the source. Support for the above contention is derived from the observation that PGE_2 does not stimulate contraction of oviduct muscles in *T. commodus* (Loher 1984) nor in cockroach (Cook et al. 1984). This led Loher (1984) to surmise that the PGs in the release of the egg-laying process may function at the level of the central nervous system. For detailed information on prostaglandins in insect reproduction the reader is referred to a recent review (Stanley-Samuelson and Loher 1986).

13 Prostaglandins and Reproduction in Fish Species

Most of the present information concerning prostaglandins and fish reproduction is related to their roles in reproduction of the female teleosts. Though information about the PG-biosynthesizing ability of the various female reproductive organs is scanty, a wealth of information exists showing the occurrence of varius PGs in fish ovaries and gonads, oocytes, follicles and follicular fluids, blood and plasma from the sexually mature fish species. There is increasing evidence that PGs are of significance in the release of gonadotrophin-releasing hormone (GRH) in stimulating ovulation and controlling female spawning behavior. Here, we will not attempt to duplicate the work of Stacey and Goetz (1982) who have written an excellent review on PG and fish reproduction. Rather, a brief chronology of findings related to PG functions in teleost reproduction will be presented.

13.1 Effect of Prostaglandins on Gonadotrophin (GRH) Release

Luteinizing hormone (LH) and follicle stimulating hormone (FSH) are gonadotrophic hormones liberated from the anterior pituitary which affect the ovary and testis. These hormones consist of a decapeptide called gonadotrophin-releasing hormone (GRH). Prostaglandin E_2 or E_1 stimulates FSH and LH release from the pituitary in rat (Spies and Norman 1973). The evidence that PGs stimulate GRH release has come from studies using PG inhibitors. If indomethacin doses are given to rats, plasma LH levels are decreased. These results strongly suggest that endogenous PGs play a role in regulating oestradiol-induced release of LH (Sato et al. 1975). In fish Kapoor and Toor (1979) have clearly demonstrated that a single injection of indomethacin blocked ovulation for up to 12 days in carp, *Cyprinus carpio*, in which under the same conditions controls ovulated spontaneously within 24 h. As clomiphene citrate, which induces GRH release in carp (Breton et al. 1975) and ovulation in intact but not hypophysectomized goldfish (Pandey et al. 1973), restored ovulation in indomethacin-treated carp (Kapoor and Toor 1979), led Stacey and Goetz (1982) to suggest that indomethacin blocked the preovulatory GRH release in the carp. The other examples of the $PGF_{2\alpha}$ or PGE_2 effects on GRH have been described in goldfish (Peter and Billard 1976), catfish, *Heteropneustes fossilis* (Singh and Singh 1976). The exact mechanism by which PGs affect GRH release in goldfish and catfish by an action on the hypothalamus or the pituitary is not known. Furthermore, the effect of prostaglandins like PGI_2, PGD_2, and TXA_2 have not yet been determined. On the basis of these results, it is not yet clear whether prostaglandins have a role to play in the physiological regulation of GRH release in teleost fish species.

13.2 Ovulation – A Role for Prostaglandins

GRH release is modulated in some species of fish by PGs and their effect on the pituitary. However, this is not the only way in which PGs may affect ovulation (follicular rupture). Indomethacin treatment of rats at the time of expected GRH surge from the pituitary blocks ovulation and this block cannot be reversed by injecting LH (Armstrong and Grimwich 1972). A comparable observation has been reported by Stacey et al. (1979) where indomethacin also inhibited ovulation in goldfish if given during the spontaneous preovulatory GRH surge. These observations in rat and fish suggest that indomethacin prevents the ability of the ovary to respond to LH. Thus, PGs made in the ovary itself may also modulate ovulation both in rat and fish species. Exactly how ovarian PGs modulate ovulation in response to LH remains a matter of debate. It could be pointed out that rat ovarian follicle cells contain fibers of smooth muscle on which prostaglandin causes "contraction", thus increasing the pressure within the follicle, making it burst open and release the ovum. Jalabert and Szollosi (1975) demonstrated that rainbow trout *Salmo gairdneri* follicles matured in vivo could be induced to ovulate in vitro upon exposure to PGF_2. This response was a result of the contractile effects of PGF_{2a} on the smooth muscle cells of the theca comparable to a similar effect in the ovulation process of rat. More examples presenting evidence for the role of prostaglandins in ovulation have been provided by various studies in a number of species.

In vitro studies have shown that PGs affect ovulation in several teleost although effective prostaglandins vary from species to species. PGF_{2a} enhanced the in vitro ovulation of brook trout *Salvelinus fontinalis* oocytes, while PGE inhibited ovulation (Goetz et al. 1982). Prostaglandins E_2, E_1, and D_2 were equally effective in restoring ovulation in goldfish when administered 11 h postgonadotropin injection. Ogata et al. (1979) reported a rise in PGF_{2a} observed following gonadotropin-induced ovulation in pond loach, *Misgurnus anguillicaudatus* where levels of PGF_{2a} increased three fold after postgonadotropin treatment, while indomethacin blocked this stimulatory effect. In fish not treated with indomethacin, ovarian PGF_2 levels remain unchanged up to 6 h postgonadotropin treatment, however, rising three- and six-fold, respectively, after 12 and 24 h postinjection before all fish ovulated. This suggests that PG synthesis in teleost ovaries increases in vivo during the preovulatory period. These results are in agreement with the earlier work of Bouffard (1979) where in goldfish levels of PGF in ovarian fluid and plasma are shown to increase only at the completion of ovulation. The high level of PGF in the ovarian fluid, which bathes the ovulated oocytes, suggests that the increase in plasma levels is of ovarian origin (Bouffard 1979). In a very recent study, Tan et al. (1987) clearly show that empty postpartum follicles and vitellogenic oocytes of the guppy, *Poecilia reticulata*, can synthesize both PGE and PGF in vitro from endo- and exogenous precursors. As the PGE and PGF synthesis was inhibited by indomethacin it may be indicative of active synthesis rather than the release of preformed prostaglandins. This observation substantiates the idea that locally formed PGs in ovaries and ovarian follicles play a role in ovulation (follicular rupture).

14 Prostaglandins and the Nervous System

PGs in mammals have been implicated in the control of many physiological functions which are also under regulatory control of the nervous system (Horrobin 1978). In nonmammalian species, evidence of PGs involvement via the nervous system is available in female sexual behavior of teleosts, goldfish (Stacey 1976), amphibians, *Rana pipiens* (Diakow and Nemiroff 1981); behavioral thermoregulation and fever in frogs, a lizzard, several fishes, and a number of invertebrate species (Stanley-Samuelson 1987).

14.1 Prostaglandins and Spawning Behavior in Fish

When ovulated oocytes are injected through the ovipore into the ovarian lumen of the nonovulated female goldfish, the presence of ovulated oocytes initiated the spawning behavior (Stacey and Liley 1974). Consequently, when nonovulated females are injected with ovulated oocytes, spawning behavior is completely inhibited by indomethacin and restored in indomethacin-treated fish by the injection of PGF_2 (Stacey 1976). Indomethacin appears to inhibit spawning behavior by reducing PG synthesis and not by hindering behavioral responsiveness to PGF. Bouffard (1979) has also implicated the role of PG in mediating the effects of ovulated oocytes in spawning behavior of goldfish. As the low doses of $PGF_{2\alpha}$ in a goldfish, where ovipore, oviduct, and posterior portions of the ovaries were removed, still induced the spawning behavior (Stacey and Peter 1979), it is safe to suggest that PGF, synthesized by the ovaries or by ovulated oocytes (Stacey and Liley 1974), enters the circulation and then acts directly on the brain to trigger spawning behavior (Stacey 1981). PGF_2 in as low a range as $10\,ng\,g^{-1}$ injected into the brain was far more effective in initiating spawning behavior than injections through other routes (Stacey and Peter 1979). This suggest that at least in goldfish PGs may act centrally as potent short-latency, time-dependent stimuli on the brain to promote the spawning behavior. Stimulation of spawning behavior in a number of oviparous fish species such as Pacific herring *Clupea hallengus pallasi*, rainbow trout *Salmo gairdneri*, pond loch, and three-spine stickleback *Gasterosteus aculeaus* has been reported by Stacey (1981) and Stacey and Goetz (1982). There is some evidence that in some teleost species at least, sexually active males may exude a pheromone-like material which may in turn induce ovulation and spawning behavior in female, or vice versa (Partridge et al. 1976; Crow and Liley 1979). The nature of such a pheromone is not yet known. In the future, if this substance is identified as PG-like material, then synchronization of ovulation and spermiation would be easy to explain and PG will also have a role in fertilization.

14.2 Mediation of Behavioral Thermoregulation and Fever

Prostaglandins synthesized within the brain do not affect the control of body temperature as PG inhibitors do not lower normal body temperature in nonfebrile mammalian organisms (Rosendorff and Cranston 1968). It is also believed that PGs in mammals play no part in the regulation of temperature in hot or cold climates. In contrast, there is considerable evidence that PGs mediate the increase in body temperature caused by bacterial infection (Milton 1982). This type of behavioral response in bacterial infection is shown by various aquatic invertebrates. Crawfish *Camborus bartoni* when inoculated with a suspension of killed bacteria *Aeromonas hydrophila* moved into a 2 °C higher temperature gradient trough (Casterlin and Reynolds 1977). Uninfected crawfish when injected with increasing doses of PGE_1 over the range of 50 to 500 μg/individual resulted in an increase of 1° to 3.5 °C temperature. This led Casterlin and Reynolds (1978) to postulate that higher temperature preference is mediated by the endogenous formation of PGE_1 in the injected crawfish. In a subsequent study Casterlin and Reynolds (1979) reported similarly increased temperature preferences by more than 4 °C after injection of 100 pg PGE_1 in three species of marine arthropods, American lobster *Homarus americanus*, the pink shrimp *Penaeus duorarum*, and the horseshoe crab *Limulus polyphemus*. Likewise in other arthropods, the scorpions *Bathus occitanus* and *Andrroctonnus australis*, the body temperature increased by 15 ° and 20 °C, respectively, after treatment with low doses of PGE (Cabanac and Le Guelte 1980). These reports suggest that prostaglandins may be a part of thermoregulatory functions at least in the invertebrate species tested. Reynolds and Casterlin (1976) reported blue gill sunfish *Leponis machrochirus* and largemouth blackbass, *Micropterus salmonoides*, to exhibit behavioral thermal preferences. Both species after inoculation with 4×10^8 killed *A. hydrophila* cells showed a mean temperature increase of 2.6°–2.7 °C for the bluefish and 2.1 °C for the bass. Like the invertebrate species, the effect of PGE treatment on the fish species is unknown. There are many unanswered questions at present. Are PGs synthesized by the above mentioned species? Does PG synthesis increase after injection and what is the effect of PG at least in fish hypothalamus, the center of behavioral thermoregulation? The role of PGs in behavioral thermoregulation can only be evaluated after answers to the above questions are found.

15 Perspectives

In mammalian systems biosynthesis, regulation, degradation, and roles of various eicosanoids are relatively well understood. In some invertebrates and fish species conversion of endogenous or exogenous substrate into PGs is well documented but studies concerning their transport and metabolism have not been attempted so far. Two important features affecting the metabolic rate of overall PG formation are the availability of nonesterified substrate and the availability of hydroperoxide activator for cyclooxygenase. Both parameters in mammals are under metabolic regulation. In nonmammalian species the availability of the PG substrate may depend upon environmental factors because of the fluctuations in the amount and type of food available. Thus, in these species it would be rewarding to study phospholipase A_2 and cyclooxygenase functions and the endoperoxide formation. The fate of the endoperoxide formed, then depends upon the relative amounts of the synthases and dehydrogenases in each type of synthesizing cell; such information from the tissues of nonmammalian species is also needed. Another area worthy of attention in nonmammalian species is the transport and metabolism of prostaglandins. In egg-laying behavior of cricket and fish spawning behavior there is sufficient evidence that PGs act through the nervous system to alter the behavior, though the mechanisms are not yet understood. Thus, at least in insects and fish species the uptake of PG from the extracellular space across the plasma membrane and into the cytoplasm of the cell via hemolymph or plasma must be studied. If the PGs act as circulating mediators of physiological functions in the species concerned, the stability of the PGs has to be established. The enzyme responsible for the biological inactivation of PGs is a specific 15-hydroxyprostaglandin dehydrogenase (15-PGDH) which has been purified to homogeneity from mammalian sources (Hansen 1976). So far this enzyme has not been reported from any of the nonmammalian sources. As reported by Stanley-Samuelson et al. (1986) that in cricket the highest egg-laying activity may be induced by 15-keto-PGE_2, an inactive compound produced from PGE_2 by the action of 15-PGDH in mammals suggests that in some nonmammalian species the PG metabolites may act as the active species. In mammalian kidney, liver, and gut tissues prostaglandins of the E, F, and D series are enzymatically interconvertible. These enzymes require NAD(H) or NADP(H) for activity (Lee and Levine 1974). The significance of these PG interconverting enzymes has been documented in the mammalian renal function (Hassid et al. 1983). As the global circulation of PGs in an organism depends on the activity of inactivating enzyme the 15-PGDH, the interconversion of PGs could be a solution to provide various PG species from a few of the PGs in a particular organ. Thus, studies concerning the interconversion of PG species in nonmammalian sources could also be of significance.

At present we are aware of certain eicosanoid functions in particular physiological processes of modulation of water and electrolytes, reproduction, and in behavioral thermoregulation in a few nonmammalian species. Using background information of mammalian PG functions we would like to conjecture a few other PG roles in nonmammalian species. Since many of the effects of prostaglandins are mediated by changes in intracellular levels of cyclic 3′,5′-adenosine monophosphate, and PG receptors in some cells at least are known to be linked to a membrane adenylate cyclase enzyme, it has been possible in some cases to correlate the binding of a PG to its receptor with its effect on adenylate cyclase. This PG effect on adenylate cyclase is highly species- and organ-specific; the increase in human platelet cAMP by PGI_2 is rapidly increased by PGE_1 but unaffected by PGD_2 (Bonne et al. 1981). The potency ratio for activating adenylate cyclase in human platelets matches the antiaggregatory potency for PGI_2, E_1, and D_2. In addition to promoting cAMP formation, PGs may also elevate cAMP by inhibiting the enzyme cAMP-phosphodiesterase which normally breaks down cAMP. Such an analogous effect in tissues of facultative anaerobes may also mediate an increase in cAMP concentration which will ensure glycogen mobilization by the activation of glycogen phosphorylase. Such a mechanism, if it occurs, in facultative animals could be highly adaptive as their energy metabolism is primarily glycogen-based. Prostacyclin has been shown to stabilize lysosomal membranes of cat liver and myocardial lysosomes during short-term acute ischemia (Araki and Lefer 1980a, b). Again, such an analogous role in facultative anaerobes will be of major significance in maintaining lysosomal integrity as these animals often experience long hypoxic periods as their way of life. Another PG-cAMP-mediated function occurs in the mammalian gut where PG acts directly on mucosal cells to stimulate intestinal secretion and increased production of cAMP. It is believed that a rise in cAMP in the villus cells lining the intestine inhibits Na^+ coupled Cl^- uptake at the brush border, while a similar rise in cAMP intestinal crypt cells opens up a Cl^- ionophore in the membrane of these cells to allow Cl^- ions to escape into lumen. Prostaglandins have been identified in the gastrointestinal tract of various fish species (Ogata and Nomura 1975; Bandyopadhyay et al. 1982). In euryhaline fish species PGs may be involved during fresh and saltwater excursions to regulate the water and electrolyte movements across the gut mucosa. In the time to come, we believe that PGs, like in the mammalian species, may also appear as topological hormones or mediators of varied physiological functions in the non-mammalian organisms. Before this is realized, general research including the PG-biosynthesizing activities, substrate preferences, endogenous PG levels, regulation, degradation, and appreciation of the possible mode of action of PG compounds at the cellular and whole organism level in the species concerned is required.

Acknowledgments. The authors wish to thank Prof. Roy E. Weber, Institute of Biology, Odense University, Odense, Denmark for his encouragement. Expert typing of the manuscript by Mrs. Inge Bøgelund and Mrs. Yrsa Kildeberg is gratefully acknowledged.

References

Anderson AA, Fletcher TC, Smith GM (1981) Prostaglandin biosynthesis in the skin of the plaice, *Pleuronectus platessa* L. Comp Biochem Physiol 70C:195–199

Änggård E, Matschinsky F, Samuelsson B (1969) Prostaglandins: enzymatic analysis. Science 163:479–480

Araki H, Lefer AM (1980a) Role of prostacyclin in the preservation of ischemic myocardial tissue in the perfused cat heart. Circul Res 47:757–763

Araki H, Lefer AM (1980b) Cytoprotective effect of prostacyclin during hypoxia in the isolated cat liver. Am J Physiol 238:H176–H178

Armstrong DT, Grimwich DI (1972) Blockade of spontaneous and LH induced ovulation in rats by indomethacin, an inhibitor of prostaglandin biosynthesis. Prostaglandins 1:21–28

Bailey JM, Bryant RW, Feinmark SJ, Makheja AN (1977) Differential separation of thromboxane from prostaglandins by one and two-dimensional thin layer chromatography. Prostaglandins 13:479–492

Bandyopadhyay GK, Datta J, Ghosh S (1982) Synthesis of diene prostaglandins in freshwater fish. Lipids 17:755–758

Bang OH (1987) More on fish oil. N E J Med 316:624

Beetens JR, Herman AG (1983) Vitamin C increases the formation of prostacyclin by aortic rings from various species and neutralizes the inhibitory effect of 15-hydroperoxy-arachidonic acid. Br J Pharmacol 80:249–254

Bergström S, Danielsson H, Samuelsson B (1964) The enzymatic formation of prostaglandin E_2 from arachidonic acid. Biochim Biophys Acta 90:207–210

Berridge MJ (1983) Rapid accumulation of inositol triphosphate reveals that agonists hydrolyse polyphosphoinositides instead of phosphatidyllinositol. Biochem J 212:849–858

Berridge MJ (1984) Inositol triphosphate and diacylglycerol as second messengers. Biochem J 220:345–360

Bhagya-Lakshami SK, Ramaiah TR (1984) Partial purification of prostaglandin synthase from silkmoth *Bombyx mori* L. Insect Biochem 14:725–728

Blackwell GJ, Flower RJ (1983) Inhibition of phospholipase. Br Med Bull 39:260–264

Blackwell GJ, Flower RJ, Nijkamp FP, Vane JR (1978) Phospholipase A_2 activity of guinea-pig isolated perfused lungs: stimulation and inhibition by anti-inflammatory steroids. Br J Pharmacol 62:79–89

Blackwell GJ, Carnuccio R, Di Rosa M, Flower RJ, Parente L, Persico P (1980) Macrocortin: a polypeptide causing the antiphospholipase effect of glucocorticoids. Nature (London) 287:147–149

Blackwell GJ, Carnuccio R, Di Rosa M, Flower RJ, Langham CS, Perente L, Persico P, Russel-Smith NC, Stone D (1982) Glucocorticoids induce the formation and release of anti-inflammatory and anti-phospholipase proteins into the peritoneal cavity of the rat. Br J Pharmacol 76:185–194

Bolger PM, Eisner GM, Ramwell PW (1978) Renal actions of prostacyclin. Nature (London) 271:467–469

Bonne C, Martin B, Watado M, Regnault F (1981) The antagonism of prostaglandins I_2, E_1 and D_2 by prostaglandin E_2 in human platelets. Throm Res 21:13–22

Borgeat P, Samuelsson B (1979) Metabolism of arachidonic acid polymorphonuclear leukocytes, structural analysis of novel hydroxylated compounds. J Biol Chem 254: 7865–7869

Bouffard RE (1979) The role of prostaglandins during sexual maturation, ovulation, and spermiation in the goldfish, *Carassius auratus*. M Sci Thesis, Univ British Columbia, Vancouver, Can

Brady UE (1983) Prostaglandins in insects. Insect Biochem 13:443–451

Breton B, Jalabert B, Foster A (1975) Induction de décharges des gonadotropes hypophysaires chez la carpe (*Cyprinus carpio*) à l'aide du citrate de cisclomiphène. Gen Comp Endocrinol 25:400–404

Brotherton AFA, Hoak JC (1983) Prostacyclin biosynthesis in cultured vascular endothelium is limited by deactivation of cyclooxygenase. J Clin Invest 72:1255–1261

Brown JA, Bucknall RM (1986) Antidiuretic and cardiovascular actions of prostaglandin E_2 in the rainbow trout *Salmo gairdneri*. Gen Comp Endocrinol 61:330–337

Bryant RW, Simon TC, Bailey JM (1982) Role of glutathione peroxidase and hexose monophosphate shunt in the platelet lipoxygenase pathway. J Biol Chem 257:14937–14943

Bryant RW, Simon TC, Bailey JM (1983) Hydroperoxy fatty acid formation in selenium deficient rat platelets: coupling of glutathione peroxidase to the lipoxygenase pathway. Biochem Biophys Res Commun 117:183–189

Bundy GL, Schneider WP, Lincoln FH, Pike JE (1972) The synthesis of prostaglandins E_2 and $F_{2\alpha}$ from (15-R)- and (15-S)-PGA_2. J Am Chem Soc 94:2123–2124

Burch RM, Halushka PV (1982) Vasopressin stimulates prostaglandin and thromboxane synthesis in toad bladder epithelial cells. Am J Physiol 243:F593–F597

Burr GO, Burr MM (1930) On the nature and role of fatty acids essential in nutrition. J Biol Chem 86:587–621

Cabanac M, Le Guelte L (1980) Temperature regulation and prostaglandin E_1 fever in scorpions. J Physiol 303:365–370

Cagen L, Qureshi Z, Nishimura H (1983) Metabolism of arachidonic acid by saline washed toadfish *Opsanus tau* red cells. Biochem Biophys Res Commun 110:250–255

Carroll KK (1986) Biological effects of fish oils in relation to chronic diseases. Lipids 21:731–732

Carsten ME, Miller JD (1977) Effects of prostaglandins and oxytocin on calcium release from a uterine microsomal fraction. J Biol Chem 252:1576–1581

Casterlin ME, Reynolds WW (1977) Behavioral fever in crayfish. Hydrobiologica 56:99–101

Casterlin ME, Reynolds WW (1978) Prostaglandin E_1 fever in the crayfish *Camborus bartoni*. Pharmacol Biochem Behav 9:593–595

Casterlin ME, Reynolds WW (1979) Fever induced in marine arthropods by prostaglandin E_1. Life Sci 25:1601–1604

Cenedella RJ, Galli C, Paoletti R (1975) Brain free fatty acid levels in rats sacrificed by decapitation versus focussed microwave irradiation. Lipids 10:290–293

Charo IF, Shak S, Karasek MA, Davidson PM, Goldstein IM (1984) Prostaglandin I_2 is not a major metabolite of arachidonic acid in cultured endothelial cells from human foreskin microvessels. J Clin Invest 74:914–919

Christ EJ, van Dorp DA (1972) Comparative aspects of prostaglandin biosynthesis in animal tissues. Biochim Biophys Acta 270:207–210

Christ-Hazelhof E, Nugteren DH (1979) Purification and characterisation of prostaglandin endoperoxide D-isomerase, a cytoplasmic, glutathione-requiring enzyme. Biochim Biophys Acta 572:43–51

Clare AS, Walker G, Holland DL, Crisp DJ (1982) Barnacle egg hatching: a novel role for a prostaglandin-like compound. Mar Biol Lett 3:113–120

Clare AS, Walker G, Holland DL, Crisp DJ (1985) The hatching substance of the barnacle, *Balanus balanoides* (L.). Proc R Soc London Ser B 224:131–147

Clare AS, van Elk R, Feyen JHM (1986) Eicosanoids: their biosynthesis in accessory sex organs of *Lymnaea stagnalis* (L). Int J Reprod Dev 10:125–131

Cook BJ, Holman GM, Meola S (1984) The oviduct musculature of the cockroach *Leucophaea maderae* and its response to various neurotransmitters and hormones. Arch Insect Biochem Physiol 1:167–178

Corey EJ, Pyne SG (1983) Synthesis of a new series of potent inhibitors of thromboxane A_2 biosynthesis. Tetrahedron Lett 24:3291–3294

Corey EJ, Albright JO, Barton AE, Hashimoto S (1980) Chemical and enzymic syntheses of 5-HPETE, a key biological precursor of slow reacting substance of anaphylaxis (SRS) and 5-HETE. J Am Chem Soc 102:1435–1436

Crawford MA (1983) Background to essential fatty acids and their prostanoid derivatives. Br Med Bull 39:210–213

Crisp DJ (1956) A substance promoting hatching and liberation of young in cirripedes. Nature (London) 178:263

Crow RT, Liley NR (1979) A sexual pheromone in the guppy, *Poecilia reticulata* (Peters). Can J Zool 57:184–188

Destephano DB, Brady UE (1977) Prostaglandin and prostaglandin synthase in the cricket, *Acheta domesticus.* J Insect Physiol 23:905–911

Destephano DB, Brady UE, Lovins RE (1974) Synthesis of prostaglandin by reproductive tissue of the male house cricket, *Acheta domesticus.* Prostaglandins 6:71–79

Destephano DB, Brady UE, Woodal LB (1976) Partial characterization of prostaglandin synthase in the reproductive tract of the male house cricket, *Acheta domesticus.* Prostaglandins 11:261–273

DeWitt DL, Smith WL (1983) Purification of prostacyclin synthase from bovine aorta by immunoaffinity chromatography. J Biol Chem 258:3285–3293

DeWitt DL, Day JS, Sonnenburg WK, Smith WL (1982) Concentration of prostaglandin endoperoxide synthase and prostaglandin I_2 synthase in endothelium and smooth muscle of bovine aorta. J Clin Invest 72:1882–1888

Diakow C, Nemiroff A (1981) Vasotocin, prostaglandin, and female reproductive behavior in the frog, *Rana pipens.* Hormon Behav 15:86–93

Diczfalusy U, Falardeau P, Hammarström S (1977) Conversion of prostaglandin endoperoxides to C17-hydroxy acids catalyzed by human platelet thromboxane synthase. FEBS Lett 84:271–274

Dietz TH, Graves SY (1981) Sodium influx in isolated gills of the freshwater mussel, *Ligumia subrostrata.* J Comp Physiol 143:185–190

Egan RW, Gale PH, Kuehl FA (1979) Reduction of hydroperoxides in the prostaglandin biosynthesic pathway by a microsomal peroxidase. J Biol Chem 254:3295–3302

Feldberg W, Kellaway CH (1938) Liberation of histamine and formation of lysocithin-like substances by cobra venom. J Physiol 94:187–226

Flower RJ, Blackwell GJ (1979) Anti-inflammatory steroids induce biosynthesis of a phospholipase A_2 inhibitor which prevents prostaglandin generation. Nature (London) 278:456–459

Flower RJ, Cheung HS, Cushman DW (1973) Quantitative determination of prostaglandins and malondialdehyde formed by the arachidonate oxygenase (prostaglandin synthase). Prostaglandins 4:325–341

Freas G, Grollman S (1980) Ionic and osmotic influences on prostaglandin release from the gill tissue of a marine bivalve, *Modiolus demissus.* J Exp Biol 84:169–185

Freas G, Grollman S (1981) Uptake and binding of prostaglandins in a marine bivalve, *Modiolus demissus.* J Exp Zool 216:225–233

German JB, Bruckner GG, Kinsella JE (1986) Lipoxygenase in trout gill tissue acting on arachidonic, eicosapentaenoic and docosahexaenoic acids. Biochim Biophys Acta 875:12–20
Goetz FW, Smith DC, Krickl SP (1982) The effects of prostaglandins, phosphodiesterase inhibitors, and cyclic AMP on ovulation of brook trout, (*Salvelinus fontinalis*) oocytes. Gen Comp Endocrinol 48:154–160
Gorman RR, Lin AH (1983) Biology and biochemistry of thromboxane synthetase inhibitors. In: Samuelsson B, Paoletti R (eds) Advances in prostaglandin thromboxane leukotriene research, vol 11. Raven, New York, pp 235–240
Gorman RR, Bunting S, Miller OV (1977) Modulation of human platelet adenyl cyclase by prostacyclin (PGX). Prostaglandins 13:377–388
Granström E, Kindahl H (1978) Radioimmunoassay of prostaglandins and thromboxanes. In: Frölich JC (ed) Advances in prostaglandin thromboxane research, vol 5. Raven, New York, pp 119–210
Graves SY, Dietz TH (1979) Prostaglandin E_2 inhibition of sodium transport in the freshwater mussel. J Exp Zool 210:195–201
Gréen K (1969) Gas chromatography-mass spectrometry of O-methyloxime derivatives of prostaglandins. Chem Phys Lipids 3:254–272
Gréen K, Samuelsson B (1964) Prostaglandins and related factors: XIX. Thin-layer chromatography of prostaglandins. J Lipid Res 5:117–120
Gréen K, Hamberg M, Samuelsson B, Smigel M, Frölich JC (1978) Measurement of prostaglandins, thromboxanes, prostacyclin and their metabolites by gas liquid chromatography-mass spectrometry. In: Frölich JC (ed) Advances in prostaglandin thromboxane research, vol 5. Raven, New York, pp 39–94
Gryglewski RJ, Panczenko B, Korbut R, Grodzinska L, Ocetkiewics A (1975) Corticosteroids inhibit prostaglandin release from perfused mesenteric blood vessels of rabbit and from perfused lungs of sensitized guinea pigs. Prostaglandins 10:343–355
Gupta OP, Lahlou B, Botella J, Porthé-Nibelle J (1985) In vivo and in vitro studies on the release of cortisol from interrenal tissue in trout. I. Effects of ACTH and prostaglandins. Exp Biol 43:201–212
Hagen DV, Brady UE (1982) Prostaglandins in the cabbage looper *Trichoplusia ni.* J Insect Physiol 28:761–765
Hajjar DP, Weksler BB, Falcone DJ, Hefton JM, Tack-Goldman K, Minick CR (1982) Prostacyclin modulates cholesteryl ester hydrolytic activity by its effect on cyclic adenosine monophosphate in rabbit aortic smooth muscle cells. J Clin Invest 70:479–488
Hamberg M, Hamberg G (1980) On the mechanism of the oxygenation of arachidonic acid by human platelet lipoxygenase. Biochem Biophys Res Commun 95:1090–1097
Hamberg M, Svensson J, Samuelsson B (1975) Thromboxanes: a new group of biologically active compounds derived from prostaglandin endoperoxides. Proc Natl Acad Sci USA 72:2994–2998
Hamilton J, Karol RJ (1982) High performance liquid chromatography (HPLC) of arachidonic acid metabolites. Prog Lipid Res 21:155–170
Hammarström S (1980) Enzymatic synthesis of 15-hydroperoxythromboxane A_2 and 12-hydroperoxy-5,8,10-heptadecatrienoic acid. J Biol Chem 255:518–521
Handler JS (1981) Vasopressin-prostaglandin interaction in the regulation of epithelial cell permeability to water. Kidney Int 19:831–838
Hanel AM, Lands WEM (1982) Modification of anti-inflammatory drug effectiveness by ambient lipid peroxides. Biochem Pharmacol 31:3307–3311
Hansen H (1976) 15-Hydroxyprostaglandin dehydrogenase – a review. Prostaglandins 12:647–679
Hansen AE, Stewart RA, Hughes G, Soderhjelm L (1962) The relation of linoleic acid to infant feeding. Acta Pediat (Stockholm) 51 (Suppl 137):1–41
Hassid A, Shebusky A, Dunn MJ (1983) Metabolism by human renal enzymes: presence of 9-hydroxyprostaglandin dehydrogenase activity in human kidney. In: Samuelsson B, Paoletti R (eds) Advances in prostaglandin thromboxane leukotriene research, vol 11. Raven, New York, pp 499–504
Hemler ME, Lands WEM (1980) Evidence for a peroxide-initiated free radical mechanism of prostaglandin biosynthesis. J Biol Chem 255:6253–6261

Hemler ME, Cook WH, Lands WEM (1979) Prostaglandin biosynthesis can be triggered by lipid peroxides. Arch Biochem Biophys 193:340–345
Herman CA, McCloskey MM, Doolittle K (1982) Prostaglandins of one, two, and three series affect blood pressure in American bullfrog, *Rana catesbeiana*. Gen Comp Endocrinol 54:491–498
Herman CA, Zimmerman PR, Doolittle KD (1984) Prostaglandin synthesis in goldfish heart, *Caraccius auratus*. Gen Comp Endocrinol 54:478–485
Higgs GA, Vane Jr (1983) Inhibition of cyclooxygenase and lipoxygenase. Br Med Bull 39:265–270
Hirata F, Schiffmann E, Venkatasubramaniam K, Salomon D, Axelrod J (1980) A phospholipase A_2 inhibitory protein in rabbit neutrophils induced by glucocorticoids. Proc Natl Acad Sci USA 77:2533–2536
Hirata F, del Carmine R, Nelson CA, Axelrod J, Schiffmann E, Warabi A, DeBlas AL, Nirenberg M, Manganiello V, Vaughman M, Kumagi S, Green I, Decker JL, Steinberg AD (1981) Presence of autoantibody for phospholipase inhibitory protein, lipomodulin in patients with rheumatic diseases. Proc Natl Acad Sci USA 78:3190–3194
Holland DL, East J, Gibson KH, Clayton E, Oldfield A (1985) Identification of the hatching factor of the barnacle *Balanus balanoides* as the novel eicosanoid 10,11,12-trihydroxy 5,8,14,17-eicosatetraenoic acid. Prostaglandins 29:1021–1029
Honn KV, Chavin W (1978) Temporal effect of prostaglandin E and F on human adrenal cAMP and cortisol output. Life Sci 22:543–552
Horrobin DF (1978) Prostaglandins: physiology, pharmacology and clinical significance. Book Press, Brattlebro, VT
Horseman ND, Meier AH (1978) Prostaglandin and the osmoregulatory role of prolactin in a teleost. Life Sci 22:1485–1490
Humes JL, Sadowski S, Galavage M, Goldenberg M, Subers E, Bonney RJ, Kuehl FA, Jr. (1982) Evidence of two sources of arachidonic acid for oxidative metabolism by mouse peritoneal macrophages. J Biol Chem 257:1591–1594
Hwang DH (1982) Characteristics of the formation of platelet lipoxygenase product from endogenous arachidonic acid. Lipids 17:845–847
Irvine RF (1982) How is the level of free arachidonic acid controlled in mammalian cells? Biochem J 204:3–16
Ito T, Ogawa K, Sakai K, Watanabe J, Satake T, Kamaya N, Hiraku S, Naito J (1983) Effects of a selective inhibitor of thromboxane synthetase (OKY-158) in human. In: Samuelsson B, Paoletti R (eds) Advances in prostaglandin thromboxane leukotriene research, vol 11. Raven, New York, pp 245–251
Jalabert B, Szollosi D (1975) In vitro ovulation of trout oocytes: effect of prostaglandins on smooth muscle-like cells of the theca. Prostaglandins 9:765–779
Johnson RA, Morton DR, Kinner JH, Gorman RR, McGuire JC, Sun FF (1976) The chemical structure of prostaglandin X (prostacyclin). Prostaglandins 12:912–928
Johnston HH, Herzog JP, Lauler DP (1967) Effect of prostaglandin E_1 on renal haemodynamics, sodium and water excretion. Am J Physiol 213:939–946
Jörg A, Henderson WR, Murphy RC, Klebanhoff SJ (1982) Leukotriene generation by eosinophils. J Exp Med 155:390–402
Kapoor K, Toor HS (1979) The effect of clomiphene citrate on ovulation and spawning of indomethacin treated carp, *Cyprinus carpio*. J Fish Biol 14:59–66
Kinoshita T, Nimura Y, Shionoya S, Goto H, Ozawa T (1987) Predominance of prostaglandin D_2 and I_2 in the rat gastric mucosa: analysis by high-performance liquid chromatography. Biochem Int 14:303–309
Korotchenko OD, Mishchenko TY, Isay V (1983) Prostaglandins of Japan sea invertebrates – I. The quantitation of group B prostaglandins in echinoderms. Comp Biochem Physiol 74C:85–88
Kulmacz RJ, Lands WEM (1985) Quantitative similarities in the several actions of cyanide on prostaglandin H synthase. Prostaglandins 29:175–190
Kunigelis SC, Saleuddin SM (1986) Reproduction in the freshwater gastropod, *Helisoma*: involvement of prostaglandin in egg production. Int J Invert Reprod Dev 10:159–167

Kurzrok R, Lieb C (1930) Biochemical studies of human semen. II. The action of semen on the human uterus. Proc Soc Exp Biol Med 18:268–272
Lands WEM, le Tellier PR, Rome LH, Vanderhoek JY (1973) Inhibition of prostaglandin biosynthesis. Adv Biosci 9:15–28
Lands WEM, Culp BR, Hirai A, Gorman R (1985) Relationship of thromboxane generation to the aggregation of platelets from humans. Effects of eicosapentaenoic acid. Prostaglandins 30:819–825
Lange AB (1984) The transfer of prostaglandin synthesizing activity during mating in *Locusta migratoria*. Insect Biochem 14:551–556
Lee SC, Levine L (1974) Prostaglandin metabolism. 1. Cytoplasmic reduced NAD dependent and microsomal reduced NADP dependent prostaglandin E 9-ketoreductase in monkey and pigeon tissues. J Biol Chem 249:1369–1375
Lefer AM, Ogletree ML, Smith JB, Silver MJ, Nicolau KC, Barnette WE, Gasic GP (1978) Prostacyclin: a potentially valuable agent for preserving myocardial tissue in acute myocardial ischemia. Science 200:52–54
Leffler CW, Hanson RC, Schneider EG (1980) Effects of prostaglandins and prostaglandin precursors on the arterial pressure of the bullfrog, *Rana catesbeiana*. Comp Biochem Physiol 66C:199–202
Lefkowitz RJ, Mullikin JD, Wood CL, Gore TB, Mukherjee C (1977) Regulation of prostaglandin receptors by prostaglandins and guanine nucleotides in frog erythrocytes. J Biol Chem 252:5295–5303
Leigh PJ, Cramp WA, MacDermot J (1984) Identification of the prostacyclin receptor by radiation inactivation. J Biol Chem 259:12431–12436
Levine L, Kobayashi T (1983) Detection of compounds immunologically related to arachidonic acid transformation products in extracts of invertebrates. Prostagland Leuk Med 12:357–369
Levine L, van Vunakis H (1970) Antigenic activity of prostaglandins. Biochem Biophys Res Commun 41:1171–1177
Lewis RA, Soter NA, Diamond PT, Austen KF, Oates JA, Roberts LJ (1982) Prostaglandin D_2 generation after activation of rat and human mast cells with anti-IgE. J Immunol 129:1627–1631
Loher W (1984) Behavioral and physiological changes in cricket-females after mating. In: Engles W, Clark WH Jr., Fischer A, Olive PJW, Nent DF (eds) Adv Invertebr Reprod 3:189–201
Loher W, Ganjian I, Kubo I, Stanley-Samuelson DW, Tobe SS (1981) Prostaglandins: their role in egg-laying of the cricket, *Teleogryllus commodus*. Proc Natl Acad Sci USA 78:7835–7838
Loher W, Ganjian I, Kubo I, Stanley-Samuelson D, Tobe SS (1985) Location and synthesis of PGE_2 an egg-laying stimulant of the cricket, *Teleogryllus commodus*. In: Lofts B, Holmes WN (eds) Current trends in endocrinology. Hongkong Univ Press, pp 391–393
Mai J, Goswami SK, Bruckner G, Kinsella JE (1981) A new prostaglandin C22-PGF_4, synthesized from docosahexaenoic acid (C22:6n3) by trout gill. Prostaglandins 21:691–698
Malik KU, Wong PY-K (1981) Leukotriene C: the major lipoxygenase metabolite of arachidonic acid in dog spleen. Biochem Biophys Res Commun 103:511–520
Marshall PJ, Kulmacz RJ, Lands WEM (1987) Constraints on prostaglandin biosynthesis in tissues. J Biol Chem 262:3510–3517
Marumo F (1986) Role of inhibitory and stimulative effects of prostaglandins on vasopressin-stimulated osmotic water flow in the toad bladder. Jpn J Physiol 36:935–944
Meijer L, Guerrier P, Maclouf J (1984) Arachidonic acid, 12 and 15-hydroxyeicosatetraenoic acids, eicosapentaenoic acid, and phospholipase A_2 induce starfish oocytes maturation. Dev Biol 106:368–378
Meijer L, Brash AR, Bryant RW, Ng K, Mclouf J, Sprecher H (1986a) Stereospecific induction of starfish oocyte maturation (8R)-hydroxyeicosatetraenoic acid. J Biol Chem. 261: 17040–17047
Meijer L, Maclouf J, Bryant RW (1986b) Arachidonic acid metabolism in starfish oocytes. Dev Biol 114:22–23
Milton AS (1982) Prostaglandins in fever and the mode of action of antipyretics. In: Handbook of experimental pharmacology, vol 60. Springer, Berlin Heidelberg New York, pp 257–304

Mizuno K, Yamamoto S, Lands WEM (1982) Effects of non-steroidal anti-inflammatory drugs on fatty acid cyclooxygenase and prostaglandin hydroperoxidase activities. Prostaglandins 23:743–757

Moncada S, Gryglewski RJ, Bunting S, Vane JR (1976) An enzyme isolated from arteries transforms prostaglandin endoperoxides to an unstable substance that inhibits platelet aggregation. Nature (London) 263:663–665

Moncada S, Herman AG, Higgs EA, Vane JR (1977) Differential formation of prostacyclin (PGX and PGI_2) by layers of the arterial wall. An explanation for the antithrombotic properties of vascular endothelium. Thromb Res 11:323–344

Moncada S, Ferreira SH, Vane JR (1978) Bioassay of prostaglandins and biologically active substances derived from arachidonic acid. In: Frölich JC (ed) Advances in prostaglandin thromboxane research, vol 5. Raven, New York, pp 211–236

Morris HR, Taylor GW, Piper PJ, Tippins JR (1980) Structure of slow-reacting substance of anaphylaxis from guinea pig lung. Nature (London) 285:104–106

Morse DE, Duncan H, Hooker N, Morse A (1977) Hydrogen peroxide induces spawning in mollusks with activation of prostaglandin endoperoxide synthase. Science 196:298–300

Morse DE, Kayne M, Tidyman M, Anderson S (1978) Capacity for biosynthesis of prostaglandin-related compounds: distribution and properties of the rate-limiting enzyme in hydrocorals, gorgonians, and other coelentrates of the Caribbean and Pacific. Biol Bull 154:440–452

Murphy RC, Hammarström S, Samuelsson B (1979) Leukotriene C: a slow-reacting substance from murine mastocytoma cells. Proc Natl Acad Sci USA 76:4275–4279

Murtaugh MP, Denlinger DL (1982) Prostaglandins E and $F_{2\alpha}$ in the house cricket and other insects. Insect Biochem 12:599–603

Nakao J, Ooyama T, Chang WC, Murota S, Orimo H (1982) Platelets stimulate aortic smooth muscle cell migration in vitro. Involvement of 12-L-hydroxy-5,8,10,14-eicosatetraenoic acid. Atherosclerosis 43:143–150

Narumiya S, Salmon JA, Cottee FH, Weatherley BC, Flower RJ (1981) Arachidonic acid 15-lipoxygenase from rabbit peritoneal polymorphonuclear leukocytes. Partial purification and properties. J Biol Chem 256:9583–9592

Needleman P, Raz A, Ferrendelli J, Minkes M (1977) Application of imidazole as a selective inhibitor of thromboxane synthetase in human platelets. Proc Natl Acad Sci USA 74:1716–1720

Needleman P, Turk J, Jakschik BA, Morrison AR, Lefkowith JB (1986) Arachidonic acid metabolism. Annu Rev Biochem 55:69–102

Nijkamp FP, Flower RJ, Moncada S, Vane JR (1976) Partial purification of rabbit aorta contracting substance-releasing factor and inhibition of its activity by antiinflammatory steroids. Nature (London) 263:479–482

Nomura T, Ogata H (1976) Distribution of prostaglandins in the animal kingdom. Biochim Biophys Acta 431:127–131

Nomura T, Ogata H, Ito M (1973) Occurrence of prostaglandins in fish testis. Tohoku J Agric Res 24:138–144

Nugteren DH (1982) Arachidonic acid-12-lipoxygenase from bovine platelets. Meth Enzymol 86:49–54

Oates JA (1982) The 1982 Nobel Prize in physiology or medicine. Science 218:765–768

Ogata H, Nomura T (1975) Isolation and identification of PGE_2 from gastrointestinal tract of shark *Triakis scullia*. Biochim Biophys Acta 388:84–91

Ogata H, Nomura T, Hata M (1979) Prostaglandin $F_{2\alpha}$ changes induced by ovulatory stimuli in the pond loach, *Misgurnus anguillicaudatus*. Bull Jpn Soc Sci Fish 45:929–931

Ogino N, Miyamoto T, Yamamoto S, Hayaishi O (1977) Prostaglandin endoperoxide E isomerase from bovine vesicular gland microsomes, a glutathione-requiring enzyme. J Biol Chem 252:890–895

Ono K, Osada M, Matsutani T, Mori K, Nomura T (1982) Gonodal prostaglandin F_2 profile during sexual maturation in the oyster, *Crassostrea gigas*. Mar Biol Lett 3:223–230

Orloff J, Handler J, Bergström S (1965) Effect of PGE_1 on the permeability response of toad bladder to vasopressin, theophylline and adenosine 3'5'-monophosphate. Nature (London) 205:397–398

Pace-Asciak C, Wolfe LS (1968) Inhibition of prostaglandin synthesis by oleic, linoleic and linolenic acids. Biochim Biophys Acta 152:784–787
Pace-Asciak C, Wolfe LS (1971) A novel prostaglandin derivative formed from arachidonic acid by rat stomach homogenates. Biochemistry 10:3657–3664
Pagels WR, Sachs RJ, Marnett LJ, DeWitt DL, Day JS, Jeffrey S, Smith WL (1983) Immunochemical evidence for the involvement of prostaglandin H synthase in hydroperoxide-dependent oxidations by ram seminal vesicle microsomes. J Biol Chem 258:6517–6523
Palm CD, Levine BA (1987) Simultaneous preparation and extraction of prostaglandins from gastric tissue with separation and quantitation by high pressure liquid chromatography. J Liquid Chromatogr 10:881–897
Pandey S, Stacey NE, Hoar WS (1973) Mode of action of clomiphene citrate in inducing ovulation of goldfish, *Carassius auratus*. Can J Zool 51:1315–1316
Partridge BL, Liley NR, Stacey NE (1976) The role of pheromones in the sexual behavior of the goldfish. Anim Behav 24:291–309
Paulsrud JR, Pensler L, Whitten CF, Stewart S, Holman RT (1972) Essential fatty acid deficiency in infants induced by fat free intravenous feeding. Am J Clin Nutrit 25:897–904
Peter RE, Billard R (1976) Effects of third ventricle injection of prostaglandins on gonadotropin secretion in goldfish, *Carassius auratus*. Gen Comp Endocrinol 30:451–456
Peyraud-Waitzenegger M, Nomura T, Peyraud C (1976) Cardiovascular effects on PGE_2 in the carp (*Cyprinus carpio*). In: Samuelsson B, Paoletti R (eds) Advances in prostaglandin thromboxane research, vol 2. Raven, New York, p 912
Piomelli D (1985) Leukotrienes in teleost fish gills. Naturwissenschaften 72:276–277
Piper J (1983) Pharmacology of leukotrienes. Br Med Bull 39:255–259
Pohlman T, Yates J, Needleman P, Klahr S (1983) Effects of prostacyclin on short-circuit current and water flow in the toad urinary bladder. Am J Physiol 244:F270–F277
Reed GA, Brooks EA, Eling TE (1984) Phenylbutazone-dependent epoxidation of 7,8-dihydroxy-7,8-dihydrobenzo(*a*)pyrene. A new mechanism for prostaglandin H synthase-catalyzed oxidations. J Biol Chem 259:5591–5595
Reynolds WW, Casterlin ME, Covert JB (1976) Behavioural fever in teleost fishes. Nature (London) 259:41–42
Rice JR, Zenser TV, Davis BB (1985) Formation of thioether conjugates of the bladder carcinogen ANFT catalyzed by prostaglandin H synthase. Carcinogenesis 6:585–590
Rittenhouse-Simmons S (1979) Production of diglyceride from phosphatidylinositol in activated human platelets. J Clin Invest 63:580–587
Rollins TE, Smith WL (1980) Subcellular localization of prostaglandin-forming cyclooxygenase in Swiss mouse 3T3 fibroblasts by electron microscopic immunocytochemistry. J Biol Chem 255:4872–4875
Rome LH, Lands WEM (1975) Structure requirements for time dependent inhibition of prostaglandin biosynthesis by anti-inflammatory drugs. Proc Natl Acad Sci USA 72:4863–4865
Rosendorff C, Cranston WI (1968) Effects of salicylate on human temperature regulation. Clin Sci 35:81–91
Roth GJ, Majerus PW (1975) The mechanism of the effect of aspirin on human platelets. I. Acetylation of a particular fraction protein. J Clin Invest 56:624–632
Roth GJ, Machuga ET, Strittmatter P (1981) The heme binding properties of prostaglandin synthetase from sheep vesicular gland. J Biol Chem 256:10018–10022
Rouzer CA, Scott WA, Hamill AL, Cohn ZA (1980) Dynamics of leukotriene C production by macrophages. J Exp Med 152:1236–1247
Saintsing DG, Hwang DH, Dietz TH (1983) Production of prostaglandins E_2 and $F_{2\alpha}$ in the freshwater mussel *Ligumia subrostrata*: relation to sodium transport. J Pharm Exp Ther 226:455–461
Salari H, Braquet P, Borgeat P (1984) Comparative effects of indomethacin, acetylenic acid, 15-HETE, nordihydroguaiaretic acid and PW755C on the metabolism of arachidonic acid in human leukocytes and platelets. Prostagland Leuk Med 13:53–60
Salmon JA, Smith DR, Flower RJ, Moncada S, Vane JR (1978) Further studies on the enzymatic conversion of prostaglandin endoperoxide into prostacyclin by porcina aorta microsomes. Biochim Biophys Acta 523:250–262

Samuelsson B, Hamberg M, Sweeley C (1970) Quantitative gas chromatography of prostaglandin E_1 at the nanogram level: use of deuterated carrier and muliple-ion analyzer. Anal Biochem 38:301–304

Samuelsson B, Granström E, Gréen K, Hamberg M (1971) Metabolism of prostaglandins. In: Ramwell P, Shaw JE (eds) Prostaglandins. Ann NY Acad Sci 180:138–163

Samuelsson B, Granström E, Gréen K, Hamberg M, Hammarström S (1975) Prostaglandins. Annu Rev Biochem 44:669–695

Samuelsson B, Goldyne M, Granström E, Hamberg M, Hammarström S, Malmsten C (1978) Prostaglandins and thromboxanes. Annu Rev Biochem 47:997–1029

Samuelsson B, Borgeat S, Hammarström S, Murphy RC (1980) Leukotrienes: A new group of biologically active compounds. In: Samuelsson B, Ramwell PW, Paoletti R (eds) Advances in prostaglandin thromboxane research, vol 6. Raven, New York, pp 1–18

Sato M, Dunn MJ (1984) Interactions of vasopressin, prostaglandins, and cAMP in rat renal papillary collecting tubule cells in culture. Annu J Physiol 247:F423–F433

Sato J, Jyujo T, Hirono M, Iesaka T (1975) Effects of indomethacin, an inhibitor of prostaglandin synthesis in rats. J Endocrinol 64:777–779

Schafer AI, Cooper B, O'Hara D, Handin RI (1979) Identification of platelet receptors for prostaglandin I_2 and D_2. J Biol Chem 254:2914–2917

Setty YBN, Ramaiah TR (1979) Isolation and identification of prostaglandins from the reproductive organs of male silkmoth, *Bombyx mori* L. Insect Biochem 9:613–617

Setty YBN, Ramaiah TR (1980) Effect of prostaglandins and inhibitors of prostaglandin biosynthesis in oviposition in the silkmoth *Bombyx mori*. Indian J Exp Biol 13:539–541

Sheng WY, Wyche A, Lysz T, Needleman P (1982) Prostaglandin endoperoxide E_2 isomerase is dissociated from prostaglandin endoperoxide synthetase in renal cortex. J Biol Chem 257:14632–14634

Sheng WY, Lysz TA, Wyche A, Needleman P (1983) Kinetic comparison and regulation of the cascade of microsomal enzymes involved in renal arachidonate and endoperoxide metabolism. J Biol Chem 258:2188–2192

Shimizu T, Yamamoto S, Hayaishi O (1979a) Purification and properties of prostaglandin D synthetase from rat brain. J Biol Chem 254:5222–5228

Shimizu T, Mizuno N, Amano T, Hayaishi O (1979b) Prostaglandin D_2 a neuromodulator. Proc Natl Acad Sci USA 76:6231–6234

Shukla SD (1982) Minireview. Phosphatidylinositol specific phospholipase C. Life Sci 30:1323–1335

Siegel MI, McConnel RT, Porter NA, Selph JL, Traux JF, Vinegar R, Cuatrecasas P (1980a) Aspirin-like drugs inhibit arachidonic acid metabolism via lipoxygenase and cyclo-oxygenase in rat neutrophils from carragenan pleural exudates. Biochem Biophys Res Commun 92:688–695

Siegel MI, McConnel RT, Porter NA, Cuatrecasas P (1980b) Arachidonate metabolism via lipoxygenase and 12 L-hydroperoxy-5,8,10,14-eicosatetraenoic acid peroxidase sensitive to anti-inflammatory drugs. Proc Natl Acad Sci USA 77:308–312

Singh AK, Singh TP (1976) Effect of clomid, sexovid, and prostaglandins on induction of ovulation and gonadotropin secretion in a freshwater catfish, *Heteropneustes fossili* (Bloch). Endokrinologie 68:129–136

Smith WL, Lands WEM (1972) Oxygenation of polyunsaturated fatty acids during prostaglandin biosynthesis by sheep vesicular glands. Biochemistry 11:3276–3285

Smith WL, Wilkin GP (1977) Immunochemistry of prostaglandin endoperoxides to C17-hydroxy acids catalyzed by human platelet thromboxane synthase. FEBS Lett 84:271–274

Smith WL, DeWitt DL, Allen ML (1983) Bimodal distribution of the prostaglandin I_2 synthase antigen in smooth muscle cells. J Biol Chem 258:5922–5926

Spector AA (1975) Fatty acid binding to plasma albumin. J Lipid Res 16:165–179

Spies HG, Norman RL (1973) Luteinising hormone release and ovulation induced by the intraventricular infusion of prostaglandin E_1 into pentobarbitone-blocked rats. Prostaglandins 4:131–141

Srivastava KC (1985a) Docosahexaenoic acid (C22 : 6 ω3) and linoleic acid are antiaggregatory, and alter arachidonic acid metabolism in human platelets. Prostagland Leuk Med 17:319–327

Srivastava KC (1985b) Transformation of exogenous arachidonic acid in human platelets in the presence of oleic- and eicosapentaenoic acids. Prostagland Leuk Med 18:31–37

Srivastava KC (1985c) Ascorbic acid enhances the formation of prostaglandin E_1 in washed human platelets and prostacyclin in rat aortic rings. Prostagland Leuk Med 18:227–233

Srivastava KC, Awasthi KK (1983a) In vitro effect of unsaturated fatty acids on the balance between thromboxane A_2 and prostacyclin in the blood vascular system. Prostagland Leuk Med 12:449–459

Srivastava KC, Awasthi KK (1983b) Separation and quantitative determination of radiolabeled prostaglandins, thromboxanes, 6-keto-prostaglandin $F_{1\alpha}$ and other arachidonic acid metabolites produced in biological material. J Chromatogr Biomed Appl 275:61–70

Srivastava KC, Mustafa T (1984) Arachidonic acid metabolism and prostaglandins in lower animals. Mol Physiol 5:53–60

Srivastava KC, Mustafa T (1985) Formation of prostaglandins and other comparable products during aerobic and anaerobic metabolism of (1-^{14}C)arachidonic acid in the tissues of sea mussels, *Mytilus edulis L.* Mol Physiol 8:101–112

Srivastava KC, Tiwari KP, Awasthi KK (1982) A convenient TLC schedule for differential separation and quantification of prostaglandins, thromboxanes, and hydroxy fatty acids formed from (1-^{14}C)arachidonic acid in human blood platelets. Microchem J 27:246–253

Stacey NE (1976) Effects of indomethacin and prostaglandins on spawning behaviour of female goldfish. Prostaglandins 12:113–126

Stacey NE (1981) Hormonal regulation of sexual behaviour in teleosts. Am Zool 21:305–316

Stacey NE, Goetz FW (1982) Role of prostaglandins in fish reproduction. Can J Fish Aqu Sci 39:92–98

Stacey NE, Liley NR (1974) Regulation of spawning behaviour in the female goldfish. Nature (London) 247:71–72

Stacey NE, Peter RE (1979) Central action of prostaglandins in spawning behaviour of female goldfish. Physiol Behav 22:1191–1196

Stacey NE, Cook AF, Peter RE (1979) Ovulatory surge of gonadotropin in the goldfish, *Carassius auratus.* Gen Comp Endocrinol 37:246–249

Stanley-Samuelson DW (1987) Physiological roles of prostaglandins and other eisosanoids in invertebrates. Biol Bull 173:92–109

Stanley-Samuelson DW, Loher W (1986) Prostaglandins in insect reproduction. Ann Entomol Soc Am 79:841–853

Stanley-Samuelson DW, Klocke JA, Kubo I, Loher W (1983) Prostaglandin and arachidonic acid and reproductive tissue from virgin and mated female crickets, *Teleogryllus commodus.* Entomol Exp Appl 34:35–39

Stanley-Samuelson DW, Peloguin JJ, Loher W (1986) Egg-laying in response to prostaglandin injections in *Teleogryllus commodus.* Physiol Entomol 11:213–219

Stokes JB (1979) Effect of prostaglandin E_2 on chloride transport across the thick ascending limb of Henle. Selective inhibition of the medullary portion. J Clin Invest 64:495–502

Tan CH, Lam TJ, Wong LY, Pang MK (1987) Prostaglandin synthesis and its inhibition by cyclic AMP and forskolin in postpartum follicles of the guppy. *Poecilia reticulata.* Prostaglandins 34:697–715

Taylor GW, Morris HR (1983) Lipoxygenase pathways. Br Med Bull 39:219–222

Terragno A, Rydzick R, Terragno NA (1981) High performance liquid chromatography and UV detection for the separation and quantitation of prostaglandins. Prostaglandins 21:101–112

Tobe SS, Loher W (1983) Properties of the prostaglandin synthase complex in the cricket *Teleogryllus commodus.* Insect Biochem 13:137–141

Tocher DR, Sargent JR (1987) The effect of calcium ionophore A23187 on the metabolism of arachidonic acid and eicosapentaenoic acid in neutrophils from a marine teleost fish rich in (n–3) polyunsaturated fatty acids. Comp Biochem Physiol 87B:733–739

Tsunamoto K, Todo S, Imasuku S (1987) Separation of prostaglandins and thromboxane by two-dimensional thin-layer chromatography. J Chromatogr Biomed Appl 417:414–419

Ullrich V, Graf H (1984) Prostacyclin and thromboxane synthase as P-450 enzymes. Trends Pharmacol Sci 5:352–355

Van den Bosch H (1980) Intracellular phospholipases A. Biochim Biophys Acta 604:191–246

Vanderhoek JY, Tare NS, Bailey JM, Goldstein AL, Pluznik DH (1982a) New role for 15-hydroxyeicosatetraenoic acid. Activator of leukotriene biosynthesis in PT-18 mast-basophil cells. J Biol Chem 257:12191–12195

Vanderhoek JY, Bryant RW, Bailey JM (1982b) Regulation of leukocyte and platelet lipoxygenases by hydroxyeicosanoids. Biochem Pharmacol 31:3463–3467

van Dorp DA (1971) Recent developments in the biosynthesis and the analysis of prostaglandins. In: Prostaglandins. Ramwell P, Saw JE (eds) Ann NY Acad Sci 180:181–199

van Dorp DA, Beerthuis RK, Nugteren DH, Vonkeman H (1964) Enzymatic conversion of all-cis-polyunsaturated fatty acids into prostaglandins. Nature (London) 203:839–841

Vane JR (1978) The mode of action of aspirin-like drugs. Agents Actions 8:430–443

Wakayama EJ, Dillwith JW, Blomquist GJ (1980) In vitro biosynthesis of prostaglandins in the reproductive tissues of the male housefly, *Musca domestica* L. Am Zool 20:904

Wakayama EJ, Dillwith JW, Blomquist GJ (1986) Occurrence and metabolism of prostaglandins in the housefly, *Musca domestica* L. Insect Biochem 16:895–902

Warso MA, Lands WEM (1983) Lipid peroxide in relation to prostacyclin and thromboxane physiology and pathophysiology. Br Med Bull 39:277–280

Watanabe T, Shimizu T, Narumiya S, Hayaishi O (1982) NADP-linked 15-hydroxyprostaglandin dehydrogenase for prostaglandin D in human blood platelets. Arch Biochem Biophys 216:372–379

Watanabe K, Yoshida R, Shimizu T, Hayaishi O (1985) Enzymatic formation of prostaglandin F_2 alpha from prostaglandin H_2 and D_2. Purification and properties of prostaglandin F synthetase from bovine lung. J Biol Chem 260:7035–7041

Weinheimer AJ, Spraggins RL (1969) The occurrence of two new prostaglandin derivatives (15-epi-PGA_2 and its acetate methyl ester) in the gorgonian *Plexaura homomalla*: chemistry of coelentrates XV. Tetrahedron Lett 59:5185–5188

Wolfe LS (1982) Eicosanoids: Prostaglandins, thromboxanes, leukotrienes, and other derivatives of carbon-20 unsaturated fatty acids. J Neurochem 38:1–9

Woodward JJ, Smith L, Casillas E (1981) Thrombocyte aggregation in rainbow trout. Comp Biochem Physiol 68A:457–466

Yokoyama C, Mizuno K, Mitachi H, Yoshimoto T, Yamamoto S, Pace-Asciak C (1983) Purification and characterization of arachidonate 12-lipoxygenase from rat lung. Biochim Biophys Acta 750:237–243

Zusman RM, Kieser HR, Handler JS (1978) Effect of adrenal steroids on vasopressin-stimulated PGE synthesis and water flow. Am J Physiol 234:F532–F540

Chapter 7

Maternal-Fetal Trophic and Respiratory Relationships in Viviparous Ectothermic Vertebrates

B. Korsgaard[1] and *R.E. Weber*[1,2]

Contents

1 Introduction

Viviparity in ectothermic vertebrates may be defined in different ways. For the present purpose we may conveniently use Wourm's (1981) definition, viz., that viviparity is "a process in which eggs are fertilized internally and are retained within the maternal reproductive system for a significant period of time during which they develop to an advanced state and then are released". The maternal organism accordingly must provide the developing embryo with those materials that it otherwise obtains from its physical surroundings to satisfy its trophic and respiratory requirements.

Viviparity appears to have evolved enumerable times in ectothermic vertebrates (almost 100 times in squamates alone; Shine 1984), and often more than once within the same genus. Although the high frequency of evolutionary origins implies a wide interspecific diversity in physiological adaptations, the trophic and respiratory exchanges between the maternal and fetal organisms have been studied

[1] Biology Institute, Odense University, Campusvej 55, 5230 Odense M, Denmark
[2] Present Address: Zoophysiology Laboratory, Aarhus University, 8000 Aarhus C, Denmark

Advances in Comparative and Environmental Physiology, Vol. 5

in only a few species and future investigations of vertebrate ectotherms are likely to unveil novel adaptation mechanisms to viviparity.

Viviparous reproduction provides an excellent model for studying the mechanisms of biochemical and physiological adaptations to modes of life. Nutritionally, viviparity ranges from exclusively yolk-dependent embryonic development (lecithotrophy) to predominating maternal-fetal nutrient exchange (matrotrophy) during gestation. It thus illustrates a diversity of trophic adaptations to facilitate maternal-fetal nutrient transfer. Surprisingly little is known of the extent of nutrient transfer in relation to the actual energetic benefit and growth of the embryo during its intramaternal development. Information is similarly sparse regarding the relationships between energetic input via transferred organic substrates, embryonic growth, respiratory rates, and the pattern of energy utilization.

A range of organismic, cellular, and molecular adaptations support fetal-maternal exchanges of oxygen and excretory products. According to Fick's law the net diffusive transfer rate, M_x of a given substance X, equals the product of the concentration difference (C_x), and the "structural transfer capacity" kA/L (cf. Webb and Brett 1972b), where A is the total exchange surface area, L the diffusion distance, and k the diffusion coefficient.

The transfer of oxygen from the maternal to the fetal circulation almost invariably depends on a higher oxygen affinity (lower half-saturation oxygen tension, P_{50}) in the fetal than in the maternal blood. This affinity difference may be due to lower concentrations in the fetus of erythrocytic allosteric factors, which decrease the oxygen affinity of hemoglobin (Hb) by binding to deoxyHb and delaying the transition to the oxygenated state. In contrast to mammals, where the main erythrocytic organic cofactor is 2, 3-diphosphoglycerate (DPG), that in vertebrate ectotherms is ATP which may be found together with guanosine triphosphate (GTP), other nucleoside triphosphates (NTP), and inositol pentaphosphate (IPP) (Weber 1982). A higher organic phosphate concentration in the adult compared to the fetal red cells will, however, also decrease blood oxygen affinity indirectly, via the Bohr effect ($\Delta \log P_{50}/\Delta pH$), since it lowers intracellular pH through alterations in the Donnan distribution of protons across the red cell membranes (cf. Wood and Johansen 1973).

An alternative mechanism that may be implicated is the occurrence of specific fetal Hbs with higher intrinsic oxygen affinities than maternal Hbs (measured under the same physicochemical conditions). In contrast to mammals like mouse, pig, and man, however, there appears to be no evidence for embryonic Hbs that precede fetal and adult ones in ectothermic vertebrates. Other strategies to facilitate fetal oxygen transfer are lower cellular Hb concentrations, and greater degrees of oxidation to metHb in the fetal blood, both of which raise oxygen affinity (Weber et al. 1976; Pough 1977; Ingermann and Terwilliger 1982).

As in adults, fetal CO_2 elimination depends on the internal to ambient diffusion gradient, and on the blood concentration of Hb, which can increase CO_2 transport directly via increased CO_2 binding (as carbamino compounds) and indirectly by mopping up protons released by HCO_3 formation.

We here review the different patterns of maternal-fetal trophic and respiratory exchange, its hormonal control, the anatomical structures involved, and the adaptations to environmental conditions and metabolic requirements in viviparous fish, amphibians, and reptiles, concentrating mainly on recent data that appear not to have been dealt with in major reviews.

2 Chondrichthyan Fish

Viviparous reproduction is widespread among the chondrichthyan fish. The gross reproductive anatomy of the viviparous species is fairly similar to that of oviparous ones, and internal fertilization, a prerequisite for viviparous reproduction, typically occurs in both reproductive forms. In viviparous species, however, part of the oviduct is modified to form an uterus, the structure of which varies with species and with various adaptations facilitating the maternal-fetal exchange of different materials.

2.1 Nutrition

Viviparity in chondrichthyans may be grouped into two different nutritional categories: lecithotrophy, in which the embryos rely on their yolk sacs as the sole nutritional source during development; and matrotrophy, which is found in species with a placental analog or a true yolk sac placenta. Oophagy, defined as consumption of unfertilized eggs by the growing embryos, or adelphophagy, which is embryonic cannibalism in utero, are less common in these fish. These different chondrichthyan reproductive patterns have been extensively reviewed by Wourms (1977, 1981). Despite the widespread occurrence of viviparity, surprisingly little appears to be known concerning the physiological adaptations of viviparity within this class of fish.

Interestingly, the nutritional profit in terms of increased embryonic organic weight does not appear to be related to the intimacy of maternal and fetal structures. In species with a placental analog nutrient transfer to the embryos primarily results from a secretory function of the modified uterine epithelium. Although secretions from the uterus may be abundant, the actual amount of nutrients transferred may be small as indicated by temporary weight losses of about 35% in some species of *Torpedo* (Amoroso 1960).

In placental sharks embryonic dry weight may increase eight-to tenfold compared to that of the egg (Hamlett 1983). The true placental forms, which have yolk sac placentas, are limited to the two families of sharks, the Carcharinidae and Sphyrnidae. However, individual genera of these two families may show markedly different nutritional forms. Thus, *Mustelus manazo* lacks a placenta and the embryo is typically lecitrotrophic, whereas *Mustelus griseus* embryos become increasingly dependent on the placenta as the yolk mass becomes reduced with growth (Teshima 1981). The placenta consists of two parts, the maternal internal uterine wall and the embryonic yolk sac wall (Teshima 1981), which show increasing interdigitation with embryonic growth. Exchanges between the maternal organism and the embryos occur across an embryonic membrane, which is found between the maternal and the embryonic portions of the placenta, and is considered to play an important role in the nutrition, respiration, and excretion of the embryos of *Mustelus griseus.* However, no actual transport studies have yet been carried out in this species.

The ultrastructure of the yolk sac placenta of different viviparous sharks has recently been described by Hamlett and Wourms (1984) in relation to its function in gas exchange and hematrophic nutrient transport. The preimplantation yolk sac is believed to secrete proteins and act as a respiratory organ and thus to be implicated in synthetic as well as in transport processes. Hamlett (1983) observed that typan blue and horseradish peroxidase are rapidly absorbed by the distal portion of the shark placenta, and hypothetized that the maternal shark liver continues to synthesize vitellogenins during gestation and that these precursor proteins are transferred to and taken up by the fetal yolk sac epithelium.

Complex ontogenetic changes as regards the source of nutrition have recently been described in the sand tiger shark, *Odontaspis taurus* (Gilmore et al. 1983), where the embryos initially are lecithotrophic, but progressively they change to adelphophagic and oophagic nutrition (Fig. 1). During early tissue differentiation before hatching, nutrition is derived from the internal coelomic yolk and subsequently from the encapsulated yolk. After hatching the yolk sac supply presumably is supplemented by uterine secretions. At about 100 mm they start to hunt

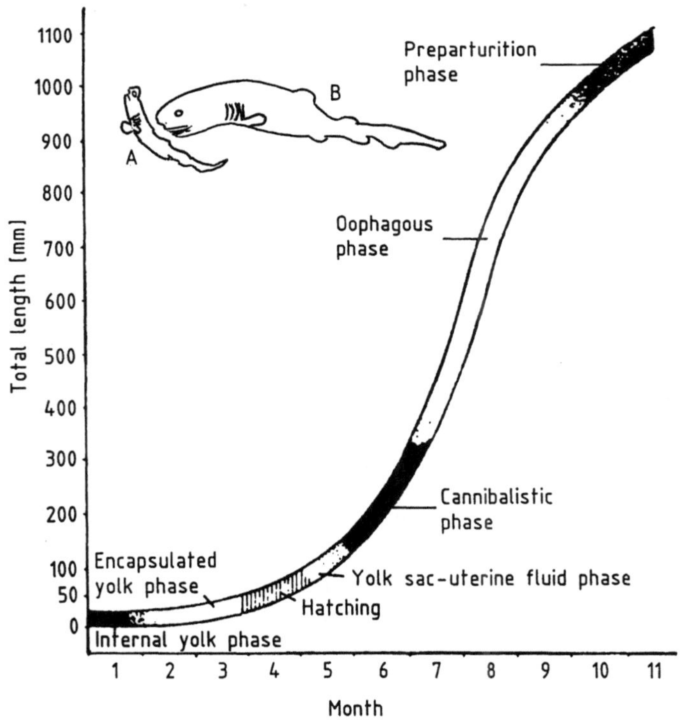

Fig. 1A, B. Embryonic growth curve and nutritional phases in the development of the teleost fish *Odontaspis taurus. Inset:* A 51-mm embryo (*A*) attacked and damaged by a male, 100-mm embryo (*B*) in utero (After Gilmore et al. 1983)

and consume other intrauterine embryos (adelphophagy). Late in gestation the single remaining embryo in each uterus obtains additional nutrition by consuming ovulated eggs from torn egg capsules in the uterus (oophagy). Oophagy-adelphophagy is considered a most efficient (albeit egoistic) way of obtaining nutrition during the prolonged gestation in sharks and results in marked intra-uterine embryonic weight increase.

In the viviparous spiny dogfish, *Squalus acanthias*, the embryos are retained in the uterus for a gestation period of about 22 months. This species is lecithotrophic and shows no signs of a nutritive secretory activity of the uterine epithelium (Hogarth 1976; Wourms 1977). A considerable weight loss of up to 55% during embryonic development is accompanied by ultrastructural changes that convert the uterine mucosa into a respiratory membrane, which may also participate in transport of water and electrolytes. During the first 6 months of gestation when the embryos are enclosed in an elongated egg case, the embryonic kidneys are still immature (Hogarth 1976), and the urea concentration of the uterine fluid is at the same level as in the maternal blood. After the 6 months the embryos hatch and are able to maintain their own uremia. Evans et al. (1982) observed that even enclosed in the presumed egg case ("candle") and incubated in sea water, the embryos were capable of regulating their plasma electrolyte and urea concentrations. Older hatched embryos in utero are capable of osmoregulation against sea water and display a reduced ionic permeability and a functional rectal gland, as well as other pathways for extrusion of NaCl.

2.2 Respiration

In some sharks gas and metabolite exchanges occur across placentae, which are modifications of the distal portions of the yolk sac that are in contact or interdigitate with vascularized areas of the uterus. In skates and rays, long thin villi (trophenemata) extend from the uterine wall, and at least in one species these enter the fetal oesophagus via the spiracles (Wood-Mason and Alcock 1891). In *Squalus acanthias*, which lacks a placental analog, a vascular plexus covers the fetal body in the initial phases of development, and the yolk sac throughout most interuterine existence, while external gill filaments, followed by internal gills, are found in the prenatal period (Manwell 1958).

Although McCutcheon (1947) observed the same oxygen affinity in dilute solutions of Hb from adult and late-gestational stages of three species of viviparous rays, his study does not exclude the presence of a fetal-adult affinity shift earlier in ontogeny (cf. Ingermann and Terwilliger 1984). There is considerable evidence for structurally and functionally unique fetal Hbs in sharks. In *S. acanthias*, higher oxygen affinities are found in whole blood, erythrocyte suspensions, and in stripped (phosphate-free) Hb from neonates compared to adults (Manwell 1963; Wells and Weber 1983; Weber et al. 1983). In addition, differences in organic phosphate interaction may be implicated. Adult *S. acanthias* erythrocytes accordingly have higher NTP concentrations than fetal and neonate erythrocytes (Bartlett 1982; Wells and Weber 1983). Interestingly GTP, the dominant cofactor in adult *Mustellus canis* red cells, is absent in the fetal erythrocytes

(Bartlett 1982). Given that GTP depresses the oxygen affinity of teleost and elasmobranch Hbs about 1.5–2 times more than ATP (Weber et al. 1975, 1983), this suggests that an ATP --> GTP switch may contribute to ontogenetic changes in blood oxygen affinity. This conversion is catalyzed by nucleoside diphosphokinase that occurs in high concentration in elasmobranch and teleost red cells (Parks and Agarwahl 1973; Weber 1982). The contribution of organic phosphates to the maternal-fetal shift in elasmobranchs may, however, be restricted by the fact that urea, which occurs at high concentrations in the red cells, reduces the allosteric interaction between Hb and ATP (Weber et al. 1983).

There is very little evidence for hormonal control of viviparity in elasmobranchs beyond the process of vitellogenesis and ovulation. Dodd et al. (1960) have shown that pituitary-gonadal interactions exist in viviparous elasmobranchs and that hypophysectomy causes degeneration of developing eggs in *Squalus canicula* and *S. acanthias*. Although it is hypothesized that corpora lutea secrete progesterone (Tsang and Callard 1987a, b), which maintains uterine functions during pregnancy, there is as yet no direct evidence for overall ovarian control of uterine function. Relaxin from ovaries of pregnant *Odontopspis taurus* has recently been characterized structurally (Gowan et al. 1981). Koob et al. (1982) found that levels of estradiol-17β in plasma were correlated to the seasonal development of the nidamental gland in *Raja erinaceae*, suggesting that estradiol induces glandular development. A more recent study (Koob et al. 1984) showed that relaxin increased the cervical cross-sectional area resulting in a premature loss of developing embryos in *S. acanthias*.

3 Bony Fish

In bony fish reproduction is typically oviparous with external fertilization. Viviparity, however, has developed independently in several teleost species within 13 different families and is also found in the coelacanths (Wourms 1981). The embryos of viviparous teleost fish are gestated within the ovary, a feature which is unique among vertebrates. Intraovarian gestation is either "follicular", with the eggs fertilized and developing within the ovarian follicles, or "ovarian", in which case the eggs, fertilized or unfertilized, or developing embryos are released from the follicles into the ovarian cavity for further development. The nutritional relationship may be defined as either mainly lecithotrophic or as matrotrophic. A number of morphological specializations have developed in embryonic as well as in maternal tissues to facilitate nutrient transfer to matrotrophic developing embryos. Thus, the embryonic epidermis, fin epithelium, gut, or even yolk sac may be specialized for nutrient absorption, while the maternal ovarian epithelium may similarly be differentiated to increase its secretory function. Often maternal and embryonic tissues combine and form a so-called follicular placenta. The adaptations to viviparity among fish have been carefully reviewed by Wourms (1981). Since then new information has been obtained both for teleosts with intrafollicular gestation, including *Anableps, Poecilia*, and *Clinus superciliosus* and for teleosts with typical ovarian gestation, including *Zoarces viviparus, Ameca splendens, Xenophorus captivus*, and different species of Scorpaenid and Embiotocid teleosts. As the experimental design and the results obtained differ in many details these findings will be reviewed family-wise below.

3.1 Family Anablepidae

The viviparous, four-eyed fish *Anableps* displays follicular gestation with two marked morphological specializations, the follicular placenta and a spectacular enlargement of the embryonic gut. Development is typically matrotrophic. Embryonic postfertilization weight increases about 3000 to 8400 times, reflecting a massive transfer of maternal metabolites across the follicular placenta (Knight et al. 1985). No experimental studies appear to have been carried out to elucidate the transport processes and the existent knowledge is thus almost entirely based on morphological studies.

3.2 Family Poeciliidae

Most species of poeciliid fish diplay follicular gestation, with a typical gestation period of approximately 1 month. Superfetation (simultaneously occurring multiple stages of embryos) is common. The expanded pericardial sac constitutes the principal fetal morphological adaptation. Nutritionally they are lecithotrophic as well as matrotrophic, the former often resulting in net dry weight loss during embryonic development, and the latter in constant or increasing embryonic dry weight during gestation (Thibault and Schultz 1978). Thus, the maternal-fetal trophic relationship may be quantified by a low rate of weight loss in the embryo. In *Poecilia reticulata* which appears to be minimally dependent on maternal nourishment, the intrafollicular exchange seems to be limited to minerals and respiratory gases (Wourms 1981). Maternal contributions to developing embryos have been demonstrated in the Gila topminnow (*P. occidentalis*), where calorific content per embryo increases by 31.6% during development (Constanz 1980). Interestingly, changes in cal/embryo appear to be correlated with changes in energy shunting between different broods within the same female. In the sailfin molly (*P. latipinna*) the embryos in three of four populations studied showed no sign of weight loss during their intraovarian development. In a fourth population the embryos lost 40% of their weight during the first year of investigation, whereas embryos from the same population did not lose weight during the following year suggesting environmental influences on interdemic differences (Trexler 1985).

Following high salinity adaptation of the pregnant female guppy *P. reticulata*, Depeche and Schoffeniels (1975) demonstrated increases in embryonic electrolyte, urea, and the amino acid concentrations. The contribution of urea to embryonic osmoregulation seems to dominate in early development, whereas the amino acid pool appears to be more important during the later stages, indicating an ontogenetic change in embryonic nitrogen metabolism. However, the mechanism underlying nitrogen accumulation in the embryos, including adaptive changes in enzymatic activities of an ornithine-urea cycle, has not yet been specified, although there is evidence for a urea cycle in oocytes and developing embryos of the guppy (Depeche et al. 1979).

Investigating the involvement of the pituitary in control of gestation in the guppy, Lam et al. (1985) found that postpartum treatment with the gonadotropin-blocking agent methallibure inhibited gestation, presumably by chemical inhibition of the secretion of the pituitary hormones in the maternal organism as suggested by Lam et al. (1985). In a study of pituitary cytology of the viviparous *P. latipinna*, Young and Ball (1982) suggested that corticosteroids, secreted in response to elevated output of ACTH may influence follicular rupture and expulsion of the embryos from the ovary at parturition. These reproductive processes also seem to be dependent on neurohypophysial peptides. Different studies indicate that thyroid hormones accelerate early embryonic development in various *Poecilia* species (Bromage and Sage 1968; Lam and Loy 1985). Studies on endocrine changes in the ovary are complicated by superfetation (see p. 215). Occasionally finding very high levels of estradiol in the maternal plasma, Kime and Groves (1986) reported a cyclic pattern for these and hypothesized a transient role for 17,20 progesterone during ovulation or parturition.

3.3 Family Clinidae

The embryos of *Clinus superciliosus* and *C. dorsalis* which develop intrafollicularly and are supplied with a small amount of yolk, absorb nutrients through various absorptive areas of the body (Veith and Cornish 1986). Nutritional transfer in *C. superciliosus* follows from a fetal weight increase of 34%. Veith (1979, 1980) found the follicular fluid to be rich in lipids and amino acids, which are believed to be actively secreted by the hypertrophied portion of the follicular epithelium. Autoradiographic studies indicate that the embryos are able to absorb these secretions, as tritiated thymidine injected into the pregnant fish was first concentrated in the follicular epithelium and subsequently absorbed by the embryos.

3.4 Family Zoarcidae

In *Zoarces viviparous* the eggs leave the ovarian follicles before fertilization. Gestation is thus ovarian and lasts approximately 5 months. After hatching the embryos develop in the ovarian fluid secreted by the ovarian epithelium. During the period between fertilization and hatching, dry weight of the eggs/embryos remains stable, indicating that the embryos are mainly nourished on the yolk. After hatching, while the embryos carry an external yolk sac, embryonic dry weights increase considerably from 4.60 mg, to 18.24 mg 1 month later, and oxygen consumption increases correspondingly from 0.106 to 0.701 μmol $embryo^{-1}$ h^{-1} (Korsgaard 1986). The embryos are now capable of assimilating and metabolizing ^{14}C-glucose that is injected into the ovary, and from in vitro media (Fig. 2; Korsgaard and Andersen 1985; Korsgaard 1986, 1987). During later stages of gestation they show linear growth and accumulate substantial amounts of different tracer compounds, i.e., glucose and amino acids. The uptake rates of these

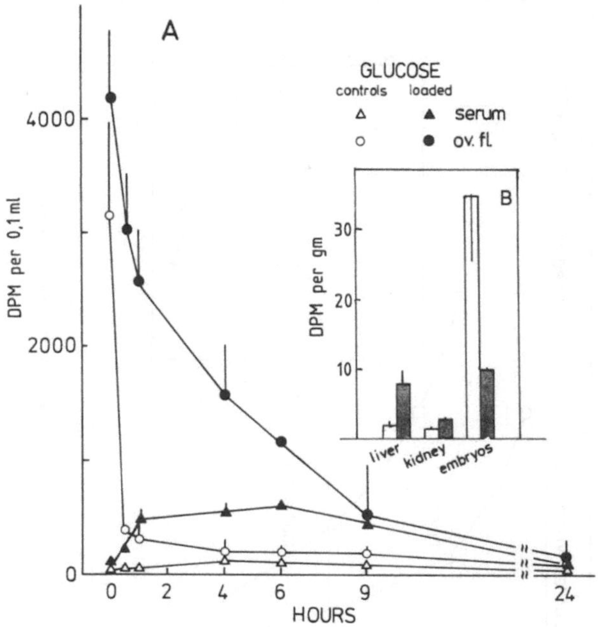

Fig. 2A, B. The time course appearance of injected radiolabeled glucose in the serum (*triangles*) and ovarian fluid (*circles*) of glucose-loaded (*solid symbols*) and non-loaded (*open symbols*) female specimens of *Zoarces viviparus*. *Inset*: The distribution of radiolabeled glucose in the liver and kidney and yolk-sac embryos 24 h after the injection of labeled glucose. *Hatched histograms* are carrier-loaded and *open histograms* carrier-unloaded groups (After Korsgaard 1987)

tracer compounds in vitro appear to be correlated with the ambient concentration of unlabeled carrier substrate within the natural range of concentrations found in the voluminous ovarian fluid. This fluid is rich in glucose, free fatty acids, and amino acids, but no proteins can be detected (Kristofferssen et al. 1973; Korsgaard 1983). The rate of clearance of intraovarian glucose or amino acid loads shows that the ovarian fluid is not a static pool, but exhibits rapid metabolite exchange with significantly lower half-life values for excess glucose (4.8 h) and excess amino acid (3.8 h) compared with the values found in the maternal serum for glucose (10.1 h) and amino acids (5.2 h). The observed growth to respiration ratio of 0.91 indicated a relative high efficiency for converting food to growth.

The impact of pregnancy on the maternal organism is reflected in considerable changes in biochemical events. Thus, during the latter part of pregnancy a marked decrease is observed in maternal liver weight, liver lipids, and glycogen, and lipid simultaneously accumulates in the maternal blood (Korsgaard and Petersen 1979). The activity levels of glutamate dehydrogenase and two transaminases of maternal liver increased significantly during midterm pregnancy suggesting changes in overall protein metabolism. These metabolic changes are consistent with increases in maternal-fetal energy transfer (Korsgaard 1982).

The osmolarity and the concentration of inorganic ions in the maternal blood and the ovarian fluid of *Z. viviparous* are approximately equal, but very different from that of the surrounding brackish water. The newborn fish adapt well to ambient osmotic stress (brackish water of 6 to 20‰ salinity) indicating development of an osmoregulatory capacity already before birth (Kristofferssen et al. 1973).

The mechanisms underlying maternal-fetal oxygen transfer in viviparous teleosts is known in only a few species. In *Zoarces viviparous* late gestational fetal

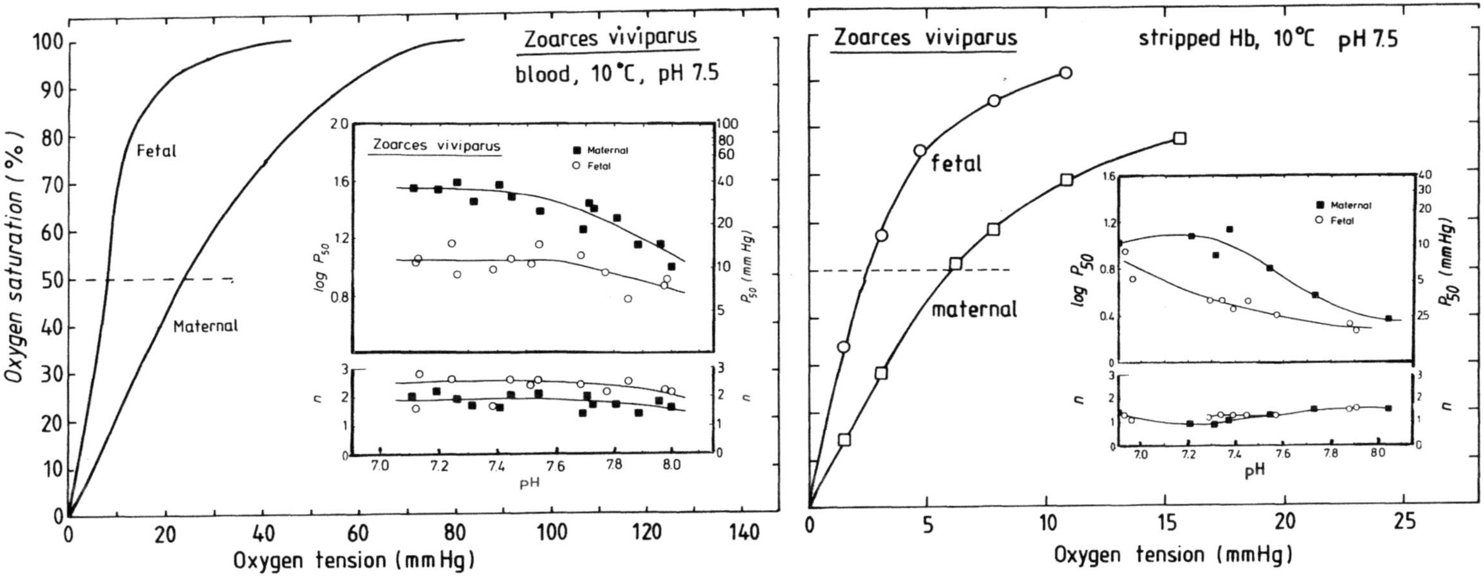

Fig. 3. Oxygen equilibria of whole blood (*left panel*) and stripped hemoglobin solutions (*right panel*) from late-gestational fetal and adult stages of *Zoarces viviparus* at 10 °C and pH 7.5. *Insets:* Variations in half-saturation oxygen tensions (P_{50}) and cooperativity (n) with pH for the fetal (*closed squares*) and adult (*open circles*) specimens (After Weber and Hartvig 1984)

blood exhibits a considerably higher oxygen affinity than maternal blood (P_{50} = 9 and 23 mmHg, respectively, at 10 °C and pH 7.5; Fig. 3). Oxygen transfer will additionally be favored by the greater cooperativity in oxygen binding in the fetal than in adult blood (Hill's cooperativity coefficient, n, which characterizes the sigmoid nature of the oxygen equilibrium curve being approximately 2.5 and 2.0, respectively), and by the larger maternal Bohr factor (Fig. 3) which increases the affinity difference under acidotic conditions (Hartvig and Weber 1984; Weber and Hartvig 1984) that occur when *Zoarces* becomes active or exposed to hypoxic conditions (Broberg and Kristoffersson 1983). The pronounced difference in oxygenation characteristics of fetal and maternal blood together with the relatively high oxygen tension found in late gestational ovarial fluid (38 ± 7 mmHg) may compensate for the low structural transfer capacity resulting from the lack of specific anatomical adaptations (placentalike structures) (Hartvig and Weber 1984). Interpolation on the blood oxygen equilibrium curves indicates that the mean ovarian oxygen tension (38 mmHg) corresponds to oxygen saturations of about 97% and 92%, respectively, in the fetus and adult, respectively, at 10 °C and pH 7.5.

Given that fetal and adult blood of *Zoarces* have similar NTP concentrations and NTP/GTP ratios, and that their Hbs exhibit similar sensitivities to ATP and GTP, the significantly higher intrinsic affinity in the fetal than the adult Hb (P_{50} = 2.8 and 7.4 mmHg, respectively) is seen as responsible for the blood-affinity difference (Weber and Hartvig 1984). This is consistent with electrophoretic differences in Hb multiplicity between adult and newly born specimens (Hjorth 1974).

The hormonal control of gestation in *Z. viviparous* is still unexplored. Kristofferrson et al. (1976), however, observed that steroid metabolism by ovarian tissue in vitro followed the same pathway as in oviparous fish, except that there is a considerably lower activity during late pregnancy compared to nonovulated fish. Also, the muscular tissue of the ovarian wall shows a clear alteration in contractile activity during different seasons (Kristoffersson et al. 1976). After ovulation the follicle can still concentrate the yolk-precursor protein vitellogenin upon stimulation with exogenous estradiol, proving functionality of the postovulatory follicle (Korsgaard 1983).

3.5 Family Goodeidae

Embryos of the Goodeidae develop within the ovarian cavity changing from lecithotrophy through progressive matrotrophy via the so-called trophotaenial placenta. Lombardi and Wourms (1985 a) have provided a detailed description of the morphology of the trophotaenial placenta in *Ameca splendens*, where the yolk reserves are sparse and consumed in early embryonic development. The placenta consists of two different components, the extraembryonic membranes or trophotaeniae, and the maternal component, the internal ovarian epithelium (Fig. 4). It is suggested that maternal nutrients are released by the internal ovarian epithelium into the ovarian cavity and taken up via the embryonic trophotaeniae, which increase embryonic dry weight by about 99% in the fully developed em-

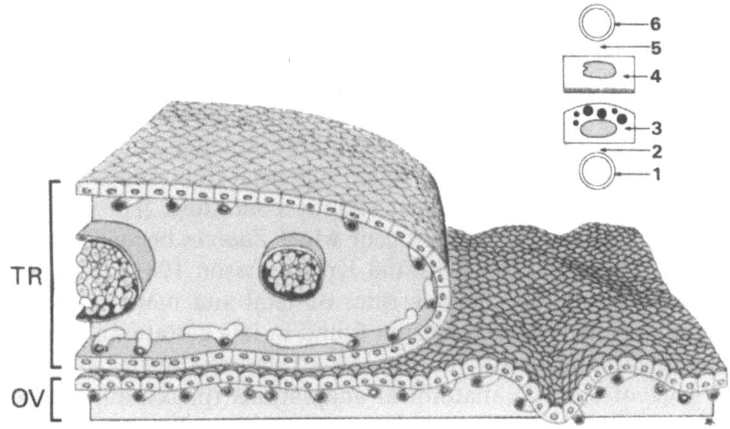

Fig. 4. Left: Schematic representation of tissue organization in the trophotaenial placenta of the goodeid teleost *Ameca splendens*, showing a section of trophotaenial ribbon in cross-section (*TR*) lying in direct apposition to the internal ovarian tissue (*OV*) that lines the ovarian lumen. *Inset*: Path of nutrient transfer from the ovarian to the trophotaenial capillaries (*1–5*) (Lombardi and Wourms 1985)

bryo. Lombardi and Wourms (1985b) recently demonstrated that the trophotaenial cells are specialized for endocytosis of exogenous proteins dissolved in the ovarian fluid and their subsequent enzymatic degradation (Fig. 4) and found that the uptake rate of horseradish peroxidase by the trophotaenial epithelium in vitro initially was 13.5 ng mg^{-1} trophotaenial protein min^{-1}, while peroxidase was degraded by the trophotaenial cells at a rate of 0.5 ng mg^{-1} trophotaenial protein min^{-1}, after transferring the embryos to a peroxidase-free medium. In embryos of another goodeid species *Xenophorus captivus*, Schindler and de Vries (1987) described a cutaneous gas-exchange pathway formed by a stratified epithelium, 2 to 4 µm in thickness, and demonstrated the presence of chloride cells in the basal part of the trophotaenea and in the abdominal wall of the embryos. The trophotaenial structure in this species seems thus also to function in gas exchange and osmoregulation, although no actual transport experiments have been carried out so far.

3.6 Family Scorpaenidae

The scorpaenid genus *Sebastes* exhibits a predominantly lecithotrophic viviparity. Recent work by Boehlert and Yoklavich (1984) on *Sebastes melanops*, however, demonstrates the occurrence of matrotrophism with a maternal contribution of approximately 70% of the catabolic energy used during embryonic development estimated on the basis of nitrogen and protein budgets and calorimetrically determined embryonic energy contents. Efficiency of yolk utilization in approximately 37 days that on average elapse between fertilization to parturition was estimated to be approx. 36%. As 81% of the initial energy investment is calculated still to be present at parturition, the difference appears to be supplied by maternal nutritional contribution (Boehlert and Yoklavich 1984). Oxygen consumption per embryo increases with development. The oxygen tension inside the ovary is unknown, but *Sebastes* species have a specialized dual arterial system supplying blood to both ovaries. The embryonic weight and caloric and protein contents gradually decreased through development, but total nitrogen remained nearly constant indicating that nitrogenous energy sources are involved in the exogenous nutrition of the embryos. This is also evident from a similar investigation of *Sebastes schlegeli* (Boehlert et al. 1986). Uptake of maternal nutrients was histologically demonstrated to occur by ingestion and adsorption of ovarian fluid by the hindgut of the embryos. These studies demonstrate a similar pattern of matrotrophic embryonic energetics in two different *Sebastes* species.

3.7 Family Embiotocidae

Most recent experimental research in the family Embiotocidae has been carried out on *Cymatogaster aggregata* and *Embiotoca lateralis*, where the eggs that are ovulated in early segmentation stages occur in the cavity of the ovary as free-floating blastulae surrounded by a thin proteinaceous ovarian fluid. Pregnancy usually lasts for several months. During early and midgestation the ovarian epithelium becomes hypertrophic and secretes ovarian fluid. The amino acids and lipids found in the ovarian fluid correspond to those of the maternal serum, but occur at lower levels (deVlaming et al. 1983). Electrophoresis experiments, however, indicate that novel peptides distinct from those of maternal serum are synthesized by the ovarian epithelium. Throughout gestation ovarian fluid contains large quantities of urea, whereas urea could not be detected in the maternal serum. The capacity to concentrate urea thus appears to be lost or reduced in the adult fish. The high level of glutamic acid observed in the ovarian fluid might reflect pronounced amino acid catabolism by the embryos, as glutamate might serve as the amino group donor in urea production. Endocrine support of gestation appears to be of limited significance since estradiol and progesterone occur at low levels throughout gestation.

Among embiotocid teleosts, the circulatory, metabolic, and respiratory adaptations associated with viviparity have been investigated in the pile perch *Rhaccochilis vacca* and striped seaperch, *Embiotoca lateralis* (Webb and Brett 1972a, b; Ingermann and Terwilliger 1981a, b, 1982, 1984). Seaperch fetuses are characterized by extensive, richly capillarized fin webs that appear to be involved in gas and ion exchange. In pile perch the ovarian fluid oxygen tension falls markedly during gestation (from about 36 to 14 mmHg as individual embryonic weight increases from 0.5 to 4 g), whereas the oxygen content of the ovarian fluid initially decreases and subsequently increases as fluid volume rises toward term (Webb and Brett 1972a, b). Ovarian tissue intrudes the fetal gill cavities in pile perch and several other embotocids, reducing the diffusion distance (L) and increasing the structural transfer capacity (Webb and Brett 1972b). Birth evidently is a traumatic event requiring a rapid shift from fin-dermal to gill breathing, and corresponding shunting of blood circulation.

In the sea perch, blood oxygen affinity is higher in midgestational than in late gestational stages, and lowest in the adult (Ingermann and Terwilliger 1984). Although the native fetal and maternal Hbs are electrophoretically indistinguishable, the fetal protein exhibits a higher intrinsic oxygen affinity and distinct peptide and amino acid compositions. The higher fetal blood affinity can additionally be attributed to lower erythrocytic NTP concentrations and lower cellular Hb concentrations, which correlate with the fact that the erythrocytes are approximately 60% larger than in the adult organism (Ingermann and Terwilliger 1981a, b, 1982). The NTP concentration difference persists even when the red cell levels of Mg^{2+}, which increases fish Hb-oxygen affinity by forming stable complexes with NTP (Weber and Lykkeboe 1978) is accounted for (Ingermann and Terwilliger 1981b). As shown earlier (Weber and Lykkeboe 1978), reductions in Hb concentration at constant Hb/NTP ratio drastically increases the oxygen affinity of fish Hb at concentrations approaching those found in the red cells (Weber and Lykkeboe 1978).

Considered in conjunction with the decrease in blood oxygen affinity during mid- to late gestational development in sea perch, the fall in ovarian fluid oxygen tension during gestational development and the increased weight-specific fetal oxygen consumption observed in the pile perch (Webb and Brett 1972), present perplexing questions concerning the regulatory mechanisms underlying fetal oxygen uptake in fish (cf. Ingermann and Terwilliger 1984b). A similar prenatal reduction in oxygen affinity (representing an adaptation to postfetal life; Ingermann and Terwilliger 1984) is, however, seen in the viviparous lizard, *Sphenomorphus guoyii* and in mammals like mouse and man.

4 The Amphibians

Most amphibians need to return to water for breeding and embryonic development. Internal fertilization and viviparity might thus represent an evolutionary step freeing the amphibians from the reproductive dependence of water. Among the three orders of extant amphibians viviparity is found most frequently in the Apoda (Caecilians). In the Urodela viviparous reproduction has been described for only a few species of salamanders, and in the anurans, oviductal viviparity appears to have been reported only in the bufonid *Nectophrynoides*. Of more curious reproductive features among the amphibians are the different forms of external incubations by the use of brood pouches and the oral cavity, both of which also occur among the teleost fish. Such extraoviductal incubation of fertilized eggs may be regarded as a highly specialized form of parental care. In truly viviparous amphibians, where embryonic development occurs in a more or less modified oviduct, the maternal trophic adaptation to embryonic nutrition most often involves secretory activity of the uterus, even though uterine structural modifications may vary from being almost nonexistent to highly specialized adaptations. According to Wake (1977), embryos of all seven known viviparous apodan genera are highly dependent on the maternal organism for nutrition. Since Wake's review on the evolutionary significance of viviparity in the apodans, most research work on this order of amphibians has been carried out on the caecilian *Typhlonectes compressicaudus.* As reproduction in general is similar among viviparous caecilians (Wake 1977; Exbrayat and Delsol 1985), the reproductive patterns of *Typhlonectes compressus* will be described here as typical.

During pregnancy, lasting approximately 6 months, the oviducts are progressively distended and the uterine epithelium displays increased secretory activity and concomitant structural changes. At ovulation ciliated cells are differentiated in the uterine epithelium and vitelline vesicles derived from degenerated unfertilized eggs are enveloped in a cellular mass deriving from the oviductal wall (Exbrayat 1983). Later the uterine epithelium displays cytoplasmic proliferations and develops highly vascularized zones.

Trophic adaptations in the embryo includes development of voluminous external gills after hatching and the aquisition of a curious rasplike tripartite dentition at an embryonic size of 12–70 mm, which disappears again after birth. The gradual disappearance of the vitelline vesicle and development of maternal-fetal trophic structures are associated with progressive changes through three different embryonic modes of nutrition. After feeding for a short period on the vitelline vesicles, the embryos develop a so-called ectotrophoblast, an absorptive structure of the ventral skin, later followed by a pair of voluminous gills that are pressed against the vascularized uterine wall forming a branchial placenta. The gills are lost before or shortly after birth (Exbrayat and Delsol 1985). There has been some debate on the nutritive function of embryonic teething (Wake 1977; Exbrayat et al. 1981). Generally, it is suggested that the embryonic teeth stimulate secretions by irritating the uterine epithelium and that they might also be used to ingest uterine secretory products and even cellular components of the oviductal wall (Wake 1977). The nutritive components of the maternal secretion and their energetic value need to be further elucidated by actual transport and assimilation studies.

Embryos of *T. compressicauda* have two extensive external gills, at least one of which makes close contact with the vascularized oviduct, suggesting that it is a site for maternal-fetal gas exchange. This tallies with a significantly higher embryonic blood oxygen affinity ($P_{50} = 11$ compared to 22 mmHg in the maternal blood) (Toews and Macintyre 1977; Garlick et al. 1979). That the difference is mediated entirely by organic phosphates follows from identical electrophoretic patterns and intrinsic Hb-oxygen affinities in both stages, and a threefold greater concentration in the adults of NTP, which consists of about 90% ATP and 10% GTP (Garlick et al. 1979). ATP drastically increases the Bohr factor of the stripped Hb ($\Delta \log P_{50}/\Delta pH$ falling from 0 to -0.45), indicating the occurrence of a larger Bohr effect in the adult blood that will increase the transfer potential under lower pH. The pH effects parallel those in oviparous frogs, where pH insensitivity in the tadpole Hb correlates with acetylation of the N-terminal amino acid of the a-chains which are a major site for binding Bohr protons in Hbs.

The viviparous urodele, *Salamandra atra*, produces about 100 eggs, but only the one most distant from each ovary actually develops in each uterus. Two live larvae are born after a gestation period of about 1 year but prolonged gestation of 2 to 4 years, depending primarily on temperature, often occurs in this species. The developing embryos obtain their nutrition in three different phases. After using up their own substantial yolk supply the embryos feed by oophagy, consuming the yolk mass from the degenerated embryotrophic eggs in the oviduct, until they measure 25 to 42 mm in length. During the remaining gestation period the embryos derive nutrition through epitheliophagy, the epithelial cells providing nourishment originating from the maternal zona trophica, a specially differentiated portion of the uterine wall (Guex and Chen 1986). The epithelial cells are released into the lumen of the uterus by partial necrosis of the underlying connective tissue, and the mere presence of the growing embryos in the uterus appears to induce the development of the zona trophica (Guex and Chen 1986).

Recent investigations on the ovoviviparous fire salamander, *Salamandra salamandra*, provides an interesting basis for comparing viviparous adaptive patterns with those occurring in alleged oviparous species (Greven 1980; Schindelmeiser and Greven 1981; Greven et al. 1983). According to Wourms' (1981) definition of viviparity, it would be more correct to consider the reproduction of *Salamandra salamandra* as true viviparity, as the fertilized developing eggs are carried for as long as a year, producing late larval stages, which are surrounded by the egg jelly at birth. The deposited eggs hatch immediately and the larvae are capable of feeding immediately after hatching. Ultrahistochemical and autoradiographic studies indicate that an active transport of ions occurs from the uterine lumen via the uterine epithelium which may be involved in the regulation of the intrauterine milieu and which may facilitate the removal of waste products from the developing embryos. The urea content in the sparse uterine fluid as well as in plasma from pregnant *Salamandra salamandra* is higher than in nonpregnant females. Nitrogen is excreted as urea from embryos excised from the uterus while extrauterine larvae excreted ammonia at slightly higher levels than urea (Schindelmeiser and Greven 1981). The capacity for urea synthesis during intrauterine development may thus be regarded as an adaptation to the prolonged stay in utero in this species.

True viviparity occurs in only one genus of the anurans, the African bufonid *Nectophrynoides*. All species within this genus live at high altitude, experiencing alternate wet and dry seasons. The reproductive patterns have been reviewed for *Necthrophynoides* by Xavier (1977) and Wake (1980). The ovulated eggs from the truly viviparous species, *Nectophrynoides occidentalis*, are very small and poor in yolk, weighing approximately 200 µg. However, the embryos show a considerable increase in size during their 9 months of intrauterine development and they are born as totally metamorphosed young weighing 30 to 60 mg at birth, showing that their nutrients are dominantly of maternal origin. Nonetheless, there are no placental structures and the embryos occur freely in the lumen of the uterus. The uterus displays continuous secretory activity and hyperemia throughout gestation and it was demonstrated that the embryos are capable of accumulating ^{14}C-lysine injected into the mother and of incorporating the radiolabelled amino acid into proteins of the digestive tract, which develops very early in development. The only other obvious embryonic morphological adaptation to viviparity is the highly vascularized, enlarged embryonic tail, which may play a role in gas exchange and as an absorptive organ.

The endocrine control of viviparity in the amphibians has been investigated in only a few species. The exact role of the corpora luteae, which are present in all species investigated, remains unknown. These ovarian structures are, however, also found in oviparous species, posing some doubt to their significance in amphibian viviparity. Exbrayat and Collenot (1983) observed that the corpora luteae of *Typhlonectes caudatus* persist throughout pregnancy, showing morphological changes parallel to the development of the embryos and then degenerating rapidly after parturition. Removal of the ovaries of pregnant *Nectophrynoides* has an effect on the course of pregnancy, accelerating embryonic development and advancing parturition by 3 or 4 months (Xavier 1977). Progesterone injected into ovariectomized pregnant females correspondingly delays embryonic development and prolongs gestation by 2 or 3 months, and Xavier (1977) accordingly suggests that the corpus luteum of *Nectophrynoides* may be important for the uterine function rather than serving primarily to maintain pregnancy.

5 Reptiles

The reptiles utilize a broad variety of reproductive modes. Most reptiles, including all Crocodilia and Chelonia, are oviparous with internal fertilization. In these species the shelled eggs are retained only briefly in the uterus and embryos do not develop beyond the gastrulation stage at oviposition. An interesting intermediate between oviparity and viviparity is encountered in the so-called egg retainers, which typically retain the eggs in utero until about half way through embryonic development. Viviparity being defined as live bearing is distinct from egg retention (Hogarth 1976; Yaron 1985). Approximately one-fifth of lizards, snakes, and amphibians are viviparous and a range of fetal and maternal structural adaptations to viviparity are known in reptiles. Uterine placental relationships have recently been decribed in a commendable review by Yaron (1985). Interestingly, some of the placental structures may derive from tissues which develop within the shelled cleidoic egg. These include the extraembryonic membranes and an inner membrane that forms the amnion and retains amniotic fluid. The vascularized chorion is derived from an outer membrane and the allantois arises as another accessory membrane. In some viviparous species these membranes that originally were formed in the cleidoic egg may be involved in the placental connnections with the maternal organism, forming the chorioallantoic placenta which is typical of placental squamates and brings maternal and fetal capillaries in intimate contact. In others the yolk sac develops in close connection with the maternal uterine tissue to form the so-called yolk sac placenta. These two forms of placental development may exhibit a great variation in interdigitation of the maternal and fetal tissues, and the various forms of placentae may serve different functions in maternal-fetal physiological exchange (Packard et al. 1977). Given the rather extensive knowledge of placental morphology, surprisingly few studies have been carried out to elucidate the physiology of the placenta of the reptiles.

Nutritionally, lecithotrophy is considered typical, and matrotrophy appears to be more sporadically distributed among viviparous reptiles. However, constant or decreasing dry mass of the embryos during their intrauterine development, which generally is considered to reflect lecithotrophy, does not exclude transplacental transfer of material. In the viviparous lizard, *Sphenomorphus guoyii*, the wet weight and the amounts of calcium, sodium, and potassium in developed embryos exceed the values found in the yolk, although the embryonic dry weight does not exceed the initial dry weight of the yolk and there is a net loss of proteins and lipids in the developing egg (Thompson 1981, 1982). Thus, while water and inorganic ions are incorporated into the developing embryos, there are no indications of an embryonic dependence on the maternal organism for organic nutrients. The functional significance and mechanisms of the electrolyte transport remain unknown. Newly born specimens of the saurian *Gerrhonotus coeruleus* and the snake *Nerodia rhombifera* (Stewart and Castillo 1984) similarly contain more water but less total lipid and dry mass than their respective initial yolk supplies, although the nitrogen levels are similar and interspecific variations are observed. However, like in *Sphenomorphus*, these species also contain a larger quantity of inorganic material than ovulated eggs and calcium similarly is a major part of the total ash content. In these cases thus an advantage of viviparity may be the additional source of water and mineral nutrients provided by the maternal organism.

The nutritive importance of placental transfer is more pronounced in the skink, *Mabuya heathi*, which has an extended gestation period of 8–12 months (Blackburn et al. 1984) and ovulates the smallest known reptilian egg with an average diameter of 1 mm and an average dry mass of 0.40 mg. The dry mass increases about 380 times at near term. The well-developed chorioallantoic placenta, which is derived from the uterine gland, is the site for nutrient transfer (Blackburn et al. 1984), and 99% of the increase in embryonic dry mass occurs after the formation of this structure. The viviparous lizard *Sceloporus aeneus bicanthalis* possesses a double placentation system and Guillette and Jones (1985) described both a yolk sac placenta and a chorioallantoic placenta. These authors nonetheless suggest that this placental arrangement is unimportant nutritionally, that it only serves in water transport, oxygen delivery, and carbon dioxide removal, and that the original yolk is sufficient for complete embryonic development.

Sceloporous aeneus exhibits reproductive bimodality, one subspecific taxon being oviparous and another viviparous, which allows for an interesting comparative study of two closely related forms. Whereas the shell membrane of the viviparous subspecies is single, very thin, and forms part of the chorioallantoic placental region until it degenerates completely during late embryonic development, that in the oviparous subspecies is thick, multilayered, and persists through oviposition. Oviductal vascularity in the incubatory region of the uterus increases significantly during embryonic development in the viviparous subspecies as a consequence of placentation, and the higher oxygen consumption in pregnant than the oviparous females is considered to reflect embryonic oxygen consumption (Guillette and Jones 1985).

Labeled inorganic ions and amino acids injected into pregnant lizards and snakes are later detected in the embryos. Although amino acids accumulated in the amniotic fluid may be incorporated into fetal proteins their metabolic fate or energetic significance is not yet known. The maternal-fetal amino acid transport is likely to provide additional nitrogen for embryonic development (Yaron 1985).

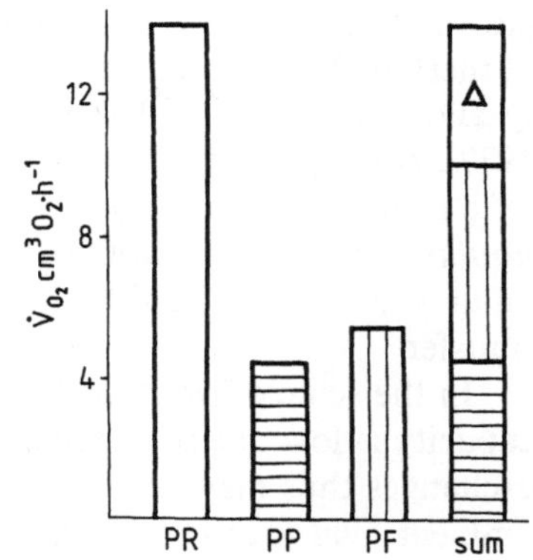

Fig. 5. Partition of oxygen consumption in a representative garter snake, *Thamnophilis sirtalis*, showing pregnant (*PR*), postpartum (*PP*), predicted fetal (*PF*) and residual (*Δ*) consumption rates (Birchard et al. 1984a)

Investigations of ionoregulation during viviparous embryonic development in reptiles are sparse, and most such information is derived indirectly from observed patterns of nitrogen excretion. Reptilian embryos generally are ureotelic and uric acid comprises only a minor part of the nitrogenous waste products. Urea may be an important osmoregulatory factor in embryonic water balance. The concentration of uric acid increased significantly in the allantoic cavity at the end of pregnancy of *Sphenomorphus guoyii* (Thompson 1977a, b). The highly correlated concentrations of urea in the amniotic and allantoic fluids suggest that urea diffuses from one compartment to the other and accumulates in the allantoic fluid, and that nitrogenous wastes are stored until birth, and not excreted via the maternal placental route. In the garter snake, *Thamnophis sirtalis*, however, urea is transported across the placenta into the maternal circulation (Clark and Sisken 1956).

A major problem of prolonged intrauterine development in reptiles is gas exchange, particularly during late development, when embryonic oxygen demand increases. In the garter snake, *Thamnophis sirtalis*, pregnancy significantly increases heart rate as well as weight-specific oxygen consumption, which is correlated with the number of embryos in the uterus, and shows that fetal oxygen consumption is scaled to fetal and not to maternal weight (Fig. 5; Birchard et al. 1984a) in accordance with observations for oviparous species (Seymour 1979). Resolving the oxygen consumption rates of the pregnant garter snake into three major components, Birchard et al. (1984a) found that the metabolic "cost of supporting pregnancy" is almost as large as either the standard metabolism of the mother or the predicted fetal metabolism, and thus that it makes an important contribution to the physiological energetics of live-bearing reptiles.

Although the blood hemoglobin, hematocrit, and oxygen-carrying capacity values in adult *T. sirtalis* do not change during pregnancy (Birchard et al. 1984a), hemoglobin and hematocrit levels are lower in late gestational fetuses and increase steadily after birth, when blood oxygen affinity decreases (P_{50} at 25 °C and pH 7.2 = 21 and 44 mmHg in newly born and adults, respectively; Pough 1977). Despite parallel changes in abundance of electrophoretically separable Hb components, there is no ontogenetic change in oxygen affinity of the stripped Hbs, nor can the affinity shift be explained by the ATP/Hb ratio, which falls during growth. Significantly, however, newly born garter snakes have high concentrations of metHb (25% compared to <2% in the adults), which raise Hb's oxygen affinity and appears to contribute to the higher fetal blood oxygen affinity (Pough 1977). A recent study (Berner and Ingermann, 1988) showed the same electrophoretic properties and oxygen affinities in stripped fetal and adult Hbs of *Thamnophis elegans*, but that in this species higher Hb and ATP concentrations in the adult erythrocytes could provide the basis for in vivo maternal-fetal oxygen transfer

In the scincid lizard, *Sphenomorphus guoyii*, in which the structural transfer capacity is low due to a weakly developed placenta, and in which maternal fetal exchanges thus may be substantially supplemented by indirect exchanges via the ambient oviductal and extraoviductal environments, the oxygen affinity of fetal blood far exceeds that in the adult (P_{50} = 30 and 70 mmHg, respectively) despite identical electrophoretic Hb patterns (Grigg and Harlow 1981). No investigations have yet identified underlying differences in red cell allosteric factors or pH. In the ovoviviparous cottonmouth snake, *Agkistrodon piscivorus*, fetal and adult Hbs similarly have identical electrophoretic properties, suggesting that the large blood fetal-maternal P_{50} shift (20 to 49 mmHg) is attributable to the significantly higher NTP/Hb ratios in the adult (Birchard et al. 1984b).

Grigg and Harlow (1981) postulated that high embryonic blood oxygen affinities in reptiles are not specifically adaptive to viviparity but represent a preadaptation to ambient oxygen paucity that is general in embryos of ovi- as well as viviparous species.

Only fragmentary information is available on hormonal control of reptilian viviparity. Active corpora lutea are observed in oviparous as well as in viviparous species of reptiles but their functional role has not yet been elucidated (Yaron 1985). In all oviparous squamates investigated corpora lutea exist until oviposition; in egg-retaining oviparous species they appear to be persistently active until oviposition. However, all reproductive modes including viviparity appear to show species differences in luteal life. High levels of progesterone have been detected in the maternal blood during pregnancy of some species (Bourne et al. 1986) and the effects of removal of the ovaries or corpora lutea during pregnancy appears similarly to be species-dependent. In several reptiles investigated this operation does not affect the normal progression of pregnancy, whereas in others the embryos die in utero (Yaron 1985). In the lizard *Tiliqua rugosa* ovariectomy does not completely eliminate circulating progesteronelike substances, indicating progesterone production in other than ovarian tissues (Bourne et al. 1986). Ovariectomy of this species does not result in abortion of the developing embryos, but parturition is delayed, as observed in other viviparous reptiles, indicating that luteal activity is related to the duration of pregnancy.

In *Sceloporus aeneus*, which exhibits reproductive bimodality (see p. 226) (Guillette and Jones 1985), both the ovi- and the viviparous subspecies have distinct corpora lutea, one for each ovulated egg, which are present immediately after ovulation and remain active until just prior to oviposition or parturition. The two subspecies moreover exhibited the same pattern of luteal regression (Guillette and Jones 1985).

6 Conclusions

In accordance to the great variability in their morphological organization, metabolic patterns, and environmental conditions, ectothermic vertebrates exhibit a much greater range of behavioral, anatomical, physiological, and molecular adaptations to live bearing, compared to homeothermic vertebrates. Although the functional morphology of viviparous ectotherms has been relatively well studied, the available knowledge on the physiology of ectothermic viviparity is cursory and fragmentary. Considering the vast diversity of adaptational patterns that have evolved independently, it might be a beneficial strategy to concentrate future studies on the physiological and molecular adaptations in a few, closely related species, which could provide a basis for greater insight into the biological significance and evolutionary implications of viviparity in ectothermic vertebrates.

References

Amoroso EC (1960) Viviparity in fishes. Symp Zool Soc (London) 1:153–181
Bartlett GR (1982) Phosphate compounds in red cells of two dogfish sharks: *Squalus acanthias* and *Mustellus canis*. Comp Biochem Physiol 73a:135–140
Berner NJ, Ingermann RL (1988) Molecular basis of the difference in oxygen affinity between maternal and foetal red blood cells in the viviparous garter snake *Thamnophis elegans*. J exp Biol 140:437–453
Birchard GF, Black CP, Schuett GW, Black V (1984a) Influence of pregnancy on oxygen consumption, heart rate and hematology in the garter snake: Implications for the "cost of reproduction" in live bearing reptiles. Comp Biochem Physiol 77A:519–523
Birchard GF, Black CP, Schuett GW, Black V (1984b) Foetal-maternal blood respiratory properties of an ovíviviparous snake the cottonmouth, *Agkistrodon piscivorus*. J Exp Biol 108:247–255
Blackburn DG, Vitt LJ, Beuchat CA (1984) Eutherian-like reproductive specializations in a viviparous reptile. Proc Natl Acad Sci USA 81:4860–4863
Boehlert GW, Yoklavich MM (1984) Reproduction, embryonic energetics, and the maternal-fetal relationship in the viviparous genus *Sebastes* (Pisces: Scorpaenidae). Biol Bull 167:354–370
Boehlert GW, Kusakari M, Shimizu M, Yamada J (1986) Energetics during embryonic development in kurosoi, *Sebastes schlegeli* Hilgendorf. J Exp Mar Biol Ecol 101:239–256
Bourne AR, Stewart BJ, Watson TG (1986) Changes in blood progesterone concentration during pregnancy in the lizard tiliqua (*Trachydosaurus*) Rugosa. Comp Biochem Physiol 84A 3:581–583
Broberg S, Kristoffersson R (1983) Oxygen consumption and lactate accumulation in the intra-ovarian embryos and young of the viviparous fish, *Zoarces viviparus* (L.) in relation to decreasing water O_2 concentration. Ann Zool Fennici 20:301–306
Bromage NR, Sage M (1968) The activity of the thyroid gland of *Poecilia* during the gestation cycle. J Endocrinol 41:303–311
Clark H, Sisken BF (1956) Nitrogenous excretion by embryos of the viviparous snake *Thamnopsis sirtalis* (L.). J Exp Biol 33:384–393
Constanz GD (1980) Energetics of viviparity in the gila topminnow. Copeia 4:876–878
Depeche J, Schoffeniels E (1975) Changes in electrolytes, urea and free amino acids of *Poecilia reticulata* embryos following high salinity adaptation of the viviparous female. Biochem Syst Ecol 3:111–119
Depeche J, Gilles R, Daufresne S, Chiapello H (1979) Urea content and urea production via the ornithine-urea cycle pathway during the ontogenic development of two teleost fishes. Comp Biochem Physiol 63A:51–56

de Vlaming V, Baltz D, Anderson S, Fitzgerald R, Delahunty G, Barkley M (1983) Aspects of embryo nutrition and excretion among viviparous embiotocid teleosts: potential endocrine involvements. Comp Biochem Physiol 76A 1:189–198

Dodd JM, Evennett PS, Godard CK (1960) Reproductive endocrinology in cyclostomes and elasmobranchs. Symp Zool Soc London 1:77–103

Evans DH, Oikari A, Kormanik GA, Mansberger L (1982) Osmoregulation by the prenatal spiny dogfish, *Squalus acanthias*. J Exp Biol 101:295–305

Exbrayat J-M (1983) Zoologie des vertebres. – Quelques observations sur l'évolution des voies génitales fémelles de *Typhlonectes compressicaudus* (Duméril et Bibron 1841), Amphibien Apode vivparé, au cours du cycle de réproduction. CR Acad Sci Paris 298 Ser III, 1, 1984

Exbrayat J-M, Collenot G (1983) Quelques aspects de l'évolution de l'ovaire de *Typhlonectes compressicaudus* (Duméril et Bibron 1841). Batracien Apode viviparé. – Etude quantitative et histochimique des corps jaunes. Nutrit Dev 23(5):889–898

Exbrayat J-M, Delsol M (1985) Reproduction and growth of *Typhlonectes compressicaudus* – a viviparous gymnophione. Copeia 4:950–955

Exbrayat J-M, Delsol M, Flatin J (1981) Zoologie des Vertebres. – Premieres remarques sur la gestation chez *Typhlonectes compressicaudus* (Duméril et Bibron 1841) Amphibien Apode viviparé. CR Acad Sci Paris 292

Garlick RL, Davis BJ, Farmer M, Fyhn HJ, Fyhn UEH, Noble RW, Powers DA, Riggs A, Weber RE (1979) A fetal-maternal shift in the oxygen equilibrium of the hemoglobin from the viviparous caecilian, *Typhlonectes compressicauda*. Comp Biochem Physiol 62A:239–244

Gilmore RG, Dodrill JW, Linley PA (1983) Reproduction and embryonic development of the sand tiger shark, *Odontaspis taurus* (Rafinesque). Fish Bull 8(2):201–225

Gowan LK, Reinig JW, Schwabe C, Bedakar S, Blundell TL (1981) On the primary and tertiary structure of relaxin from the sand tiger shark (*Odontaspis taurus*). FEBS Lett 129:80–82

Greven H (1980) Ultrahistochemical and autoradiographic evidence for epithelial transport in the uterus of the ovoviviparous salamander, *Salamandra salamandra* (L.) (Amphibia, Urodela). Cell Tissue Res 212:147–162

Greven H, Schindelmeiser J, Straub H (1983) The innervation of the uterus in *Salamandra salamandra* (L.) (Amphibia, Urodela). A morphological and biochemical study. Cell Tissue Res 232:421–431

Grigg GC, Harlow P (1981) A fetal-maternal shift of blood oxygen affinity in an Australian viviparous lizard, *Sphenomorphus guoyii* (Reptilia, Scincidae). J Comp Physiol 142:495–499

Guex GD, Chen PS (1986) Epitheliophagy: intrauterine cell nourishment in the viviparous alpine salamander, *Salamandra atra* (Laur.). Experientia 42:1205–1218

Guillette LJ Jr, Jones RE (1985) Ovarian, oviductal, and placental morphology of the reproductively bimodal lizard, *Sceloporus aeneus*. J Morphol 184:85–98

Hamlett WC (1983) Maternal-fetal relations in elasmobranch fishes. Ph D Diss Clemson Univ

Hamlett WC, Wourms JP (1984) Ultrastructure of the pre-implantation shark yolk sac placenta. Tissue Cell 16(4):613–625

Hartvig M, Weber RE (1984) Blood adaptations for maternal-fetal oxygen transfer in the viviparous teleost, *Zoarces viviparus*. In: Seymour RS (ed) Respiration and metabolism of embryonic vertebrates. Junk, Dordrecht, Neth, pp 17–30

Hjorth JP (1974) Genetics of *Zoarces* populations. VII. Fetal and adult hemoglobins and a polymorphism common to both. Hereditas 78:69–72

Hogarth PS (1976) Viviparity. Arnold, London

Ingermann RL, Terwilliger RC (1981a) Oxygen affinities of maternal and fetal hemoglobins of the viviparous seaperch, *Embiotoca lateralis*. J Comp Physiol 142:523–531

Ingermann RL, Terwilliger RC (1981b) Intraerythrocytic organic phosphates of fetal and adult seaperch *Embiotoca lateralis*: their role in maternal-fetal oxygen transport. J Comp Physiol 144:253–259

Ingermann RL, Terwilliger RC (1982) Blood parameters and facilitation of maternal-fetal oxygen transfer in a viviparous fish *Embiotoca lateralis*. Comp Biochem Physiol 73A:497–501

Ingermann RL, Terwilliger RC (1984) Facilitation of maternal-fetal oxygen transfer in fishes: anatomical and molecular specializations. In: Seymour RS (ed) Respiration and metabolism of embryonic vertebrates. Junk, Dordrecht, Neth, pp 1–15

Ingermann RL, Terwilliger RC, Roberts MS (1984) Foetal and adult blood oxygen affinities of the viviparous seaperch, *Embiotoca lateralis*. J Exp Biol 108:453–457
Kime DE, Groves DJ (1986) Steroidogenesis by gonads of a viviparous teleost, the sailfin molly *Poecilia latipinna*, in vitro and in vivo. Gen Comp Endocrinol 63:125–133
Knight FM, Lombardi J, Wourms JP, Burns JR (1985) Follicular placenta and embryonic growth of the viviparous four-eyed fish *Anableps*. J Morphol 185:131–142
Koob TS, Tsang P, Laffan SS, Callard IP (1982) Possible role for estradiol in nidamental gland function. Bull Mount Desert Isl Biol Lab 22:97–99
Koob TS, Laffan JJ, Callard IP (1984) Effects of relaxin and insulin on reproductive tract size and early fetal loss in *Squalus acanthias*. Biol Reprod 31:231–238
Korsgaard B (1982) Changes in activity levels of glutamate dehydrogenase and two transaminases of maternal liver during reproduction and after administration of oestradiol and progesterone in *Zoarces viviparus* (L.). J Fish Biol 20:445–454
Korsgaard B (1983) The chemical composition of follicular and ovarian fluids of the pregnant blenny *Zoarces viviparus* (L.). Can J Zool 61(5):1101–1108
Korsgaard B (1986) Trophic adaptations during early intraovarian development of embryos of *Zoarces viviparus* (L.). J Exp Mar Biol Ecol 98:141–152
Korsgaard B (1987) Assimilation and metabolism of exogenous glucose by yolk-sac embryos after intraovarian preincubation and in vitro in *Zoarces viviparus*. Can J Zool 65:1201–1205
Korsgaard B, Andersen FO (1985) Embryonic nutrition, growth and energetics in *Zoarces viviparous* L. as indication of a maternal-fetal trophic relationship. J Comp Physiol B 155:437–444
Korsgaard B, Petersen IM (1979) Vitellogenin, lipid and carbohydrate metabolism during vitellogenesis and pregnancy, and after hormonal induction in the blenny *Zoarces viviparus* (L). Comp Biochem Physiol 63B:245–251
Kristofferssen R, Broberg S, Pekkarinen M (1973) Histology and physiology of embryotrophic formation, embryonic nutrition and growth in the eel-pout, *Zoarces viviparus* (L.). Ann Zool Fennici 10:467–477
Kristofferssen R, Pesonen S, Terarainen T (1976) Patterns of steroid metabolism in the ovarian tissue of a viviparous teleost fish, *Zoarces viviparus* (L.). Ann Zool Fennici 13:189–194
Lam TJ, Loy GL (1985) Effect of L-thyroxine on ovarian development and gestation in the viviparous guppy, *Poecilia reticulata*. Gen Comp Endocrinol 60:324–330
Lam TJ, Pavri SC, Ng KL (1985) Effect of methallibure on gestation in the guppy, *Poecilia reticulata*. Gen Comp Endocrinol 60:315–323
Lombardi J, Wourms JP (1985a) The trophotaenial placenta of a viviparous goodeid fish. I. Ultrastructure of the internal ovarian epithelium, the maternal component. J Morphol 184:277–292
Lombardi J, Wourms JP (1985b) The trophotaenial placenta of a viviparous goodeid fish. III: Protein uptake by Trophotaeniae, the embryonic component. J Exp Zool 236:165–179
Lombardi J, Wourms JP (1985c) The trophotaenial placenta of a viviparous goodeid fish. II. Ultrastructure of trophotaeniae, the embryonic component. J Morphology 184:293–309
Lykkeboe G, Weber RE (1978) Changes in the respiratory properties of the blood in the carp, *Cyprinus carpio*, induced by diurnal variation in ambient oxygen tension. J Comp Physiol 128:117–125
Manwell C (1958) A "fetal-maternal shift" in the ovoviviparous spiny dogfish *Squalus suckleyi* (Girard). Physiol Zool 31:93–100
Manwell C (1963) Fetal and adult hemoglobins of the spiny dogfish *Squalus acanthias*. Arch Biochem Biophys 101:505–511
McCutcheon FH (1947) Specific oxygen affinity of hemoglobin in elasmobranchs and turtles. J Cell Comp Physiol 29:333–344
Packard GC, Tracy CR, Roth JJ (1977) The physiological ecology of reptilian eggs and embryos, and the evolution of viviparity within the class Reptilia. Biol Rev 52:71–105
Parks RE, Agarwal RP (1973) Nucleoside diphosphokinases. In: Boyer PD (ed) The enzymes, vol 8. Academic Press, New York, pp 307–333
Pough FH (1977) Ontogenetic change in molecular and functional properties of blood of garter snakes, *Thamnophis sirtalis*. J Exp Zool 201:47–56

Schindelmeiser J, Greven H (1981) Nitrogen excretion in intra- and extrauterine larvae of the ovoviviparous salamander, *Salamandra salamandra* (L.) (Amphibia, Urodela). Comp Biochem Physiol 70A:563–565

Schindler JF, deVries U (1987) Maternal-embryonic relationships in the goodeid teleost, *Xenophorus captivus*. Embryonic structural adaptations to viviparity. Cell Tissue Res 247:325–338

Seymour RS (1979) Dinosaur eggs. Gas conductance through the shell, water loss during incubation and clutch size. Palaeobiology 5:1–11

Shine R (1983) Reptilian reproductive modes: the oviparity-viviparity continuum. Herpetologica 39:1–8

Shine R (1984) Physiological and ecological questions on the evolution of reptilian viviparity. Perspect Vertebr Sci 3:147–154

Steward JR, Castillo RE (1984) Nutritional provision of the yolk of two species of viviparous reptiles. Physiol Zool 57(4):377–383

Teshima K (1981) Studies on the reproduction of Japanese smooth dogfishes, *Mustelus manazo* and *M. griseus*. J Shimonoseki Univ Fish 29(2):113–199

Thibault RE, Schultz RJ (1978) Reproductive adaptations among viviparous fishes (Cyprinodontiformes: Poeciliidae). Evolution 32(2):320–333

Thompson J (1977a) Embryo-maternal relationships in a viviparous skink *Sphenomorphis guoyi* (Lacertilia: Scincidae). In: Calaby JH, Tyndale CH (eds) Reproduction and evalution. Aust Acad Science, Canberra, pp 279–280

Thompson J (1977b) Embryo-maternal relationships in a viviparous lizard, *Sphenomorphus guoyii*. Comp Biochem Physiol 71A:107–112

Thompson J (1981) A study of the sources of nutrients for embryonic development in a viviparous lizard, *Sphenomorphus guoyii*. Comp Biochem Physiol 70A:509–518

Thompson J (1982) Uptake of inorganic ions from the maternal circulation during development of the embryo of a viviparous lizard, *Sphenomorphus guoyii*. Comp Biochem Physiol 71A:107–112

Toews D, Macintyre D (1977) Blood respiratory properties of a viviparous amphibian. Nature (London) 266:464–465

Trexler JC (1985) Variation in the degree of viviparity in the sailfin molly, *Poecilia latipinna*. Copeia 4:999–1004

Tsang P, Callard IP (1987a) Luteal progesterone production and regulation in the viviparous dogfish, *Squalus acanthias*. J Exp Zool 241:377–382

Tsang PCW, Callard IP (1987b) Morphological and endocrine correlates of the reproductive cycle of the aplacental viviparous dogfish, *Squalus acanthias*. Gen Comp Endocrinol 66:182–189

Veith WJ (1979) The chemical composition of the follicular fluid of the viviparous teleost *Clinus superciliosus*. Comp Biochem Physiol 63A:37–40.

Veith WJ (1980) Viviparity and embryonic adaptations in the teleost *Clinus superciliosus*. Can J Zool 58(1):1–12

Veith WJ, Cornish DA (1986) Ovarian adaptations in the viviparous teleosts *Clinus superciliosus* and *Clinus dorsalis* (Perciformes: Clinidae). South African J Zool 21(4):343–347

Wake MH (1977) Fetal maintenance and its evolutionary significance in the Amphibia: Gymnophiona. J Herpetol 11(4):379–387

Wake MH (1980) The reproductive biology of *Nectophrynoides malcolmi* (Amphibia: Bufonidae), with comments on the evolution of reproductive modes in the genus *Nectophrynoides*. Copeia 2:193–209

Webb PW, Brett JR (1972a) Respiratory adaptations of prenatal young in the ovary of two species of viviparous seaperch *Rhacochilus vacca* and *Embiotoca lateralis*. J Fish Res Board Can 29:1525–1542

Webb PW, Brett JR (1972b) Oxygen consumption of embryos and parents, and oxygen transfer characteristics within the ovary of two species of viviparous seaperch, *Rhacochilus vacca* and *Embiotica lateralis*. J Fish Res Board Can 29:1543–1553

Weber RE (1982) Intraspecific adaptation of hemoglobin function in fish to oxygen availability. In: Addink ADF, Spronk N (eds) Exogenous and endogenous influences on metabolic and neural control. Pergamon, Oxford, pp 87–102

Weber RE, Hartvig M (1984) Specific fetal hemoglobin underlies the fetal-maternal shift in blood oxygen affinity in a viviparous teleost. Mol Physiol 6:27–32
Weber RE, Lykkeboe G (1978) Respiratory adaptations in carp blood. Influences of hypoxia, red cell organic phosphates, divalent cations, and CO_2 on hemoglobin-oxygen affinity. J Comp Physiol 128:127–137
Weber RE, Lykkeboe G, Johansen K (1975) Biochemical adaptation of hemoglobin-oxygen affinity of eel to hypoxia. Life Sci 17:1345–1350
Weber RE, Wood SC, Lomholt JP (1976) Temperature acclimation and oxygen-binding properties of blood and multiple haemoglobins of rainbow trout. J Exp Biol 65:333–345
Weber RE, Wells RMG, Rossetti JE (1983) Allosteric interactions governing oxygen equilibria in the hemoglobin system of the spiny dogfish, *Squalus acanthias.* J Exp Biol 103:109–120
Weber RE, Lykkeboe G, Johansen K (1985) Biochemical aspects of the adaptation of hemoglobin-oxygen affinity to hypoxia. Life Sci 17:1345–1350
Wells RMG, Weber RE (1983) Oxygenational properties and phosphorylated intermediates in blood and erythrocytes of the dogfish, *Squalus acanthias.* J Exp Biol 103:95–108
Wood SC, Johansen K (1973) Organic phosphate metabolism in nucleated red cells: influence of hypoxia on eel HbO_2 affinity. Neth J Sea Res 7:328–338
Wood-Mason J, Alcock A (1891) On the uterine villiform papillae of Pteroplatea micrura and their relation to the embryo. Proc R Soc London Ser B 49:359–367
Wourms JP (1977) Reproduction and development of chondrichthyan fishes. Am Zool 17:379–410
Wourms JP (1981) Viviparity: the maternal-fetal relationship in fishes. Am Zool 21:473–515
Xavier F (1977) An exceptional reproductive strategy in anura: *Nectophrynoides occidentalis* angel (Bufonidae), an example of adaptation to terrestrial life by viviparity. In: Hecht MK, Goody PC, Hecht BM (eds) Major patterns in vertebrate evolutions. Plenum, New York, pp 545–552
Yaron Z (1985) Reptilian placentation and gestation: structure, function and endocrine control. In: Gans C (ed) Biology of the reptilia, vol 15B. John Wiley & Sons, New York, pp 527–604
Young G, Ball JN (1982) Ultrastructural changes in the adenohypophysis during the ovarian cycle of the viviparous teleost *Poecilia latipinna.* III. The growth hormone, adrenocorticotrophic and prolactin cells and the pars intermedia. Gen Comp Endocrinol 52:86–101

Subject Index

MIX
Papier aus verantwortungsvollen Quellen
Paper from responsible sources
FSC® C105338

If you have any concerns about our products,
you can contact us on
ProductSafety@springernature.com

In case Publisher is established outside the EU,
the EU authorized representative is:
Springer Nature Customer Service Center GmbH
Europaplatz 3, 69115 Heidelberg, Germany

Printed by Libri Plureos GmbH
in Hamburg, Germany